DATE DUE			

Mitochondria in Higher Plants

Structure, Function, and Biogenesis

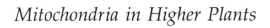

AMERICAN SOCIETY OF PLANT PHYSIOLOGISTS MONOGRAPH SERIES

Anthony H. C. Huang, Richard N. Trelease, and Thomas S. Moore, Jr. *PLANT PEROXISOMES, 1983.*

Roland Douce *MITOCHONDRIA IN HIGHER PLANTS: STRUCTURE, FUNCTION, AND BIOGENESIS, 1985.*

Mitochondria in Higher Plants

Structure, Function, and Biogenesis

Roland Douce

*Centre d'Études Nucléaires
and Université Scientifique et Médicale
Grenoble, France*

1985

ACADEMIC PRESS, INC.
(Harcourt Brace Jovanovich, Publishers)
Orlando   San Diego   New York   London
Toronto   Montreal   Sydney   Tokyo

ACADEMIC PRESS, INC.
Orlando, Florida 32887

United Kingdom Edition published by
ACADEMIC PRESS INC. (LONDON) LTD.
24–28 Oval Road, London NW1 7DX

Library of Congress Cataloging in Publication Data

Douce, Roland.
    Mitochondria in higher plants.

    (American Society of Plant Physiologists monograph
series)
    Bibliography: p.
    Includes index.
    1. Mitochondria. 2. Plant cells and tissues. I. Title.
II. Series.
QK725.D66   1985   582'.087342   84-20398
ISBN 0–12–221280–0 (alk. paper)

PRINTED IN THE UNITED STATES OF AMERICA

85 86 87 88   9 8 7 6 5 4 3 2 1

I am grateful to my teacher and friend
Walter D. Bonner
for the guidance and training he gave me
during the early part of my research career.

Contents

Foreword

From the earliest days of plant science there always have been a few plant physiologists studying respiration, commonly with respect to the activities of fruits, seeds, and roots. It was soon evident that the growth and development of plant tissues required metabolic energy derived from the oxidation ("dissimilation") of reduced carbon, but the chemistry of the process was poorly understood. Even after the essentials of glycolysis and fermentation were uncovered, there remained an exasperating void in the enzymology of the aerobic phase of respiration. Following World War II, however, there was an explosive growth in biological science based in large part on the new techniques of chromatography, radioisotope tracing, and electron microscopy. Among the disclosures was the role of the mitochondrion as the locus of the Krebs cycle and ATP formation in animal cells.

Mitochondria (sometimes called "chondriosomes") were long known as cytoplasmic inclusions in plant cells, and were suggested to have an enzymatic role in starch degradation due to an apparent clustering around amyloplasts. By drawing on the techniques used with animal mitochondria, researchers established in the early 1950s that mitochondria from plants and animals were similar in structure and function, but not entirely. As investigations proceeded, it became evident that in certain details there were important differences in cytological relationships, substrate oxidation rates, substrates utilized, electron transport chains, ion transport, etc. Many distinctions were uncovered by plant physiologists with physiological questions in mind (rarely were plant mitochondria used as material for primary studies of mitochondrial function), and failure to meet the rat liver standard was sometimes attributed to damage during isolation. Over the next two decades, however, it became clear that while all mitochondria have a common basic pattern for oxidative phosphorylation and participation in intermediary metabolism, those from plants have some special adaptations for autotrophic metabolism. Other characteristics of plant mitochondria are shared with lower organisms. In certain cases phenomena are observed which as yet have no adequate physiological or biochemical explanation [e.g., external NAD(P)H oxidation and cyanide-resistant respiration].

Enough research has now been done on the special properties of plant mitochondria to justify the effort of bringing the results together in one reference volume. Professor Douce has done this, and done it in the rigorous and scholarly fashion that characterizes the research from his laboratory. The material is logically organized into five chapters beginning with morphological and cytological observations, and proceeding through membrane and matrix functions to participation in metabolism and biogenesis. Each section presents the unique properties of plant mitochondria within the framework of general mitochondrial structure and function; the student using this book as a text will have little need to supplement it from other sources. It covers the relevant literature very well indeed, and can serve as the standard reference for plant mitochondria for some years to come. Of special value is the recurring effort to place mitochondrial activities in the larger context of biochemical and physiological functions.

Lastly, a challenge appears in this comprehensive summary of what is known about plant mitochondria: dozens of interesting research problems present themselves. Perhaps these stimuli will be the most valuable contribution of Professor Douce's book.

J. B. Hanson
Botany Department
University of Illinois
Urbana, Illinois

Preface

Plants are capable of utilizing light energy for the photosynthetic reduction of the atmospheric carbon dioxide plus water into carbohydrates and oxygen. This is a complex process that is performed in chloroplasts of land plants and the corresponding brown, red, and green plastids of the aquatic algae, including the phytoplankton of the sea. It is this process that perhaps more than anything else distinguishes plants from animals. The total amount of organic compounds formed each year by plants has been estimated at more than 100 billion tons. A biochemical operation of this magnitude would quickly deplete the earth's atmosphere of its carbon dioxide if there were no compensating process. This compensating process known as respiration (i.e., the complex oxidation of carbohydrate to carbon dioxide and water, molecular oxygen serving as the ultimate electron acceptor) occurs not only in animal cells but also in plant cells. Oxygen uptake and carbon dioxide output are the external manifestations of this process. In plants, as in other organisms, respiration takes place in mitochondria, where the energy freed in oxidation of cellular organic substrates is converted into the phosphate bond energy of adenosine triphosphate.

Flemming (1882) and Altmann (1890) discovered mitochondria in animal cells. The evidence that mitochondria also occur in plant cells (in tapetum cells of the anthers of Nymphae) first dates from the beginning of this century (Meves, 1904), and the mitochondria of a cell were collectively designated by the term chondriome (Guillermond *et al.*, 1933). In the three decades since the pioneering work of Millerd *et al.* (1951), the experimental literature on the dynamics of plant mitochondrial reactions has been growing enormously in both volume and complexity. Although work on the activities of higher plant mitochondria has lagged behind that conducted on the animal counterpart, particularly mammalian mitochondria (largely because of technical difficulties in handling the plant material), it is becoming more and more obvious that the basic enzymology of the biological oxidations and phosphorylations of higher plant mitochondria is remarkably similar to that found in animal systems. It is also obvious, however, that in addition to this highly conser-

vative basic system, developed at an early stage of evolution, the mitochondria of higher plants possess several additional features that are
absent in animal systems.

The aim of this book, then, is to collect and interpret the rapidly
growing experimental information on plant mitochondria, not only the
basic enzymology of ATP synthesis coupled to electron transport that
appears to constitute the major activity of the mitochondria but also
many other aspects that make plant mitochondria rather more diverse
than their animal counterparts. Finally, we have tried to emphasize the
important problem of mitochondria functioning under the wide variety
of metabolic conditions encountered in the cytoplasm of plant cells.

This book is intended not only for research workers and students
interested in the enzymology of plant mitochondria respiration but also
for graduate and undergraduate students in the field of plant biochemistry, cell physiology, and molecular biology. We also hope that this book
may be useful as a starting point for those students wishing to pursue
special studies in this field.

Roland Douce

Acknowledgments

The author is greatly indebted to Dr. Michel Neuburger for criticism and advice. He is also grateful to Dr. David Day, who read the entire manuscript carefully and provided numerous helpful suggestions. Errors of fact and interpretation that remain are, of course, my own responsibility.

I owe major debts to Dr. Richard Bligny, who did some illustrations with artistic skill. Mme. Françoise Bucharles has given valuable assistance by typing innumerable drafts. As for the technical assistance, I particularly thank Agnès Jourdain. Finally I am indebted to Jacques Joyard, Etienne Journet, and research worker friends who gave me the idea that the book should be written.

Mitochondria in Higher Plants

Structure, Function, and Biogenesis

1

General Organization of Plant Mitochondria

I.   MITOCHONDRIA IN THE INTACT CELL

The technique of electron microscopy which has provided the most information on plant mitochondrial structure is thin sectioning. Specimens are fixed, dehydrated, and then embedded in epoxy resins or polyester resins. Reagents widely used as fixatives include glutaraldehyde, osmium tetroxide, and potassium permanganate. Frequently, combinations of these reagents either sequentially or together are employed. Staining is carried out during fixation with osmium tetroxide or potassium permanganate or subsequently by treating the fixed specimen with a solution of heavy-metal salt (uranyl acetate, lead citrate). Osmium tetroxide, which is a strong oxidant, can cross-link and stabilize molecules such as proteins and polar lipids. A primary fixation with glutaraldehyde that reacts with amino groups of polypeptide chains assures a rapid stabilization of proteins. Under these conditions, the subcellular structures containing proteins and nucleoproteins are well

1

preserved. Potassium permanganate is also a strong oxidant which gives outstanding contrast to mitochondrial membranes.

A. Gross Features of Mitochondria

1. Morphology

In higher plant cells, Janus green B–stained particles called mitochondria (Greek mitos=thread and chondrion=granule) have a variety of shapes ranging from spheres to filamentous forms (Fig. 1.1). However, the most common shape assumed by plant mitochondria is that of a rod with hemispherical ends about 0.5 μm in diameter and up to 2 μm in length (Fig. 1.1b). The facts that branched mitochondria are only occasionally detected in individual sections and that mitochondria of various origin are mobile inside plant cells by passive and/or active movements within the streaming cytoplasm (Honda et al., 1966) strongly suggest that formation of single reticular mitochondrion in plant cells (i.e., the formation of extensive systems interconnected in three dimensions) is rare (see Vartapetian et al., 1977; Fig. 1.2). This is in contrast with Chlorella, Chlamydomonas, Euglena, and yeast, for which three-dimensional reconstructions indicate the existence of a single reticular mitochondrion in the cell (Gunning and Steer, 1975). Mitochondria in tissue from dormant tubers of the Jerusalem artichoke (Helianthus tuberosus) characteristically contain large numbers of mitochondria shaped like cups (or bells). These mitochondria are frequently sectioned to reveal an annular profile (Bagshaw et al., 1969). It is clear, however, that great care must be taken in the analysis of three-dimensional shapes from single sections, and three-dimensional reconstructions are required in order to visualize the exact shape of the mitochondria. Unfortunately, in higher plant cells very few studies employing serial sectioning have been undertaken (Gunning and Steer, 1975).

Many cells seem to contain mixtures of mitochondria of different shapes. This heterogeneity may arise as a coexistence of separate and independent populations of mitochondria within the same cell. However, electron micrographs tend to leave a static impression of rigidity of

---

Fig. 1.1.   Electron micrographs of mitochondria in sections of (A) mesophyll cell of *Hordeum albostrians* leaf (×102,000), (B) stomatal guard cell of *H. albostrians* leaf (×62,500), and (C) glandular hair of *Arctium minus* leaf (×75,000). Abbreviations: im, inner mitochondrial membrane; om, outer mitochondrial membrane; t, tonoplast; and pm, plasma membrane. (Courtesy of J. P. Carde.)

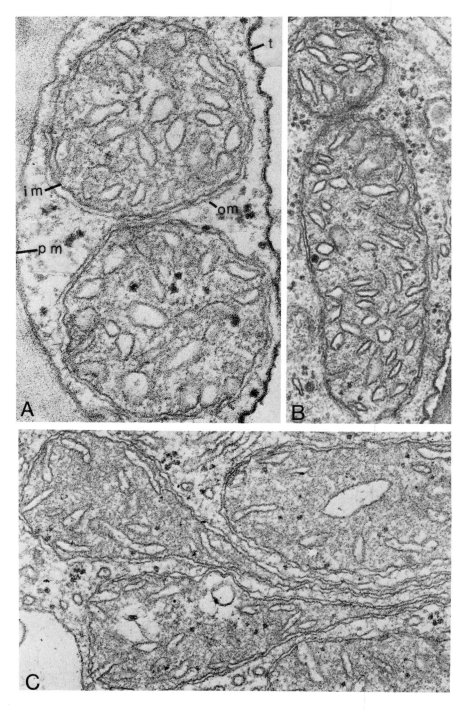

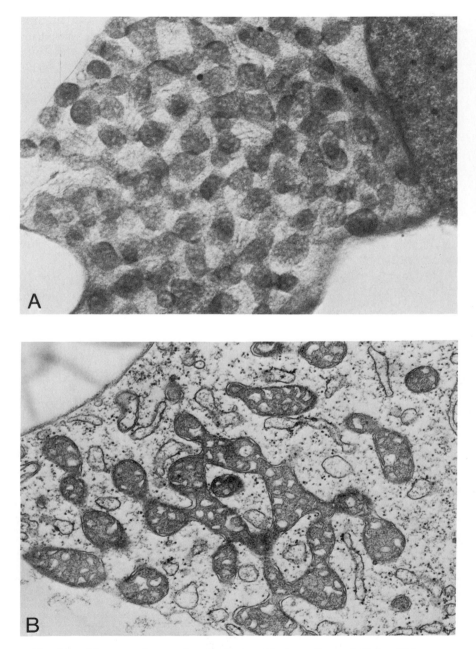

Fig. 1.2.   Electron micrographs of mitochondria in sections of *Ginkgo biloba* mega-gametophyte. (A) Thick section (×25,000) and (B) ultrathin section (×50,000). Note in this cell the existence of a single reticular mitochondrion. (Courtesy of J. P. Carde and R. Rohr.)

structure, while in reality mitochondria are highly dynamic structures. Thus, a careful examination of living plant cells by phase-contrast microscopy and time-lapse cinematography demonstrates both the diversity of form and the dynamic characteristics of mitochondria. These cell organelles can be seen continuously changing their shape, that is, becoming threadlike, branched, or globular, sometimes fusing with one another, or splitting up into smaller organelles (Frédéric and Chèvremont, 1952; Honda *et al.*, 1966). Since the mechanism and physiological significance of these movements remain largely unexplained, we now need to establish more precisely how the controls of mitochondrial shape or fusion operate. We believe that the mitochondrial micromorphology is very delicate and that mitochondria are the organelles that are the most sensitive indicators of outside influence. The observations that 2,4-dinitrophenol and ATP cause pronounced changes in mitochondrial forms and movements, as does anaerobiosis, strongly suggest that the shape and probably the volume of mitochondria *in situ* are linked to respiration and phosphorylation (Frédéric, 1958). In addition, it is possible that the shape of the mitochondria is partially attributable to local cytoplasmic constraints because isolated mitochondria are spherical.

2. Numbers

The number of mitochondria per cell in higher plants is likely to be in the hundreds of even thousands depending on the size and type of cell and the extent of cell differentiation. In stereological analysis of electron micrographs of sycamore cells during their exponential phase of growth, in which mitochondria are spheres or small rods, we have found 250 mitochondria per cell. These mitochondria contain from 6 to 7% of the total protein of the cell and occupy, on the average, 0.7% of the total volume of the cell (including the vacuole space) and some 7% of the total cytoplasmic volume. In very active plant cells, like secretory cells in nectaries (Gunning and Steer, 1975) or anther stalks of grasses (Ledbetter and Porter, 1970), the proportion is much higher and can match that of animal cells. The cytoplasm of very active cells, like transfer cells and companion cells (i.e., cells that have the greater sustained demands for energy transduction in connection with the secretory process or the transport of solutes at the cell surface), is richly endowed with mitochondria (Gunning and Steer, 1975). Likewise the abundance of large mitochondria in bundle sheath cells of NAD-malic enzyme-type C_4 plants, which show elaborate development of the inner membrane (Hatch *et al.*, 1975), could be related to the demands of metabolite transport. Mitochondria are very numerous in animal cells, and it has been estimated that mitochondria occupy about 20% of the volume of mammalian cells. Circumstantial evidence suggests that plant cells contain fewer mitochondria per cell but have a higher rate of mitochondrial O_2 consumption than animal cells (Table 1.1).

Guillermond (1924) was the first to realize that fundamental events during meiosis were not restricted solely to the chromosomes. With a series of co-workers, he demonstrated that changes in the "chondriosome" accompanied the male reduction division in several plants and interpreted them as evidence of the continuity of chondriosome elements through to the cytoplasm of the male gametophyte (see Heslop-Harrison's review, 1971). Likewise, detailed studies by Chesnoy (1974), Tourte (1975), Dickinson and Potter (1978), and Medina *et al.* (1981) demonstrated the continuity of mitochondria through to the cytoplasm of the female gametophyte. During the course of female gametogenesis, mitochondria undergo a large number of changes in a short time (from late zygotene to the end of meiosis), implying a preparation of the cytoplasm of the female germinal cell for the expression of the zygote genome. In fact, during the fertilization process it is the female cytoplasm, not the male, which is the zygote cytoplasm. Thus mitochondria could theoretically be traced back through generations of cells and organisms, maintaining uninterruptedly some structural principles for millions of years.

TABLE 1.1

Uptake Rates (State III) for Mitochondria Isolated from Various Tissues or Cells

Organism	Succinate[a]	α-Ketoglutarate[a]	NADH[a]
Guinea pig liver[b]	91	29	0
Beef heart[b]	117	85	0
Potato tubers[c]	505	257	328
Mung bean hypocotyls[d]	451	—	510
Spinach leaves[e]	238	—	315
Neurospora crassa[f]	260	140	450
Saccharomyces cerevisiae[f]	195	140	525

[a] Units: nmoles of O_2/min/mg protein.
[b] Wainio, 1970.
[c] Neuburger *et al.*, 1982.
[d] Douce *et al.*, 1972.
[e] Nishimura *et al.*, 1982.
[f] Lloyd, 1974.

In very young cells (meristematic cells) containing large amounts of 80 S ribosomes, dumbbell shaped mitochondria (dividing mitochondria?) are frequently encountered (Fig. 1.3). In these cells the divisions maintain the level of the mitochondrial populations. In differentiating cells, it will be interesting to see what quantitative relationships may develop between cell surface area, tonoplast area, and rate of respiration on the one hand and the number of mitochondria (or mitochondrial volume) on the other. It is possible that the number of mitochondria per unit volume of cytoplasm may be maintained while the cell enlarges (Clowes and Juniper, 1968).

3. Intracellular Location and Interaction with Various Membrane Systems

Observations of animal cells (for a review see Munn, 1974) indicate that mitochondria are frequently located near a supply of substrate, near points in the cytoplasm known to require ATP generated by the mitochondria, or near cell membranes involved in transcellular active transport (see, for example, the close association between mitochondria and the contractile portion of myofibril in striated muscle cells, between mitochondria and the plasma membrane of the proximal tubules of kidney, between mitochondria and lipid droplets in pancreatic acinar cells, and between mitochondria and the midpiece of the spermatozoon). This permits ATP or substrates to be utilized for cell function through a very short diffusion path.

In sharp contrast, observations of plant cells indicate that mitochondria are distributed randomly throughout the thin film of the cytoplasm, thus favoring all cell regions equally. For instance, they do not lie closely juxtaposed to the plasma membrane or to the tonoplast where massive energy requirements arise in connection with the transport of anions and cations. However, an aggregation of filamentous structures in a paracrystalline array associated with a stack of mitochondria has been observed in the tapetal cells of *Hyacinthoides non-scripta* (Jordan and Luck, 1980). The possibility that the position of mitochondia within cells is controlled somewhat by specific elements is suggested by this observation.

At one time or another, plant mitochondria have been reported as being associated with nearly all the other cellular organelles. In many such cases, these apparent associations are probably casual. Several reports have described close association of mitochondria with the chloroplast outer envelope (for a review see Douce and Joyard, 1979). A report based on light microscopy of living cells by Wildman *et al.* (1966)

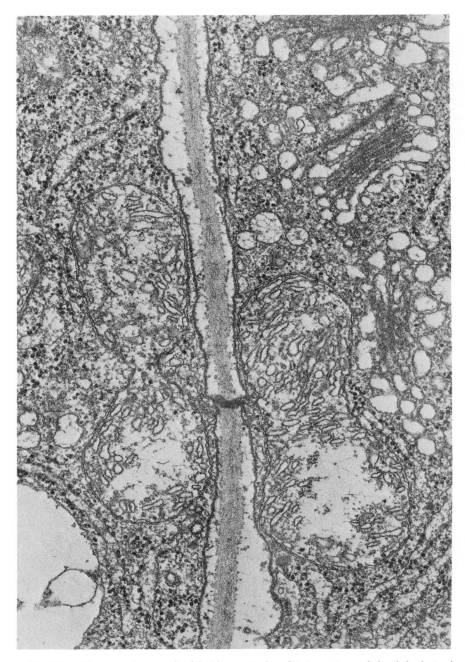

Fig. 1.3.   Electron micrograph of dividing mitochondria in sections of glandular hair of *Convolvulus arvens* leaf (×47,500). (Courtesy of J. P. Carde.)

suggested that long envelope-bound protuberances could detach from the chloroplasts to form mitochondria. However, this concept has been justly criticized (E. A. Chapman *et al.*, 1975). More recently Wildman *et al.* (1974) have proposed that mitochondria may fuse completely with the chloroplast envelope. It seems likely that at least some of the chloroplast–mitochondria interactions reported by Wildman's group may be easily explained by the formation and loss of deep mitochondria-containing invaginations in the chloroplast envelope (Montes and Bradbeer, 1976; Fig. 1.4). In *Panicum* species of the Laxa group some mitochondria in leaf bundle sheath cells are surrounded by chloroplasts when viewed in profile. Complete enclosure rather than invagination is indicated by the absence of two concentric chloroplast membranes surrounding the mitochondrial profiles (Brown *et al.*, 1983). It is clear that this "enclosure" of mitochondria raises many questions about their origin and function. Do these mitochondria form inside chloroplasts? If so, is the formation concurrent with or after plastid formation? Do enclosed mito-

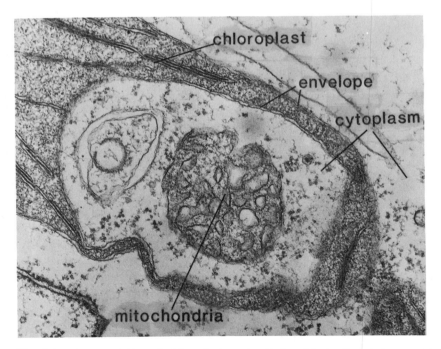

Fig. 1.4. An association of chloroplasts and mitochondria in *Funaria hygrometrica* (×40,000). Mitochondria are sometimes found in deep invaginations situated near the periphery of chloroplasts. (Courtesy of F. Nurit.)

chondria perform the same metabolic role as those outside the chloroplasts? Is their role related to the low photorespiratory loss of CO_2 in some species of the Laxa group?

In leaves, the peroxisomes are usually found closely associated in a transient fashion, with the outer mitochondrial membrane (Gruber *et al.*, 1970; Frederick and Newcomb, 1969; Black *et al.*, 1976; Fig. 1.5). Although inference of a physiological relationship from results of this sort may be premature in the absence of more conclusive information, it is possible that the transfer of metabolites between organelles could be greatly facilitated by the proximity of boundary membranes. Furthermore, the close association between mitochondria and various cellular organelles observed in most electron micrographs of green leaf cells may indicate the cooperative action of enzymes in chloroplasts, peroxisomes, mitochondria, and the cytoplasm in which they lie, which would allow sufficiently high rates of carbon flow for photorespiration (Tolbert, 1971; Chollet, 1977).

In a careful study of cell membranes in the fern *Pteris vittata*, Crotty and Ledbetter (1973) found membrane continuities between the outer membrane of the mitochondrion and the smooth endoplasmic reticulum. Morré *et al.* (1971) described similar observations in both plant and animal cells. However, the infrequency with which direct endoplasmic reticulum–mitochondrial outer membrane associations have been observed leads us to question the significance of membrane continuities between mitochondria and endoplasmic reticulum. Since the outer membrane of the mitochondrion and the endoplasmic reticulum are quite dissimilar, some local specialization at the molecular level would be needed to permit the fusion of the two. A further complication is the possibility of artifactual membrane fusions in the course of fixation for electron microscopy (Haussmann, 1977). This clouds the issue of whether the endoplasmic reticulum cisternae can fuse, even temporarily, with the outer mitochondrial membrane in order to allow the transfer of membrane components from its cytoplasm to the mitochondria and *vice versa*. Permanent junctions between mitochondria and a specific intracellular membrane system are even more unlikely, because plant mitochondria are mobile *in vivo* by passive and/or active movements within the streaming cytoplasm.

In some plants specific association of virus particles with the outer mitochondrial membrane have been noted. For example, in the palisade and spongy mesophyll of leaf tips of *Nicotiana levelandi* examined 5 days after inoculation with tobacco rattle virus, 80% of the mitochondria had large numbers (100–500 per mitochondrion) of virus particles associated

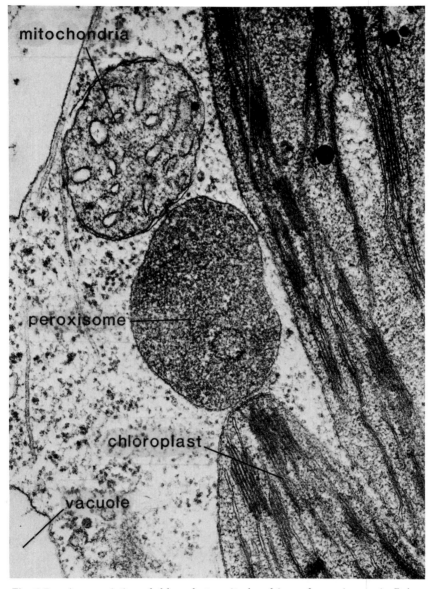

Fig. 1.5. An association of chloroplasts, mitochondria, and peroxisomes in *F. hygrometrica* (×54,000). Peroxisomes are very often appressed to the outer membrane of the mitochondrion. (Courtesy of F. Nurit.)

with them. The particles were attached to the outer membrane and radiated from it (Harrison and Woods, 1966; Harrison and Roberts, 1968). In tomato (*Lycopersicum esculentum*) pollen grains infected with pepper ring spot virus, the particles form ordered aggregates aligned perpendicularly to the mitochondrial surface (Camargo *et al.*, 1969). Likewise, monolayers of beet mosaic virus have been noted surrounding mitochondria in infected *Gomphrena globosa* seedlings (Russo and Martelli, 1969). It is possible that the outer membrane is a site of replication for some viruses.

B. Ultrastructure of Plant Mitochondria

All plant mitochondria examined so far correspond to the basic pattern first deduced by Palade (1952, 1953) and Sjöstrand (1953). Each plant mitochondrion possesses two membranes and two membrane-bound compartments: a smooth outer membrane surrounds and completely encloses a peripheral inner membrane that in turns encloses a protein-rich inner space called the matrix. Lying in the matrix are a variable number of membranous structures called cristae (or cristae mitochondriales; Palade, 1952) that appear either to be associated with the inner membrane or free. Despite the fact that cristae may be abundant, it is only rarely that a section through a plant mitochondrion shows an actual point of continuity between the membrane of a crista and the inner element of the two surrounding membranes (Fig. 1.1). A count of 50 mitochondrial profiles from a variety of plants made by Öpik (1974) yielded 1029 cristae, of which only 9% showed the junction. The inference is that the invaginations must join to the rest of the inner membrane only at narrow necks. There are no continuities between outer and inner mitochondrial membranes; in many places, however, there are electron dense areas that seem to cement both membranes. It is possible that the intermembrane space may be maintained by electrostatic repulsions between the opposed electronegative surfaces of the two mitochondrial membranes (Hackenbrock and Miller, 1975).

The pattern of membrane infoldings varies between tissues and plant species. On the basis of their appearance in plant mitochondria fixed *in situ*, cristae may be broadly classified as saccular (dilated cristae) or tubular. The dilated cristae form, irregularly oriented sacs and folds, is the most common (Fig. 1.1). In contrast, animal mitochondria infoldings are mainly lamellar (i.e., platelike). In plant cells the cristae show no preferred orientation (Fig. 1.1). As pointed out judiciously by Öpik (1968), the two most commonly employed fixatives, potassium permanganate and osmium tetroxide, can give somewhat differing pictures of plant mitochondrial structure. After permanganate fixation, membranes are very smooth and regular and the intracristal spaces are narrow, whereas with osmium the membranes often appear more wavy and diffuse. In addition, in osmium-fixed mitochondria the width of the intracristal space varies; it may be practically as narrow as with permanganate fixation or dilated to varying degrees. Consequently, interpretation of these observations must be undertaken with extreme caution because of the danger of fixation artifacts. Hackenbrock (1968) has shown that rat liver mitochondria exhibit several distinct morphologies under the electron microscope. The actively respiring mitochondria (state 3 respiring) have a condensed matrix with a volume approximately one-half that of the matrix of nonrespiring mitochondria (state 4 resting). It is thus possible that the swollen-cristae dark matrix form could represent *in vivo* actively respiring mitochondria.

Since the enzymes of electron transport and oxidative phosphorylation are embedded in the membranes of the cristae, it is clear that plant mitochondria with profuse cristae are capable of higher rates of respiration than mitochondria that have relatively sparse cristae. It is also obvious that the large surface area presented by the cristae may also be related to the activity of the anion and proton transport mechanisms of the inner mitochondrial membrane. Finally, it is possible that an active development of a number of cristae allows a more complete extraction of traces of O_2 from an environment depleted in the gas. In support of this last suggestion, Vartapetian *et al.* (1974) have shown that the high resistance of rice coleoptile cells to anaerobiosis is reflected not only in their ability to preserve mitochondrial apparatus when growing in strictly anaerobic conditions but also in an especially active development of a number of mitochondrial cristae. It is interesting to note that in this material a considerable degree of elaboration of mitochondrial structure occurs under anaerobiosis, contrasting with the situation for yeast in which mitochondria of normal structure are formed only in aerobic conditions (Öpik, 1973). It is interesting to make an estimate of the surface area of the mitochondrial membranes of a plant cell in relation to the surface area of the plasma membrane. In the case of cultured sycamore cells, we can assume for the sake of simplicity that the cell is a smooth sphere with a diameter of 30 μm containing 250 cylindrical mitochondria of average dimension 2 μm × 1.0 μm. The cellular surface area is thus about 2800 μm^2 and the total mitochondrial outer surface area in one cell is 2000 μm^2. Since in these mitochondria the area of the inner membrane is about three times that of the outer, the total mitochondrial inner membrane surface area in one cell is 6000 μm^2, or more than two times the surface of the plasma membrane.

1. Structure of Mitochondrial Membranes

The mitochondrial outer membrane is approximately as thick as the inner membrane, which is thinner than the tonoplast (6 nm) or plasma membrane (7 nm) (Fig. 1.1). The thickness of mitochondrial membranes is comparable to that of other organelles, including plastids, peroxisomes, and endoplasmic reticula. When osmium tetroxide is used as the sole fixative, the classical tripartite dark–light–dark structure of mitochondrial membranes (which may reflect the bimolecular layers of lipids in the membranes) is often not easily discernible. This is especially true for the lipid-rich outer membrane, which very often appears faint and even exhibits local breaking when the inner membrane still appears intact. Addition of 1% (w/w) tannic acid to the specimens just before dehydration, however, considerably enhances the contrast of mitochondrial membranes (Carde *et al.*, 1982; Fig. 1.6). By its complementary fixative action on membrane proteins (Futaesaku *et al.*, 1972) and phosphatidylcholine molecules (Kalina and Pease, 1977), tannic acid seems to stabilize the mitochondrial membranes against massive lipid extraction during dehydration and subsequent processing. Unfortunately, metal impregnation is destructive to proteins, and the thin-section method, therefore, cannot be used for detailed investigations of macromolecular structure. In fact, the continuous lipid bilayer backbone proposed in the unit membrane hypothesis (Robertson, 1960) may result from artifactual modifaction of the membrane structure by extensive denaturation of membrane proteins (Sjöstrand and Barajas, 1968). It is very likely that the inner mitochondrial membrane is composed of two regions of macromolecular distinction, the cristal membrane and the inner boundary membrane. These morphologically distinct membranes differ in their relative binding affinities for polycationic ferritin and thus in their anionic macromolecular structure (Hackenbrock, 1975).

Freeze-etching reveals detailed ultrastructure that is not recognized by the usual procedures of electron microscopy, including a granular fracture face (interior) of both inner and outer membranes. Unlike most other electron microscopic preparation techniques, freeze-etching does not require the use of chemical fixatives and dehydrating agents. In this method, tissues are rapidly frozen and then fractured with a microtome knife under vacuum; some of the frozen water is sublimed to a depth of

Fig. 1.6. Electron micrographs of mitochondria in sections of (A) *Portulaca oleracea* leaf (×71,500) and (B) secretary duct of *Foeniculum dulca* hypocotyl (×97,000). After a sequential fixation by glutaraldehyde, osmium tetroxide, and tannic acid, the dark–light–dark construction (arrows) of both mitochondrial membranes is well preserved. (Courtesy of J. P. Carde.)

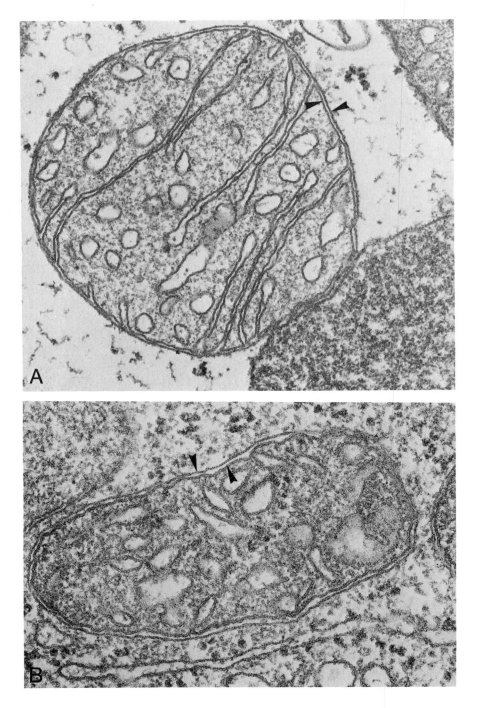

a few hundred angstroms (etching). The etched surface is then shadowed (platinum deposition), replicated, and examined by the usual electron microscopic techniques. By this method, the topography of the cleaved surface is revealed. Its main advantage is that surface views, as well as cross sections of chemically unaltered membranes in their natural surroundings, can be obtained (Staehelin and Probine, 1970). Using this method, an asymmetric distribution of particles had been observed in both outer and inner membranes, each with characteristic particle densities. The total number of particles in the outer membrane is approximately 2500 per μm^2, while that in the inner membrane is 6400 per μm^2 (Branton, 1969; Packer et al., 1973). The highest concentrations of particles are located on the sides of the membrane that face the protein-rich, soluble compartments, either the cytoplasm for the outer membrane or the matrix for the inner membrane. Conversely, the sides of the two membranes that face the intermembrane space contain fewer particles, suggesting that the transverse organization of both membranes is asymmetric. The particles observed in both mitochondrial membranes probably represent the integral proteins described by Singer and Nicholson (1972). It is extremely likely that the differences present in the ultrastructure of inner and outer membranes of the plant mitochondria as revealed by freeze-fracturing reflect functional and compositional differences, since it is assumed that membrane particles represent enzyme complexes of various sorts associated with the functioning of the membrane. The presence, however, of such particles in replicas prepared from protein-free systems underlines the fact that membrane-associated particles observed in freeze-fracture replicas of cell membranes cannot a priori be regarded as arising from intrinsic membrane proteins (Gounaris et al., 1983; for review see Quinn and Williams, 1983).

Another electron microscopic technique that has provided information on plant mitochondrial structure is negative staining. This method serves as a valuable adjunct to the classical electron microscopic method of ultrathin sectioning. A suspension of mitochondria that has been sufficiently disrupted to permit the stain to enter is mixed with a solution of potassium phosphotungstate ($P_2O_5 \cdot 24WO_3 \cdot nH_2O$) at neutral pH, and a thin film of the mixture is dried onto the electron microscope specimen grid. Since at neutral pH potassium phosphotungstate has little affinity for proteins or lipids, membrane structures are unstained and remain transparent to electrons, while the space between the structures is filled with a film of electron opaque phosphotungstate (Parsons, 1967). Using the negative staining technique, complex subunits can be seen on the cristae of all mitochondria isolated so far (Wainio, 1970; Munn, 1974). In higher plant mitochondria (Parsons et al., 1965), the subunits have a "head" 90–110 Å in diameter and a "stem" measuring 35–40 by 45 Å. The center to center spacing of the particles is about 100 Å, and the inner membrane is completely covered with 90- to 110-Å subunits. In addition, it has been clearly shown that the surface bearing the subunits is the matrix-facing surface. It is interesting to note, however, that Wriggelsworth *et al.* (1970) found no evidence by freeze-etching for the existence of stalked knobs on mitochondrial inner membranes that are observed with negatively stained preparations. Malhotra and Eakin (1967) have reported that in mitochondria isolated from wild-type *Neurospora crassa*, the particles are not firmly bound to the inner membrane since they are readily removed by 1 mM EDTA. It is obvious now that these inner membrane particles do not represent the morphological substrate of the electron transport chain [electron transport particle (ETP) or oxysomes], as originally claimed (Fernández-Morán *et al.*, 1964). Rather, each complex subunit probably represents an energy-transducing ATPase molecule (ATPaseF$_1$) associated with the electron transport chain. This ATPaseF$_1$ is postulated to function in the terminal transphosphorylation step of oxidative phosphorylation by catalyzing the synthesis of ATP from ADP and phosphate (P$_1$).

Finally, electron microscopic observations of negatively stained outer membranes of plant mitochondria by Parsons *et al.* (1965) suggested a physical basis for the high small-molecule permeability of this membrane. The surfaces of outer membranes of plant mitochondria contain dense (but apparently random) arrays of negative stain–accumulating sites ("pits," each ~3 nm in diameter), which could represent hydrophilic openings or channels through the membranes. Mannella and Bonner (1975) and Mannella (1981) subsequently found that outer membranes isolated from plant mitochondria display X-ray diffraction patterns consistent with random arrays of subunits having in-plane dimensions like those of the subunits seen in electron micrographs.

2. Mitochondrial Matrix

The plant mitochondrial matrix, which contains closely packed protein molecules (0.4 g protein/ml), can exhibit several distinct morphologies under the electron microscope. After osmium fixation, particularly when preceded by glutaraldehyde, the matrix appears electron dense and finely granular with electron-lucent interstices (Figs. 1.1 and 1.7). When permanganate is used as the fixative, matrix structure is generally lost. This heterogeneity may arise either (a) as an osmotically-induced artifact during processing (swelling involves the unfolding of the inner membrane and a decrease in the amorphous matrix, while

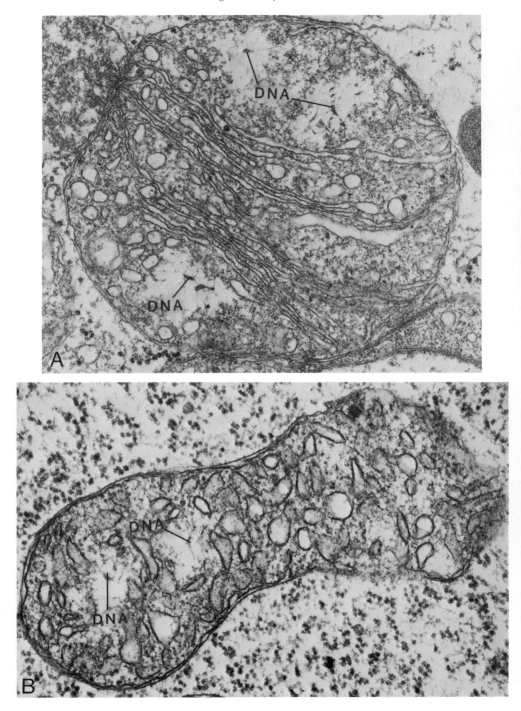

contraction involves the infolding of the inner membrane and an increase in matrix density) or (b) as a consequence of different metabolic states of individual mitochondrion at the time of fixation. Thus, mitochondria in bundle sheath cells of NAD^+-malic enzyme C_4 species are generally large and densely stained with very translucent cristae (E. A. Chapman *et al.*, 1975). They do not resemble mitochondria of mesophyll cells. In spinach leaf mitochondria a relatively higher proportion of the protein is in the matrix space compared to petiole mitochondria; consequently, leaf mitochondria have a somewhat lower lipid–protein ratio than petiole mitochondria (Gardeström *et al.*, 1983). Heterogeneity may also arise as a consequence of mitochondria being at different phases in a growth and division cycle. For example, mitochondria from fast-growing cells tend to have fewer cristae and more "open space" in their matrix than organelles from functionally differentiated and active tissues such as leaves (Fig. 1.8). The former mitochondria could contain less protein than the latter, which would be reflected in lower density values. Mitochondria of young etiolated maize shoots occur in a morphological continuum ranging from a dilute matrix with few cristae (Type I) to those with a dense matrix and extensive cristae (Type III) (Malone *et al.*, 1974; Chrispeels *et al.*, 1966). Matrix density development in plant mitochondria parallels cell development (Fig. 1.8).

In thin sections, DNA is usually localized in several electron transparent areas of the organelle matrix (Nass *et al.*, 1965; Kislev *et al.*, 1965; Swift *et al.*, 1968; Mikulska *et al.*, 1970) and appears as a network of finely dispersed fibrils (nucleoids) (Fig. 1.7), an observation that correlates with the biochemical evidence that mitochondria can contain multiple copies of their DNA. Multiple nucleoids are, however, not always present; mitochondria in *Beta* leaves contain just one (Kowallik and Hermann, 1972). As a general rule, certain very large mitochondria (up to 10 μm) contain several groups of DNA filaments in one section, whereas smaller mitochondria contain only one DNA filament. The structure of the filamentous component is dependent on the fixation procedure, being partially clumped into an opaque mass from which finer strands may radiate with osmium tetroxide but finely filamentous with Kellenberger's fixative (Kellenberger *et al.*, 1958). The filaments (highly twisted DNA molecules?) are electron dense when the specimens are treated with uranyl acetate and can be extracted with DNase after formalin fixation (Kislev *et al.*, 1965). The DNA filaments are more evident in the mitochondria of rapidly dividing cells than in the mitochondria of ma-

Fig. 1.7. Electron micrographs of mitochondria in sections of (A) *Portulaca oleracea* leaf (×71,500) and (B) spore of *Funaria hygrometrica* (×70,000) showing the prominent DNA filaments. (Courtesy of J. P. Carde and F. Nurit.)

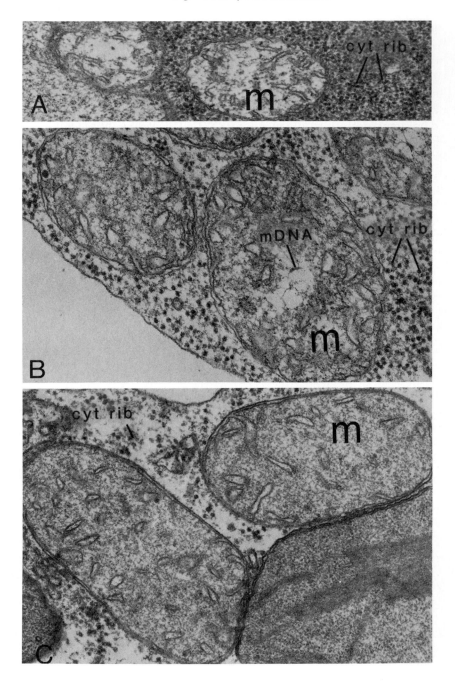

ture plant tissues, such as the differentiated zone of elongation (Swift *et al.*, 1968; Swift and Wolstenholme, 1969). Several authors have suggested that the DNA is in some way attached to cristal membranes (Nass, 1969; Swift *et al.*, 1968; Öpik, 1974). This is an attractive hypothesis because distribution of DNA could result from splitting of the attachment point (where replication can occur) by intercalating inner membrane growth in a way similar to that postulated for bacteria (Jacob *et al.*, 1963). It is clear that a great deal of research is needed before the notion that mitochondrial DNA is associated with the inner membrane can be asserted with any degree of confidence. The present lack of knowledge regarding DNA synthesis in plant mitochondria does not help the situation. Furthermore, the mechanical aspects of DNA replication in plant mitochondria (initiation points, active or passive unwinding mechanism, and swivel points) may remain obscure for some time, and, if they exist, the nature of the DNA–membrane attachments remains to be elucidated.

Ribosomelike particles about 150 Å in diameter are present in low numbers in the matrices of higher plant mitochondria (Kislev *et al.*, 1965; Öpik, 1974; Gunning and Steer, 1975; Fig. 1.9). These particles are not seen after treatment of the sections with RNase. The arrangements may appear random or they can form small groups, possibly polysomes. Sometimes the mitoribosomes appear associated with the inner membranes or cristae (Öpik, 1974) and could be involved in the synthesis of the inner membrane. However, the conditions for detecting these bound ribosomes, if they exist, may be difficult to control, since ribosome association with the inner membrane may be restricted in both time and space. In higher plants, the mitochondrial ribosomes are smaller than their cytoplasmic counterparts (Fig. 1.9) and apparently account for less than 1% of the total cellular ribosome population (Leaver, 1979).

Other structures are occasionally seen in the matrix of plant mitochondria. Protein crystals like those developed in peroxisomes of certain plant cells have been seen in meristematic cells of *Pisum sativum* mitochondria (Leak, 1968) and in lentil (*Lens culinaris*) leaf epidermal cell mitochondria (Lance-Nougarède, 1966). In the ungerminated rice coleoptile and in radicle-meristem cells of ungerminated bean seeds, mito-

---

Fig. 1.8. Electron micrographs of mitochondria in sections of mesophyll cell of spinach leaf showing the morphological appearance of mitochondria during the evolution from young leaves to mature leaves. (A) Young cell (×45,000); (B) greening cell (×56,000); and (C) mature cell (×56,000). Note the high density of 80 S ribosomes in the cytoplasm (cyt rib) of young cells. Matrix (m) density development parallels cell development. (Courtesy of J. P. Carde.)

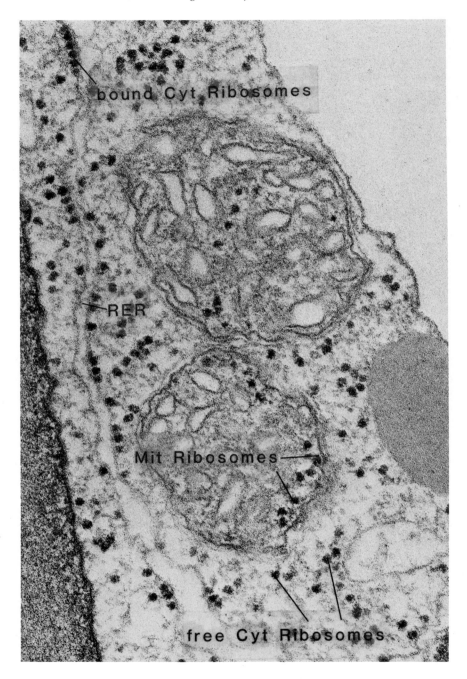

chondrial profiles often contain a compact electron-dense area, up to about 0.3 μm in diameter, surrounded by small granules that are probably mitoribosomes (Öpik, 1974). The physiological significance of these electron-opaque inclusions (storage proteins?) is entirely unknown. Numerous dense osmiophilic granules of insoluble phosphate, probably hydroxyapatite, originally observed in the matrix of some animal mitochondria are occasionally encountered in plant mitochondria observed *in situ*. In this case, energy for Ca^{2+} plus P_i uptake may be provided by oxidizable substrates or ATP (for review see R. H. Wilson and Graesser, 1976). Likewise, mitochondria from beans accumulate strontium, with the concomitant formation of many dense granules (Ramirez-Mitchell *et al.*, 1973).

II. THE STRUCTURE OF ISOLATED PLANT MITOCHONDRIA

A. Isolation of Intact Mitochondria

In plant cells, the active cytoplasm occupies a peripheral shell between the central vacuole and a rigid cell wall. The vacuole is a large watery compartment surrounded by a thick membrane (tonoplast) and containing a wide variety of substances harmful to other organelles (flavonoids, various colorless phenolic compounds, etc.) and hydrolases (lipases, proteinases, etc.). In addition, terpenes, resin acids, and other secondary products will strongly interfere with specific tissues during the isolation of organelles. When plant tissue is homogenized in order to isolate mitochondria, cellular compartmentalization is destroyed and the "secondary products" are released into the medium with effects ranging from undesirable to devasting depending on the tissue and on the isolation technique. For example, once released, the various phenolic compounds and their oxidation products (*o*- and *p*-quinones) interact immediately with the membrane-bound proteins (hydrophobic interactions, hydrogen bonding, ionic interactions, and oxidation of protein functional groups by the quinones; see Loomis, 1974) giving rise either to an uncoupling of oxidative phosphorylation (Van Sumere *et al.*, 1972; Stenlid, 1970; Weinbach and Garbus, 1966; Verhaeren, 1980; Ravanel *et al.*,

Fig. 1.9. Electron micrograph of mitochondria in section of xylem companion cell of *Hordeum albostrians* leaf (×102,000) showing numerous ribosomes. Note that the mitochondrial ribosomes are slightly smaller than their cytoplasmic counterparts. Abbreviation: RER, rough endoplasmic reticulum. (Courtesy of J. P. Carde.)

1981, 1982) or to an inhibition of the electron transport chain (Ravanel *et al.*, 1982, 1984). Likewise, during the course of the preparation of pine needle mitochondria, β-pinene (a monoterpene) released from the resin canals adheres to the mitochondrial membranes and inhibits, in a non-reversible fashion, electron flow at the level of the quinone pool (Pauly *et al.*, 1981). 4-Deoxyphorbol triester, a poisonous constituent of the latex sap of *Euphorbia biglandulosa*, was found to be a potent inhibitor of oxidative phosphorylation in mitochondria (Noack *et al.*, 1980). Enzymes that liberate free fatty acids such as linoleic and linolenic acids, the two major fatty acids of plant cell membranes (Stumpf, 1980), are potentially the most troublesome contaminants in mitochondrial preparations from plants. In most cases, these fatty acids are produced by the actions of lipolytic acyl hydrolases (vacuolar enzymes?) released during the course of the tissue grinding. These enzymes, which are strongly activated by linoleic acid, cause a very rapid hydrolysis of mitochondrial phospholipids, leading to functional impairment (uncoupling of oxidative phosphorylation; Douce and Lance, 1972; Bligny and Douce, 1980). In addition, unsaturated fatty acids are the best natural substrates for lipoxygenase (an enzyme that catalyzes the addition of O_2 to linoleic acid), producing harmful fatty acid hydroperoxides. In the presence of O_2 these hydroperoxides react strongly with all membrane proteins, including cytochromes, leading to a complete destruction of the protein (Haurowitz *et al.*, 1941; Dupont *et al.*, 1982; Dupont, 1983). Since these enzymes do not have metal ion requirements and do not show sensitivities to inhibitors suitable for use in the preparation of mitochondria, inhibition of these enzymes is difficult during the grinding of tissue. Apart from problems associated with the presence of inhibiting substances in the vacuole, there is a need for strong shearing forces to disrupt the rigid cell wall. These forces are also detrimental to the mitochondria. For example, according to the interesting suggestion of Pradet (1982), the occurrence of injury upon isolation of mitochondria from seeds during early germination might be the cause for the poor oxidative phosphorylation capacity of isolated mitochondria from dry seeds (Nawa and Asahi, 1973; Morohashi *et al.*, 1981; Sato and Asahi, 1975). Pradet's concern led us to doubt whether the apparent improvement of mitochondrial properties during germination of seeds represents a true formation of mitochondria or protein synthesis or whether it should be considered as an artifact.

Consequently, considerable expertise is required in order to prepare mitochondria displaying the same biochemical and morphological characteristics as mitochondria *in vivo*. It is difficult, if not impossible, at present to prepare 100% intact and fully functional mitochondria; it is a simple matter to prepare a "mitochondrial fraction" (i.e., a subcellular fraction containing mitochondria). Although, many methods that are currently employed yield preparations containing an appreciable proportion of intact mitochondria, they also result in substantial contamination by extra mitochondrial components and may lead, in consequence, to ambiguous and possibly misleading results. Thus, plant mitochondria have been reported to differ from animal mitochondria in characteristic ways, such as being less tightly coupled (Ikuma, 1972). Packer *et al.* (1970) concluded that these differences are due, in every instance, to damage to the plant mitochondria during the isolation procedure. These problems are aggravated by the diversity of plant material from which mitochondria are isolated and the lack of an accepted standard method of isolation. Finally, we must point out that all mitochondrial preparations from plant organs, such as roots or leaves, containing several cell types will contain a mixed population of mitochondria from the different cell types (Öpik, 1968).

1. Preparation of Washed Mitochondria

If the investigations to be undertaken do not demand a particular species, difficulties can usually be minimized by careful choice of plant material (cauliflower buds, white and sweet potato tubers, castor bean endosperm, spinach and pea leaves, etc.). Fresh material (0.5–1 kg) is cut into 3 liters of chilled medium containing 0.3 M mannitol, 4 mM cysteine, 1 mM EDTA, 30 mM 3-(N-morpholino)propane sulfonic acid (Mops) buffer (pH 7.5), 0.1% (w/v) defatted bovine serum albumin, and 0.6% (w/v) insoluble polyvinyl pyrrolidone (PVP). Mitochondria require an osmoticum for the maintenance of their structure; sucrose, sorbitol, or mannitol are the most commonly used. Since sorbitol or mannitol are not usually metabolized, they have an advantage over sucrose. Interestingly, osmotic inhibition of respiratory activity was found to be reversible on transfer of mitochondria from a hypertonic to an isotonic medium (Olabiyi *et al.*, 1983). In addition, sugar stabilized isolated mitochondria during freezing (Thebud and Santarius, 1981). Acidity, phenolic compounds, tannin formation, and oxidation products lead to the rapid inactivation of mitochondria. For example, the isolation of functional mitochondria from crassulacean acid metabolism (CAM) plants has two major problems: large amounts of phenolic substances are normally found in succulent leaves, and high concentrations of organic acids are associated with all CAM plants. These difficulties are usually overcome by adding alkaline buffers, insoluble PVP (Loomis and Battaile, 1966; Hulme *et al.*, 1964), cysteine (Bonner, 1967; Ikuma and Bonner, 1967a; Storey and Bahr, 1969), or ascorbate to the isolating medium.

Honda *et al.* (1966) recommended the use of media containing dextran, albumin, and Ficoll (Pharmacia) for the isolation of intact, predominantly rodlike mitochondria which retain the pleomorphism they show *in vivo.* It appears that insoluble PVP binds hydrogen ions as well as phenols. Other strategies to minimize the effects of vacuoles on plant mitochondria are to keep a low ratio of tissue to breakage medium (Bonner, 1967) and to separate the organelles from soluble elements of the cellular brei as rapidly as possible (Palmer and Kirk, 1974). In principle, the simplest way to prevent oxidation of plant phenolic compounds or fatty acids such as linoleic acid is to bubble an inert gas heavier than air, such as argon, into the medium and to avoid the use of blenders which whip air into the homogenates. The uncoupling of oxidative phosphorylation by free fatty acids (Pressman and Lardy, 1955) is usually overcome by adding relatively high concentrations of defatted bovine serum albumin (BSA) [0.1–1% (w/v)]. Fatty acids, such as oleic acid, inhibit ATP synthesis by releasing respiratory control without significantly affecting ATPase (Matsuoka and Nakamura, 1979). The beneficial effect of BSA, which is not fully understood [for example, Ducet (1978) believes that BSA acts at the inner membrane level to decrease proton permeability insofar as the outer membrane is damaged], is probably due to its ability to bind free fatty acids (Chen, 1967) formed by the action of acyl hydrolase enzymes during extractions (Dalgarno and Birt, 1963; Bligny and Douce, 1978; Diolez and Moreau, 1983). The well-known capacity of BSA to bind anions and lipids (the latter by hydrophobic forces) is consistent with its high content of hydrophobic amino acids and lysine (Spahr and Edsall, 1964). Bovine serum albumin appears also to be an effective phenol and quinone scavenger (Loomis, 1974). Addition of EDTA to extraction media inhibits the activity of Ca^{2+}-dependent phospholipase D and may reduce the activity of some lipases.

   Three different techniques are commonly employed for breaking open tissue cells: (a) macerating for a short period, at a moderate speed, in a motor-driven blender (e.g., a Waring blender, a Polytron, or a Moulinex), (b) mortar and pestle grinding, and (c) razor blade chopping. Invariably, the more violent the homogenization, the lower the quality of the isolated mitochondria. The shear forces must be great enough to break the cells without stripping the outer membrane from the mitochondria. In other words, tissue disruption should be carried out very gently (see, for example, Ikuma, 1970). A rapid and convenient method is to disrupt the tissue at low speed for 2 sec in a 1-gal Waring blender. The brei is rapidly squeezed through 8 layers of muslin and 50-μm nylon netting. After an initial low-speed centrifugation, the mitochondria are sedimented at 10,000 *g*, and, after washing, resedimented at 10,000 *g*. This method, when followed carefully, will produce mitochondria with a high degree of intactness (Fig. 1.10). The crucial point in the whole procedure is the grinding of the tissue. In order to obtain intact mito-

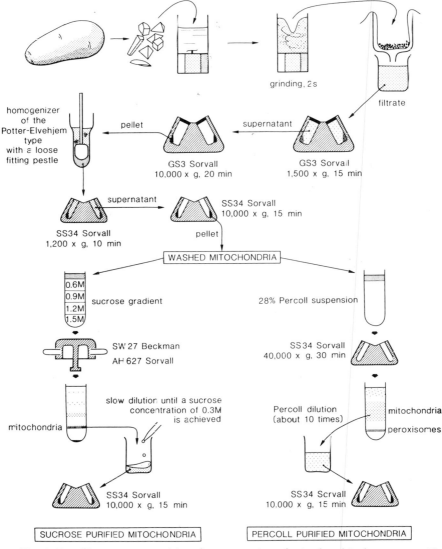

Fig. 1.10. Diagram summarizing the preparation of mitochondria from potato (*Solanum tuberosum*) tubers.

chondria, it is absolutely necessary to restrict the grinding procedure to a minimum. Longer blending improves the yield of recovered mitochondria but considerably increases the percentage of envelope-free mitochondria and of mitochondria that have resealed following rupture and the loss of matrix content. The latter derive from the filamentous mitochondria present in the cytoplasm of all plant cells (Fig. 1.1). Under these conditions, it stands to reason that variable amounts of harmful vacuolar enzymes and nuclear material might have become confined within the matrix space. It is clear, therefore, that the final yield of intact mitochondria capable of rapid substrate-dependent O_2 uptake with pronounced respiratory control and ADP/O ratios approaching "theoretical" limits is very low. For example, when the excellent classical method of Bonner, (1967) is used, we have shown that the maximum yield of intact mitochondria is roughly 1–3% of the total tissue mitochondria. A number of recent investigations have been encouraging, however, in demonstrating the potential usefulness of plant protoplasts for physiological research (Nishimura *et al.*, 1976; Edwards *et al.*, 1978; Hampp and Ziegler, 1980; Nishimura and Beevers, 1978; Robinson and Walker, 1979; W. Wirtz *et al.*, 1980). Cell wall materials are enzymatically degraded during the preparation of protoplasts, and subsequent gentle disruption of the protoplast plasmalemma is ideal for isolating intact mitochondria with a high yield (Nishimura *et al.*, 1982).

The criteria for the assessement of mitochondrial integrity, including (a) the respiratory rate in the presence of added ADP compared to the rate obtained following its expenditure (Chance and Williams, 1955), (b) the ADP/O ratio (Chance and Williams, 1955), (c) the permeability of the outer membrane to cytochrome *c* (Wojtczak and Zaluska, 1969; Douce *et al.*, 1972), (d) the latency of matrix enzymes such as malate dehydrogenase (Douce *et al.*, 1972) and fumarate hydratase, and (e) direct examination in the electron microscope, are frequently used as sensitive indicators of the success of the isolation. In our experience the best indicators of damage are succinate–cytochrome *c* oxidoreductase and cytochrome *c* oxidase (Douce *et al.*, 1972). The first assay is based on the reduction of added soluble cytochrome *c* by the mitochondrial cytochrome *c*, which is localized on the external side of the inner membrane. The reaction occurs after initiating electron transport with succinate in the presence of ATP (an activator of succinodehydrogenase) and cyanide (which prevents the oxidation of soluble cytochrome *c* by cytochrome oxidase). The outer mitochondrial membrane presents an impenetrable barrier to the added soluble cytochrome *c*. When intact, the outer membranes prevent externally added cytochrome *c* from interacting with the respiratory chain. With increasing outer membrane damage, however, there will be increasing succinate–cytochrome *c* oxidoreductase activity as the added soluble cytochrome *c* gains access to the outer surface of the inner membrane. The second assay is based on the measurement of KCN-sensitive ascorbate cytochrome *c*-dependent O_2 uptake. The reaction occurs after initiating O_2 uptake with soluble cytochrome *c* in the presence of ascorbate. If the mitochondria are fully intact, there will be no O_2 uptake. With increasing outer membrane damage, however, exogenous reduced cytochrome *c* has access to the outer surface of the inner membrane, where it is oxidized, triggering O_2 uptake (Neuburger *et al.*, 1982) (Fig. 1.11). The observations of Palmer and Kirk (1974) that the succinate–

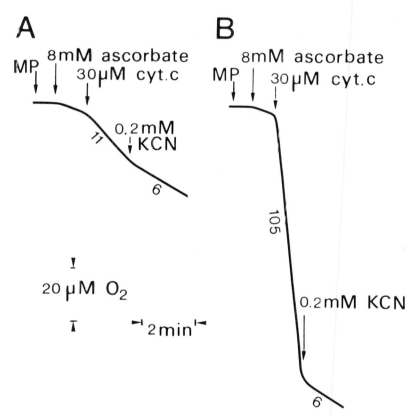

Fig.1.11. Ascorbate/cytochrome *c*-dependent O_2 uptake of Percoll-purified mitochondria. (A) Intact mitochondria and (B) burst mitochondria. The numbers on the traces refer to nmoles O_2 consumed/min (protein: 0.20 mg; final volume: 1 ml). Note that the rate of O_2 uptake by the intact mitochondria is 5% of that recorded for osmotically shocked mitochondria, giving an apparent percentage intactness of 95%. Abbreviations: MP, purified mitochondria; and cyt. c, cytochrome *c*. [From Neuburger *et al.* (1982).]

cytochrome *c* oxidoreductase activity of plant mitochondria in 0.4 *M* sucrose is ATP-activated and that the reduction kinetics of exogenous cytochrome *c* closely parallel the rate of O_2 consumption during succinate and NADH oxidation by these mitochondria do not imply penetration of intact outer mitochondrial membranes by cytochrome *c*. They demonstrate, instead, the well-known fact that most mitochondrial preparations contain a significant fraction of mitochondria with broken (or missing) outer membranes. It is the respiratory chains of this mitochondrial population that can interact with exogenous cytochrome *c*.

Even though the final preparation often contains more than 95% intact mitochondria, the material is not necessarily devoid of various extramitochondrial membranes and soluble enzymes (vacuolar enzymes, glycolytic enzymes, catalase, lipoxygenase, etc.). For example, a careful examination by electron microscopy of the washed mitochondrial pellets obtained by differential centrifugation shows that the mitochondria are always contaminated by smooth vesicles of unknown origin, intact and broken plastids, and peroxisomes (Fig. 1.12). The worst contamination encountered is most certainly vacuolar enzymes (proteinases, lipolytic acyl hydrolases, etc.) (Dahlhem *et al.*, 1982; Bligny and Douce, 1978). A number of investigators have provided evidence that the major portion of the soluble proteolytic activity present in enzymatically isolated mature plant tissue protoplasts can be recovered in vacuoles isolated from the protoplasts (Boller and Kende, 1979; Nishimura and Beevers, 1979; Wagner *et al.*, 1981; Wittenbach *et al.*, 1982). The vacuoles in higher plants function similarly to lysosomes in animal cells, as hypothesized by Matile (1978). Tomomatsu and Asahi (1980) have described an enzyme ("organelle-damaging enzyme") that probably originates from the vacuole and very rapidly disrupts the limiting border of various cellular organelles after isolation. In addition, Hasson and Laties (1976) have clearly shown that the phospholipids of intact potato tuber mitochondria are highly susceptible to degradation by potato lipolytic acyl hydrolase. They suggested, therefore, that this enzyme is involved in the extensive lipid breakdown that occurs in fresh potato slices following cutting and in the deterioration of mitochondria during their preparation and aging. In order to avoid the danger of long and fruitless controversy, the intact mitochondria obtained by differential centrifugation must be purified by isopycnic centrifugation in a sucrose (Ducet, 1960; Baker *et al.*, 1968; Moreau and Lance, 1972; Douce *et al.*, 1972) or silicasol gradient (Jackson *et al.* 1979a; Nishimura *et al.*, 1982; Neuburger *et al.*, 1982; Moreau and Romani, 1982; Schwitzguebel and Siegenthaler, 1984) or by partition in aqueous polymer two-phase systems (Larsson and Anderson, 1979; Gardeström *et al.*, 1978).

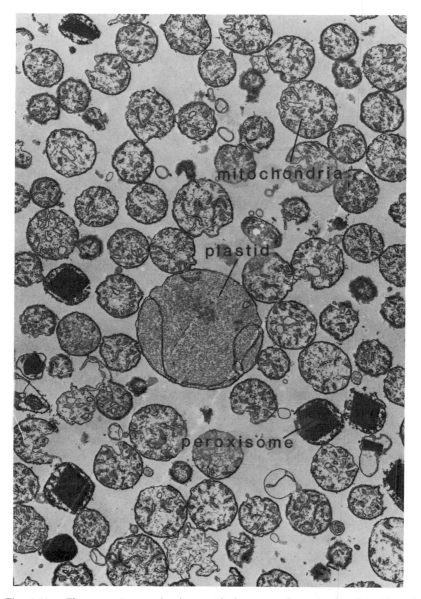

Fig. 1.12.   Electron micrograph of unpurified potato tuber mitochondria. Note that mitochondria are contaminated by plastids and peroxisomes. (Courtesy of J. P. Carde.)

2.   Purification of Plant Mitochondria

The purification procedure by centrifugation on a sucrose gradient is applicable to mitochondria from nongreen tissues such as potato tubers and mung bean hypocotyls (Douce *et al.*, 1972), but we have not been able to purify leaf mitochondria by this method. The method used consists of layering washed mitochondria on top of discontinuous sucrose gradients and centrifuging in a swinging bucket rotor for 45 min at 40,000 g. The gradients are prepared by layering sucrose solution containing 0.1% BSA and 10 mM phosphate buffer into centrifuge tubes in the following sequence of concentrations: 0.6 M, 0.9 M, 1.2 M, and 1.5 M (Fig. 1.10). Following centrifugation,* the bulk of the mitochondria are found at the boundary between the 1.2 and 1.45 M gradients. Alternatively, a simpler method consists of centrifuging the washed mitochondria through a layer of 0.6 M sucrose at 10,000 g ("cushion centrifugation"; Day and Hanson, 1977b). Two critical experimental conditions are required to maintain physiological integrity during the purification process. First, the mitochondria, after removal from the sucrose gradient, are under conditions of high osmolarity and are, therefore, extremely condensed. The volume of the matrix is greatly reduced, and the outer membrane is separated from the inner membrane by a greatly increased sucrose space. Consequently, dilution to iso-osmolar conditions must proceed very slowly to avoid the damage resulting from a too rapidly expanding inner membrane (Fig. 1.13). The experimental evidence indicates that sucrose slowly enters plant mitochondria during centrifugation. Consequently, inattention to the altered isotonic requirements could lead to uncontrolled bursting and resealing of organelles (Douce *et al.*, 1972). It is interesting to note that Sitaraman and Janardana Sarma (1981) have demonstrated that gravitational field enhances the permeability of biological membranes to sucrose and that isotonic conditions for the integrity of subcellular organelles are shown to be remarkably influenced by the concentration of sucrose present during their isolation by centrifugation. The second critical condition is a careful control of centrifugal forces. Thus, Collot *et al.* (1975) have demonstrated that mitochondria show noticeable deterioration during centrifugation on discontinuous sucrose gradients. This deterioration was caused by high water pressures resulting from large centrifugal forces (200,000 g), and such pressures cause dissociation of inner membrane lipoproteins.

---

*We have observed the adverse effects of rapidly starting a large tube swinging bucket rotor: serious changes occurred in step gradients unless there was careful speed control of acceleration and deceleration (rate controller) through the critical area of bucket reorientation.

Furthermore, according to Wattiaux-De Coninck *et al.* (1977) pressures of several hundred kg/cm^2 are readily developed during centrifugation in conventional rotors. Thus, it is very probably that under such conditions lateral phase separations take place in mitochondrial membranes. This leads to an increase in the inner mitochondrial membrane permeability to sucrose and perhaps various matrix coenzymes such as thiamine pyrophosphate and NAD$^+$. The centrifugal force (40,000 g) used on plant mitochondria during the course of purification (Douce *et al.*, 1972), however, is not sufficient to induce any irreversible membrane deterioration. The fact that plant mitochondria survive this purification procedure, in contrast with animal mitochondria, emphasizes the strength of their membranes. Evidence suggests that plant mitochondria are relatively tough structures capable of withstanding far greater osmotic stress than are their mammalian counterparts. Since the outer membrane of a mitochondrion is freely permeable to sucrose, this could lead to an osmotically induced change of volume and organelle density during centrifugation. Accordingly, the buoyant density of organelles measured in sucrose gradients may not represent their real density. Unfortunately, sucrose density gradient centrifugation has generally yielded poor-quality leaf mitochondria with inadequate separation of the mitochondria from contamination material (Douce *et al.*, 1972; Jackson and Moore, 1979; Arron *et al.*, 1979b), although, Nash and Wiskich (1983) have prepared pea leaf mitochondria on a linear sucrose density gradient. These mitochondria were substantially free of contamination by chlorophyll and peroxisomes. They showed high respiratory rates and good respiratory control and ADP/O ratios.

The recent introduction of a nontoxic silica-sol gradient material, Percoll™ [by Pharmacia Fine Chemicals Ltd., Uppsala, Sweden (Pertoft and Laurent, 1977)], has permitted the development of a rapid purification procedure utilizing iso-osmotic and low-viscosity conditions. Consequently, Percoll, which consists of colloidal silica particles of 15–30 nm in diameter which have been coated with PVP, is nearly an ideal gradient material for the separation of mitochondria from higher plants. Unlike sucrose, Percoll does not penetrate biological membranes and does not exert large osmotic effects. Subcellular organelles, therefore, band at densities lower than on sucrose gradients, and densities from 1.04 to 1.10 g/ml have been recorded for different types of mitochondria (Pharmacia Fine Chemicals, 1980; Neuburger *et al.*, 1982; Neuburger and Douce, 1983). The general method is to layer a suspension of washed mitochondria over a density gradient of the silica sol and centrifuge to equilibrium. Mitochondria from material such as potato tubers can be separated on a continuous, self-generated gradient. A centrifuge tube is

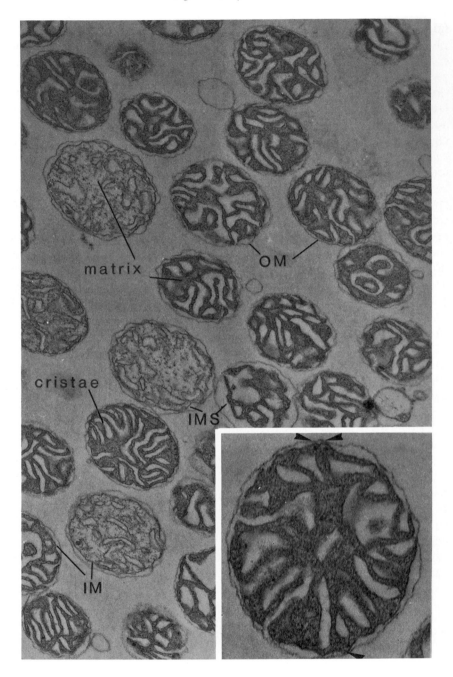

filled with Percoll medium [28% (v/v) Percoll; 300 mM sucrose; 10 mM phosphate buffer (pH 7.2); 1 mM EDTA, and 0.1% (w/v) BSA]. (The final concentration of Percoll and sucrose should be carefully determined according to the origin of the mitochondria and the nature of contaminations present in the preparation.) Mitochondria are then carefully layered onto the gradient and fractionation is undertaken for 30 min at 40,000 g (the automatic rate controller must be used). Most of the mitochondria appeared as a broad band located beneath a yellow layer at the top of the gradient (Fig. 1.10). The mitochondria are highly pleomorphic and sometimes filamentous, and they differ in the relative proportions of internal membranes and remaining matrices. Discontinuous Percoll gradients can also be used successfully to separate mitochondria from other plant organelles (Jackson *et al.*, 1979; Schwitzguebel and Siegenthaler, 1984). The passage of mitochondria from etiolated tissues through a Percoll density gradient is an advantageous step because it not only removes extramitochondrial membranes such as microbodies and amyloplast membranes containing carotenoids (Douce and Joyard, 1979) (Figs. 1.14, 1.15), but also removes contaminating hydrolases such as lipolytic acyl hydrolase (Fig. 1.16; Table 1.2). Under these conditions, the physiological integrity of isolated mitochondria can be maintained at least for 3 days (Neuburger and Douce, 1983). It seems likely, therefore, that colloidal silica particles coated with PVP may act beneficially by binding these hydrolases. In support of this suggestion, the purification of potato mitochondria by centrifugation through a discontinuous sucrose density gradient does not completely remove these contaminating hydrolases (Neuburger *et al.*, 1982; Fig. 1.16).

Unfortunately, the Percoll gradients described earlier have not been found useful for the separation of leaf mitochondria from contaminating structures, notably broken thylakoids. Attempts to reduce the chloroplast contamination have been made by continuous preparative free-flow electrophoresis (Rustin *et al.*, 1980a), but little information on mitochondrial activities have been reported. This electrophoretic method exploits differences in cell membrane surface charges (see Hannig and Heidrich, 1974). To our knowledge, the best method, which permits a clean separation of the mitochondria from broken thylakoids, is the

---

Fig. 1.13. Electron micrographs of sucrose-purified mung bean hypocotyl mitochondria. Note that mitochondria exhibit intact membranes and a dense matrix. In this medium (slightly hypertonic), the outer membrane appears to be loosely attached to the inner membrane (contact points at arrows) with large empty spaces in between. Abbreviations: IMS, intermembrane space; IM, inner mitochondrial membrane; and OM, outer mitochondrial membrane.

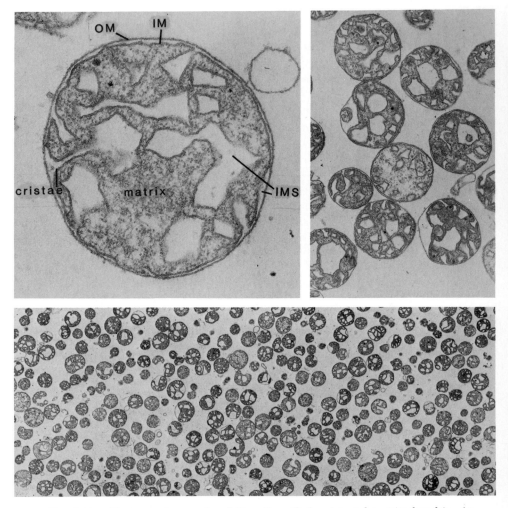

Fig. 1.14.   Electron micrographs of Percoll-purified potato tuber mitochondria. At high magnification, the dark–light–dark construction of both membranes is clearly visible. Abbreviations: IMS, intermembrane space; IM, inner mitochondrial membrane; and OM, outer mitochondrial membrane. [From Neuburger *et al.* (1982).]

procedure using partition in an aqueous dextran–poly(ethylene glycol) two-phase system and centrifugation in a silica-sol gradient. By this technique, membrane vesicles are separated according to differences in membrane surface properties. Such differences can be related both to membrane charge and to membrane hydrophobicity (Albertsson, 1971; Larsson and Anderson, 1979). Phase systems without organic solvents

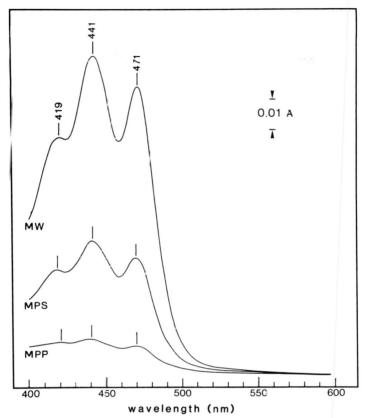

Fig. 1.15.   Absorption spectra of ethanol extracts of potato tuber mitochondria. Note that Percoll-purified mitochondria are almost devoid of carotenoids. Abbreviations: MW, washed mitochondria (3.2 mg protein; 0.32 μg carotenoid); MPS, mitochondria purified on a sucrose gradient (3.2 mg protein; 0.14 μg carotenoid); and MPP, mitochondria purified on a Percoll gradient (3.2. mg protein; 0.03 μg carotenoid). [From Neuburger *et al.* (1982).]

can be easily obtained by mixing aqueous solutions of two different polymers above certain concentrations. As an example, a mixture of 6.1% (w/w) dextran 500 (mol.wt. 500,000), 6.1% (w/w) poly(ethylene glycol) 4000 (PEG; mol.wt. 4,000), 4 mM phosphate buffer (pH 7.8); 2 mM KCl, and 0.3 M sucrose forms a two-phase system with a dextran-rich lower phase and a PEG-rich upper phase. In this phase system, 80% of the chloroplast lamellae partition in the upper phase, while 80% of the mitochondria partition at the interface and in the lower phase. Most of the upper phase is then removed, a new upper phase is added, and the partition procedure is repeated twice. The final lower phase is diluted

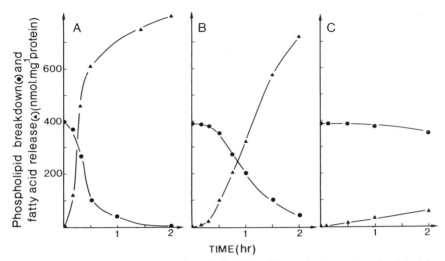

Fig. 1.16. Rate of endogenous phospholipid breakdown during aging of potato tuber mitochondria induced by 5 mM Ca^{2+}. (A) washed mitochondria; (B) mitochondria purified on sucrose gradient; and (C) mitochondria purified on Percoll gradient. Note that sucrose density gradient centrifugation did not appear to be as effective as Percoll gradient centrifugation in isolating mitochondria free from endogenous lipolytic acyl hydrolase. [From Neuburger *et al.* (1982).]

TABLE 1.2

α-Ketoglutarate Oxidation and Measurements of Various Extramitochondrial Markers in Washed and Purified Mitochondria Isolated from Potato Tubers

Extramitochondrial markers	Washed mitochondria[a]	Purified mitochondria[b]
State-III rate of α-ketoglutarate oxidation (nmoles of O$_2$/min/mg of protein)	153	216
Catalase (μmoles of O$_2$/min/mg of protein)	12	0.2
Lipoxygenase (μmoles of O$_2$/min/mg of protein)	3.5	ND
Carotenoids (μg/mg of protein)	0.1	0.008

[a] Washed mitochondria were purified by centrifugation (40,000 *g*; 30 min; rotor SS 34 Sorvall) in a Percoll gradient [28% (v/v) Percoll; 0.3 *M* sucrose; 10 m*M* phosphate buffer (pH 7.2); 1 m*M* EDTA; and 0.1% (w/v) bovine serum albumin].

[b] If the purification is repeated once (i.e., after pelleting mitochondria and layering the suspending pellet on top of a new Percoll gradient [28% (v/v) Percoll]; 0.3 *M* sorbitol; and 10 m*M* phosphate buffer (pH 7.2), mitochondrial catalase activity can be reduced to less than 0.02 μmoles of O$_2$/min/mg of protein. ND, not detected.

with seven volumes of a suitable medium and the intact mitochondria are obtained by centrifugation. The yield in a typical preparation from 50 g of leaves is about 2 mg of protein, giving a protein/chlorophyll ratio of approximately 100 (Gardeström *et al.*, 1978). About 90% of the mitochondria in the final pellet are intact according to the cytochrome *c* test described previously. The mitochondrial preparation thus obtained, however, is still contaminated by peroxisomes, as judged by catalase activity measurement. A further purification of the mitochondria by centrifugation on a discontinuous Percoll density gradient completely eliminates residual thylakoid from the final mitochondrial pellet (Bergman *et al.*, 1980; Gardeström, 1981). Consequently, Bergman *et al.* (1980) have thereby been able to completely remove chlorophyll from the mitochondrial preparation while retaining the function and intactness of the mitochondria. Likewise, Day *et al.* (1985) have developed a a technique suitable for the large-scale preparation of intact, active, and chlorophyll-free mitochondria from pea leaves. The method employs a self-generating gradient of Percoll in combination with a linear gradient of PVP [0–10% (w/v), top to bottom], in a single step. Using this simple method, a clean separation of mitochondria from contaminating chlorophyll was achieved. When PVP was omitted from the Percoll solution, the distribution of chlorophyll and glycolate oxidase remained the same, but cytochrome *c* oxidase was spread throughout the tube. Thus, the inclusion of PVP serves to narrow the distribution of the mitochondria without affecting that of the other organelles (Fig. 1.17).

B. Structure of Isolated Mitochondria

Interesting experiments with isolated plant mitochondria have demonstrated that the two membranes surrounding the organelle delineate two compartments differing in their accessibility to compounds of different molecular weights. These two compartments are the matrix space and the intermembrane space, which includes the intracristal space. The space situated between the inner and the outer membrane is found to be nonspecifically permeable to molecules, both charged and uncharged, up to a molecular weight of about 10,000 (Pfaff *et al.*, 1968). Although the possibility of membrane damage during isolation could not be completely ruled out, the outer membrane of the mitochondrion isolated by rather gentle procedures was still permeable to sucrose, nucleotides, and NAD^+ but not to cytochrome *c* (Douce *et al.*, 1972). This nonspecific permeability of the outer membrane is obviously needed for the rapid exchange of metabolic intermediates, nucleotides, and P_i between the

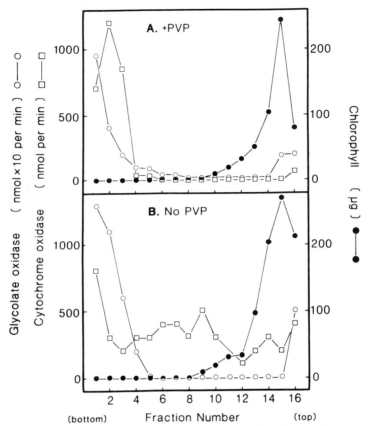

Fig. 1.17.   Distribution of markers in a Percoll–PVP gradient during the course of pea leaf mitochondria purification. The input of crude mitochondria was 0.45 mg chlorophyll (A) and 0.54 mg chlorophyll (B). Note that including PVP serves to narrow the distribution of the mitochondria without affecting that of the other organelles.

cytoplasm and the mitochondrial matrix. However, since phospholipid bilayers are essentially impermeable to hydrophylic solutes, the outer membrane must contain a component that produces transmembrane diffusion channels. Zalman *et al.* (1980) have inserted outer membrane fragments from mung bean hypocotyl mitochondria into the phospholipid bilayer of liposome vesicles. Using this system, they established that the mitochondrial outer membrane is nonspecifically permeable to hydrophilic solutes of up to several thousand daltons, and they identified a channel-forming protein with an apparent molecular weight of 30,000 through differential detergent extraction and centrifugation in a sucrose density gradient. This protein is similar to the porin of the outer membrane of a gram-negative bacterium (Schnaitman, 1971). In contrast, the inner membrane surrounding the matrix space, which is practically impermeable to sucrose, is selectively permeable to a limited number of anions. Consequently, the size of the sucrose permeable space depends on the tonicity of the medium; it is very small when the mitochondria are kept in an isotonic medium (0.2 M mannitol or sucrose). The sucrose permeable space is important when the mitochondria are kept in a hypertonic medium. The matrix space is usually very condensed in hypertonic media and expanded (i.e., less electron dense) in isotonic solution. Under extreme hypotonic conditions, water passes through the inner membranes into the matrix space, which enlarges. The inner membrane becomes extended with some material derived from the unfolding cristae and presses against the outer membrane, which having limited elasticity becomes ruptured (Fig. 1.18). Nonetheless, the bulk of the matrix enzymes such as fumarase or malate dehydrogenase are not readily released into the medium, providing that Mg^{2+} ions are present in the medium. Consequently, plant mitochondria, like their mammalian counterparts, were found to behave as simple osmometers undergoing rapid, apparently reversible changes (insofar as the outer membrane is still intact) in matrix volume V, in response to changes in the external osmotic pressure Π (Baker *et al.*, 1968; Guillot-Salomon, 1972; Fig. 1.19). The solute content of the matrix is not accurately known and probably fluctuates according to the origin of the mitochondria and the procedure used for their isolation. This could strongly influence the osmotic properties of isolated mitochondria (see Malone *et al.*, 1974). Consequently, it is obvious that the structure of isolated plant mitochondria depends on the tissue from which they are isolated and the nature of the suspension medium. Very often plant mitochondria isolated in 0.3 M mannitol or sucrose media are rounded, the peripheral space is large, the intracristal spaces are enlarged, and the matrix in sections appears electron dense (see, for example, Baker *et al.*, 1968; Douce *et al.*, 1972; Figs. 1.12, 1.13, and 1.14). The rod-shaped or filamentous mitochondria frequently observed *in vivo* (Fig. 1.1) are seldom observed in mitochondrial preparations. Laties and Treffry (1969) obtained rod-shaped mitochondria from potato tubers by using an isolation medium containing sucrose and high-molecular-weight polymers such as Ficoll (sucrose–epichlorhydrin polymer) or dextran (polyglucose). We believe that the change from an elongated to a rounded form depends on the composition of the medium rather than hypothetical membrane damage during the course of mitochondrial preparations. Whether or not isolated plant mitochondria that morphologically resemble mitochondria *in vivo* differ functionally in any significant way

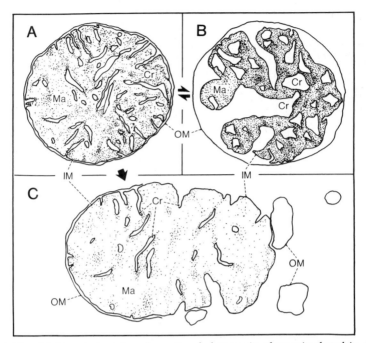

Fig. 1.18.   Medium-induced conformational changes in plant mitochondria. (A) iso-osmotic medium; (B) high-osmolarity medium; and (C) low-osmolarity medium. Note that in a high-osmolarity medium (0.6 *M* sucrose) the outer membrane appears to be loosely attached (contact sites) to the inner membrane with large empty spaces in between. In low-osmolarity medium the inner membrane becomes extended and presses the outer membrane, which having limiting elasticity becomes ruptured. Very often the outer membrane, partially disrupted by swelling, remains attached to the inner membrane at the contact sites as inverted vesicles. Abbreviations: Cr, crista; Ma, matrix; OM, outer mitochondrial membrane; and IM, inner mitochondrial membrane.

from isolated, undamaged, spherical mitochondria, remains to be determined.

The morphological responses of chloroplasts to changes in the osmolarity of suspending media are also well documented (Heldt *et al.,* 1972; Douce and Joyard, 1979). In this case, the size of the sucrose permeable space (i.e., the space situated between the inner and the outer membrane of the chloroplast envelope) also depends on the tonicity of the medium. It is very small when the chloroplasts are kept in

Fig. 1.19.   Electron micrographs of sections of mitochondria isolated from cauliflower buds in (A) 0.6 *M*-sucrose medium and (B) 0.1 *M*-sucrose medium. [From Guillot-Salomon (1972).]

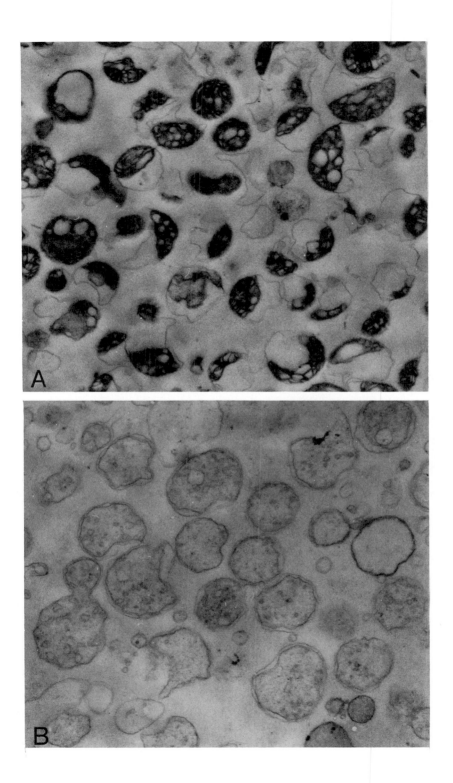

slight hypotonic medium (0.2 *M* sorbitol), whereas in a hypertonic medium (0.6 *M* sorbitol) the outer envelope membrane appears to be loosely attached to the inner envelope membrane with large empty spaces in between. In contrast to the inner mitochondrial membrane, however, the stretching of the envelope cannot originate from the addition of materials coming from the thylakoids, since demonstrable connections between the inner membrane of the envelope and the thylakoid system in mature chloroplasts have never been reported. Consequently, once the shape of a sphere is reached, further swelling is limited by the low plasticity of the membranes and bursting occurs. This in turn leads to total detachment of the envelope membranes and to liberation of the stroma material.

The large-amplitude swelling–contraction cycle that seems to follow the metabolic state of the mitochondria (Hackenbrock, 1968) has never been reported in the case of isolated intact plant mitochondria. Nevertheless, in the case of isolated corn mitochondria, Hanson and co-workers (Bertagnolli and Hanson, 1973; Hensley and Hanson, 1975; Day and Hanson, 1977b) have observed a fast but discrete change of light scattering measured at 520 nm induced by ADP during the oxidation of succinate. According to Knight *et al.* (1981) and Ducet (1980) changes in the intensity of scattered light are not reliable indices of changes in the volume of mitochondria. These authors think that changes in light scattering are often due to changes in refractive indices which are brought about by generation or discharge of the energized state without changes in volume. In other words, the inner membrane potential could influence, in a discrete and rapid way, the light scattering properties of the mitochondria. In support of this suggestion, small-angle X-ray scattering data suggest that major but reversible rearrangements of the mitochondrial inner membrane structure are induced by the uncoupler (Mannella and Parsons, 1977).

Besides the large-amplitude volume changes induced by the osmolarity of the medium, it was found that swelling of plant mitochondria can also occur in an isotonic medium and is promoted by various cations (Stoner and Hanson, 1966; Miller *et al.*, 1974) and anions (Pomeroy, 1977). This mitochondrial swelling, which concerns primarily the inner membrane, is partly related to the movements of water across the membrane which accompany the uptake and release of anions by way of specific carriers (Chappel and Crofts, 1966; Overman *et al.*, 1970; Phillips and Williams, 1973; Kirk and Hanson, 1973). These observations have served as an important and powerful tool in discovering and characterizing some of the inner membrane translocators in plant mitochondria (Phillips and Williams, 1973; Wiskich, 1974).

2

Composition and Function of Plant Mitochondrial Membranes

I. COMPOSITION OF MITOCHONDRIAL MEMBRANES

A. Methods for the Separation of the Outer and Inner Mitochondrial Membranes

Only a few attempts have been made to separately isolate the outer and the inner membranes of plant mitochondria, and most have used the techniques of hypotonic swelling (Parsons *et al.*, 1966), mechanical disruption (Douce *et al.*, 1973b), or digitonin treatment (Lévy *et al.*, 1966; Schnaitman and Greenawalt, 1968) to selectively detach the outer membrane (Meunier *et al.*, 1971; Moreau and Lance, 1972; Douce *et al.*, 1973b; Day and Wiskich, 1975). Since the buoyant density of the outer membrane proved sufficiently different from that of the inner membrane, the two membranes are then usually separated by either gradient or rate (differential) centrifugation in a single or multistep process. In most cases, the inner membrane plus matrix-containing fraction is treated further, prior to isolation of the membrane, to remove matrix proteins. One of the main problems encountered in the studies was that of contamination of the outer membrane fraction by the inner membrane, matrix enzymes, plastid and peroxisome membranes, and the endoplasmic reticulum. In fact, the preparation of mitochondrial membranes is extremely difficult for the following reasons: (a) it is difficult if not impossible at present to prepare large amounts of intact plant mitochondria devoid of various extramitochondrial membranes; (b) the outer membrane proteins represent a small proportion of the total mitochondrial protein [Bligny and Douce (1980) have shown that in the case of the mung bean hypocotyl mitochondrion, the outer membrane proteins represent only 6.8% of the total mitochondrial protein]; and (c) several authors (Hackenbrock, 1968; Hackenbrock and Miller, 1975) have clearly demonstrated that a number of contact sites occur between the two mitochondrial membranes (Fig. 1.13). Hackenbrock and Miller (1975) have also shown that the outer mitochondrial membrane, partially disrupted by treatment with digitonin, remains attached to the inner membrane at the contact sites as inverted vesicles. On the evidence of these simple morphological criteria, we can therefore conclude that some outer membrane material cosediments with the main bulk of the inner membrane during the centrifugation process.

1. Swelling Treatment

Douce *et al.*, (1972) have demonstrated that exposure of potato tuber or mung bean hypocotyl mitochondria to successively greater hypotonic shocks increases succinate–cytochrome *c* oxidoreductase activity. Just as this activity monitors the rupture of the outer membrane, the release of malate dehydrogenase into the supernatant after centrifugation of hypotonically treated mitochondria can be used to follow the bursting of the inner mitochondrial membrane (Fig. 2.1). It is interesting to note that the lowering of the osmolarity enhanced the activity of malate dehydrogenase measured in the pellet by increasing the ability of NADH to permeate the inner membrane. Consequently, the ability of NADH to cross the inner membrane and reach the malate dehydrogenase situated in the matrix is more sensitive to changes in osmolarity than is the release of the malate dehydrogenase to the medium (J. M. Palmer and Kirk, 1974). It is clear that the unnatural process of hypotonic shock markedly affects inner membrane permeability. In context, the release of this enzyme from the matrix space was used to help define the osmolarity below which it would be unsafe, in terms of inner membrane contamination, to swell the mitochondrion for the purpose of isolating the outer membrane. The succinate–cytochrome *c* oxidoreductase activity is nearly

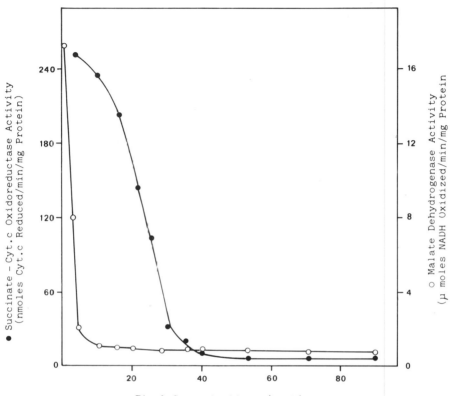

Fig. 2.1. Effect of final osmolarity of suspending media on mitochondrial enzyme activities. Aliquots of purified mitochondria (20 μl) were added to 5 ml of appropriate sucrose solutions at room temperature. After 2 min, the succinate–cyt c oxidoreductase (●) activities of the suspensions were monitored. The mitochondria were then pelleted (8000 g, 10 min) and the supernatants were assayed for malate dehydrogenase activity (○). [From Douce *et al.* (1973b).]

maximum at an osmolarity higher than that which causes the release of the bulk of malate dehydrogenase (Fig. 2.1). This implies that the outer membrane of a plant mitochondrion can be ruptured almost completely by osmotic swelling with little concomitant inner membrane breakage. Hence, by carefully controlling the final osmolarity of the swelling medium, it should be possible to isolate a good percentage of the outer mitochondrial membranes free of inner membrane contamination. Interestingly, the presence of EDTA in the suspending medium makes the inner membrane more susceptible to osmotic damage (J. M. Palmer and

Kirk, 1974). On the contrary, Mg^{2+} makes the inner membrane less susceptible to osmotic damage (R. Douce, unpublished data).

A concentrated suspension of gradient-purified mitochondria (about 100 mg/ml) is, therefore, rapidly pipetted into 50 volumes of a sucrose solution, the osmolarity of which (Cs) is determined by the equation $Cs = 51(Cf) - Co/50$, in which Cf is the desired final osmolarity and Co is the initial osmolarity of the mitochondrial suspension, which is carefully adjusted to 0.3 Osm after density gradient purification. For example, in the case of mung bean hypocotyl mitochondria, outer membrane rupture is optimal at 10 mOsm (Douce *et al.*, 1972, 1973b; Mannella, 1974), so the concentration (Cs) of sucrose used was 4 mM. This dilute suspension of mitochondria is vigorously stirred for 15 min, after which enough 1.8 M sucrose is added to the suspension to raise the final osmolarity to 0.3 Osm. The mitochondrial suspension is then layered on top of discontinuous sucrose gradients (0.6; 0.9; and 1.25 M) containing 10 mM potassium phosphate buffer (pH 7.2) and spun at 40,000 g for 60 min using a swinging bucket rotor. Based on the antimycin-insensitive NADH-cytochrome c oxidoreductase activity (a marker enzyme for the outer membrane), swelling–contraction treatment of mung bean hypocotyl mitochondria results in the recovery of 40% of this outer membrane activity in light fractions (the 0.3/0.6 M and 0.6/0.9 M sucrose interfaces). Unfortunately, by this method more than 50% of the outer membrane activity is still associated with the mitoplast fraction (inner membrane plus matrix) present in the pellet. In addition, the final yield of outer membrane is largely dependent upon the origin of the mitochondria; for example, it is very low with potato tuber mitochondria and high with mung bean hypocotyl mitochondria. The outer mitochondrial membrane appears in the electron microscope as a small, smooth vesicle without the projected 9-nm particles (ATPase molecules) that mark the inner membrane (Moreau and Lance, 1972; Douce *et al.*, 1973b).

In order to facilitate total outer membrane detachment and thereby increase the yield of outer membrane, the mitochondrial suspension maintained in a hypertonic medium was forced very slowly through a small aperture under high-pressure nitrogen gas. With mung bean hypocotyl mitochondria, pressing at 1800 psi appears to be optimal for stripping the outer membrane. The extrusion flow rate was controlled to about 0.5 ml/min by adjusting the aperture diameter (Douce *et al.*, 1973b). The resulting outer membrane vesicles and mitoplasts were recovered essentially as described above. This mechanical disruption considerably decreased the outer membrane activity remaining in the mitoplast pellet (by about 50% in the case of mung bean hypocotyl).

Another marked effect of this fragmentation procedure was to shift outer membrane marker accumulations to lighter densities. Furthermore, the percentage of antimycin-insensitive NADH-cytochrome c oxidoreductase activity and its specific activity in the soluble fraction was unusually high. This was not due simply to detachment of the protein from the outer membrane, because antimycin-insensitive NADH-cytochrome c oxidoreductase activity was still membrane bound. In fact, we have shown that the very small outer membrane vesicles are incompletely sedimented under the conditions described here but can be precipitated completely with prolonged centrifugation.

2. Digitonin Treatment

The rationale of this method is that the outer membrane is rich in sterol (Dupéron et al., 1975), which binds digitonin. Dupéron et al. (1975) measured the sterol content of cauliflower mitochondria and found that the inner membrane contained 28 μg sterol/mg protein and the outer membrane contained 77 μg sterol/mg protein. This enrichment in the outer membrane is also found in animal mitochondria. Consequently, the outer membrane is released from plant mitochondria at a lower digitonin–protein ratio than that at which the inner membrane is broken (as judged by the release of soluble malate dehydrogenase from the matrix). It is essential that the optimal ratio of digitonin/mg mitochondrial protein be determined for each specific kind of mitochondria. By "titrating" the liberation of the outer membrane and its associated marker enzymes with increasing concentrations of digitonin, these optimal ratios can be determined. For example, in the case of turnip the mitochondrial suspension is incubated with a digitonin solution, the concentration of which has been adjusted to give a digitonin–protein ratio of 2 : 10, for 30 min at 0°C with continuous stirring (Day and Wiskich, 1975). The two membranes are then separated by differential centrifugation.

Douce et al. (1973b), Moreau (1978), and Day and Wiskich (1975) noticed that outer membrane fractions obtained by digitonin or swelling treatment of plant mitochondria have a high specific activity of malate dehydrogenase. It is likely, however, that this activity arises due to the considerable amount of soluble malate dehydrogenase accessible to exogenous NADH in the intact mitochondrion (almost 1 μmole NADH oxidized/min/mg mitochondrial protein; see Douce et al., 1973b). This possibility is confirmed by the observation that 50–70% of the malate dehydrogenase activity of the outer membrane fractions remain in the supernatant after pelleting (60,000 g, 90 min) of the membranes (Mannella, 1974; Day and Wiskich, 1975).

According to Day *et al.* (1979), only a small amount of the total adenylate kinase activity is released at the low digitonin concentration that breaks the outer membrane. Yet, when adenylate kinase activity is measured in intact plant mitochondria, its activity is not affected by the addition of Triton X-100, showing that there is no permeability barrier to the reagents used in the assay, including $NADP^+$. They logically interpret these results to mean that adenylate kinase is tightly bound to the inner membrane of the plant mitochondrion, facing the intermembrane space. This is in contrast to the situation in liver (Sottocasa *et al.*, 1967) and yeast (Bandlow, 1972) in which the adenylate kinase seems to be soluble and associated with the intermembrane space.

In Jerusalem artichoke tubers (Le Floc'h and Lafleuriel, 1983) and probably in *Catharanthus roseus* cells (Hirose and Ashihara, 1982), the mitochondria possess all the enzymatic equipment necessary for recycling adenine into adenine nucleotides. Phosphorybosyl pyrophosphate synthetase (EC 3.7.6.1) and adenine phosphorybosyltransferase (EC 3.4.2.7) are both probably localized in the intermembrane space of the mitochondria. Likewise, the presence of an AMP amino hydrolase, probably bound to the outer surface of the inner membrane, suggests that mitochondria might also play an important role in the recycling of adenine into guanine nucleotides.

Finally, Dieter and Marmé (1984) have shown that most of the NAD kinase from etiolated corn coleoptiles is membrane-bound. Differential and density gradient centrifugation reveal that this NAD kinase, which is shown to be under the control of Ca^{2+} and calmodulin, is associated with mitochondria. Submitochondrial localization experiments clearly indicate that NAD kinase is located at the outer mitochondrial membrane. Consequently, in the intact plant cell this NAD kinase can be regulated via cytoplasmic calmodulin by changes in the cytoplasmic free-calcium concentration.

B. Lipid Composition

Isolated mitochondria from a variety of plant tissues are rich in phospholipids (23–27% lipid by weight), which may account for as much as 90% of the total lipids (Moreau *et al.*, 1974; Bligny and Douce, 1980) (Fig. 2.2). Phosphatidylcholine and phosphatidylethanolamine are the most abundant phospholipids. Relative to the whole cell of origin, plant mito-

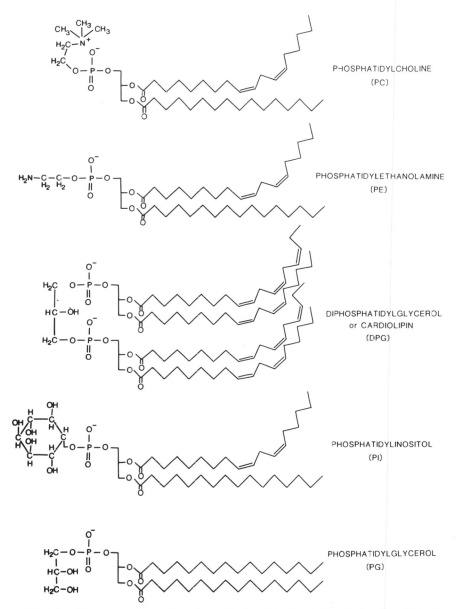

Fig. 2.2. Structure of the main polar lipids found in the plant mitochondria. Note the specific position of the main fatty acids on each of the polar lipids. In the case of phosphatidylcholine, the major polar lipid of plant mitochondrial membranes, the charge on the phosphate group is balanced, in an internal salt linkage, by the quaternary amine of the choline residue.

TABLE 2.1

Phospholipid Composition (Percentage by Weight) of Sycamore Cells and Sycamore Cell Mitochondria

	Percentage of total phospholipids[a]	
Phospholipid	Cells	Mitochondria
Phosphatidylcholine	47	43
Phosphatidylethanolamine	29	35
Diphosphatidylglycerol	1.8	13
Phosphatidylinositol	16	6
Phosphatidylglycerol	5	3

[a] Phospholipid content of the cells: 0.16 mg/mg cell protein; phospholipid content of the mitochondria: 0.3 mg/mg mitochondrial protein.

chondria generally contain less phosphatidylcholine and more phosphatidylethanolamine when the values are expressed in terms of the percentage contribution of each component in the total phospholipids (Bligny and Douce, 1980; Table 2.1). The salient feature of plant mitochondrial phospholipids is the high concentration of diphosphatidylglycerol (cardiolipin) that is not found in other cellular organelles (Douce, 1965; Coulon-Morelec and Douce, 1968; Douce *et al.*, 1968). Indeed, the number of molecules of cardiolipin per molecule of cytochrome oxidase (a specific marker of mitochondria) is practically the same in the whole sycamore cell as in isolated mitochondria: in both cases there are 80 ± 5 molecules of cardiolipin per molecule of cytochrome oxidase (Bligny and Douce, 1980; Table 2.2). The cardiolipin from plant tissues has been shown by Douce (1970) to possess the diphosphatidylglycerol structure. In addition, phosphatidylinositol, a minor component, accounts for about 5% of the phospholipids. Phosphatidylglycerol is found, if at all, only in small amounts. Phosphatidic acid also occurs only in trace amounts in plant mitochondria, and its presence in significant amounts (see, for example, Hartmann *et al.*, 1981) is almost certainly a result of phospholipase D activity during extraction (Douce, 1965; Clermont and Douce, 1970). Galactolipids and sulfolipids are missing in highly purified plant mitochondria (McCarty *et al.*, 1973). Indeed, by using a specific cytochemical marker (silver proteinate) for galactolipids, Carde *et al.* (1982) have shown that the galactolipids are exclusively localized in plastid membranes of plant cells (Fig. 2.3). Like-

TABLE 2.2

Cytochrome Oxidase and Cardiolipin Content of Sycamore Cells and Sycamore Cell Mitochondria[a]

	Cells	Mitochondria
Cytochrome aa_3[b]	0.026 ± 0.002 (6)	0.37 ± 0.02 (10)
Cardiolipin[b]	$2.1 \quad \pm 0.2$ (6)	$29 \quad \pm 2$ (10)
Cardiolipin/cytochrome aa_3[c]	$81 \quad \pm 5$	$78 \quad \pm 5$

[a] The values are given per milligram of cell protein for the cells and per milligram of mitochondrial protein for the mitochondria. The S.D. is given for 95%, the number of experiments is shown in parentheses.

[b] Units: nmoles substance per mg protein.

[c] Units: molecules per molecule.

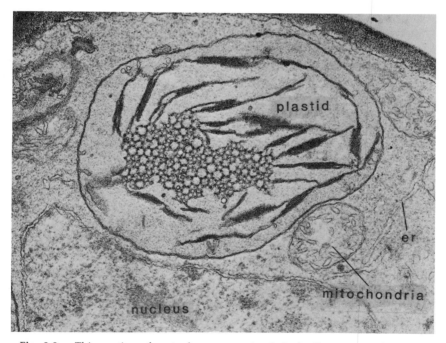

Fig. 2.3. Thin section of part of a young spinach leaf cell post-stained with silver proteinate. Note that the plastid membranes (outer and inner envelope membranes, prolamellar body, thylakoids) containing galactolipids are highly contrasted; on the contrary, the other cell membranes [nuclear and mitochondrial membranes, endoplasmic reticulum (er)] are poorly contrasted (×30,000). [From Carde *et al.* (1982).]

wise, Haas *et al.* (1979) using mesophyll protoplasts from oat primary leaves demonstrated that sulfolipid molecules are also exclusively associated with plastid membranes. On the other hand, phosphatidylethanolamine, which is a major constituent of all the mitochondria examined so far, is not detectable in plastids. Consequently, the polar lipid composition of plastids almost resembles that of blue-green algae, such as *Anacystis nidulans,* and does not resemble that of mitochondria. Unlike their energy-transducing counterparts the mitochondria, plastids are dominated by galactolipids rather than phospholipids (Table 2.3). The composition of the lipids from crown gall of scorsonerae (Douce, 1965), cauliflower inflorescence (Douce *et al.,* 1968), apple fruit (Mazliak and Ben Abdelkader, 1971), potato tuber (Ben Abdelkader and Mazliak, 1970), castor bean endosperm (Donaldson and Beevers, 1977), avocado fruit (Schwertner and Biale, 1973), mung bean hypocotyl (McCarty *et al.,* 1973), sycamore cells (Bligny and Douce, 1980), oat mesophyll cells (Fuchs *et al.,* 1981), and spinach leaf mitochondria (Gardeström *et al.,* 1981) is similar to that of animal mitochondria (McMurray, 1973).

The phospholipid composition of isolated mitochondrial membranes has been studied by several groups (Meunier and Mazliak, 1972; Moreau *et al.,* 1974; McCarty *et al.,* 1973; Bligny and Douce, 1980) using various plant tissues and is similar to that of mammalian (McMurray, 1973) and *Neurospora crassa* mitochondria (Hallermayer and Neupert, 1974); for ex-

TABLE 2.3

Polar Lipid Composition (Percentage by Weight) of the Different Membranes Isolated from Chloroplasts and Mitochondria[a]

Source	MGDG	DGDG	SL	PC	PG	PI	PE	DPG	Lipid/ protein
Chloroplast[b]									
Envelope									
Outer membrane	17	29	6	32	10	2	0	0	3
Inner membrane	49	30	5	6	8	1	0	0	0.8
Thylakoides	52	26	6.5	4.5	9.5	1.5	0	0	0.4
Mitochondria[c]									
Outer membrane	0	0	0	68	2	5	24	0	0.8
Inner membrane	0	0	0	29	1	2	50	17	0.4

[a] Abbreviations: MGDG, monogalactosyldiacylglycerol; DGDG; digalactosyldiacylglycerol; SL, sulfolipid; PC, phosphatidylcholine; PG, phosphatidylglycerol; PI, phosphatidylinositol; PE, phosphatidylethanolamine; and DPG, diphosphatidylglycerol.

[b] Spinach chloroplasts, Block *et al.,* 1983.

[c] Mung bean, Bligny and Douce, 1980.

TABLE 2.4

Phospholipid Composition (Percentage by Weight) of Sycamore Cell and Mung Bean Inner and Outer Mitochondrial Membranes[a]

Phospholipid	Sycamore cells[b]		Mung bean[c]		
	I.M.	O.M.	I.M.	O.M.	Intact mitochondria
Phosphatidylcholine	41	54	29	68	36
Phosphatidylethanolamine	37	30	50	24	46
Diphosphatidylglycerol	14.5	n.d.	17	n.d.	14
Phosphatidylinositol	5	11	2	5	2.5
Phosphatidylglycerol	2.5	4.5	1	2	1

[a] Abbreviations: n.d., not determined; I.M., inner mitochondrial membrane; and O.M., outer mitochondrial membrane.

[b] For 1 mg sycamore cell mitochondrial protein there are: matrix, 0.34 mg protein; I.M., 0.60 mg protein, 0.24 mg phospholipid; and O.M., 0.06 mg protein, 0.05 mg phospholipid.

[c] For 1 mg mung bean mitochondrial protein there are: matrix, 0.30 mg protein; I.M., 0.65 mg protein, 0.25 mg phospholipid; and O.M., 0.05 mg protein, 0.04 mg phospholipid.

ample, the relative proportions of cardiolipin are nearly equal in the phylogenetically distant organisms. Each type of membrane contains characteristic polar lipids in fixed molar ratios that are probably determined genetically. In the inner membrane, the two main phospholipids, phosphatidylcholine and phosphatidylethanolamine, represent 75–80% of the phospholipids, whereas cardiolipin accounts for 17–20% of the phospholipids. In the outer membrane, phosphatidylcholine and phosphatidylethanolamine represent up to 90–95% of the phospholipids, whereas cardiolipin is barely detectable. The ratio of phosphatidylcholine–phosphatidylethanolamine for the outer membrane is generally much higher than that found for the inner membrane (Moreau *et al.,* 1974; Bligny and Douce, 1980; Tables 2.3 and 2.4). The relative concentration of phosphatidylinositol is somewhat higher in the outer membrane than in the inner membrane (Moreau *et al.,* 1974). Since the convoluted inner membrane has a much larger surface area as compared to the outer membrane, it is not surprising that the phospholipid composition of the whole mitochondrion more closely resembles that of the inner membrane than of the outer membrane (Bligny and Douce, 1980). The outer membrane is twofold richer in lipids than the inner membrane on a protein basis (Bligny and Douce, 1980), which accounts for the different sedimentation behavior of the membranes in density gradients.

Bligny and Douce (1980) investigated the fatty acid composition of mitochondrial phospholipids and concluded that the patterns are basically similar for inner and outer membranes from mung bean hypocotyls and sycamore cells (Table 2.5). The fatty acids of the outer membrane phospholipids were slightly more saturated than those of the inner membrane (Bligny and Douce, 1980), but the difference was not as dramatic as that reported by Moreau *et al.* (1974). The most highly unsaturated phospholipid is cardiolipin, which contains nearly 85% of its total fatty acid as linoleate and linolenate. Phosphatidylcholine and phosphatidylethanolamine generally show similar fatty acid compositions. Phosphatidylinositol and phosphatidylglycerol contain a higher proportion of palmitic acid than phosphatidylcholine and phosphatidylethanolamine. The distribution of fatty acids between the 1- and 2-positions of the plant mitochondrial phospholipids shows a general pattern similar to that in animal mitochondrial phospholipids. Palmitic acid (C16 : 0) is primarily located at the 1-position, whereas the 2-position contains predominantly unsaturated fatty acids (Douce and Lance, 1972; Fuchs *et al.*, 1981). Analysis of the major polar lipids from chloroplasts

TABLE 2.5

Fatty Acid Composition (Percentage by Weight) of Sycamore Cell and Mung Bean Inner and Outer Mitochondrial Membranes[a]

Fatty acids	Total phospholipids		PC		PE		DPG		PI		PG	
	I.M.	O.M.	I.M.	O.M.	I.M.	O.M.	I.M.	O.M.	I.M.	O.M.	I.M.	O.M.
Sycamore cells												
$C_{16:0}$	25	34	24	29	28	31	7	—	49	56	59	67
$C_{18:0}$	1.6	2.5	1.6	2	1.5	2	1	—	2	3	4	5
$C_{18:1}$	3.7	4	4	5	3	4	5	—	2.5	3	5	6
$C_{18:2}$	49	44	46	45	51	50	62	—	39	33	26	18
$C_{18:3}$	20	15	24	19	16	12	25	—	7.5	5	6	4
Mung bean												
$C_{16:0}$	12	21	9	17	18	29	3	—	30	49	61	71
$C_{18:0}$	1.5	3	2	4	1.5	2.5	0.5	—	3	5	4	7
$C_{18:1}$	4.5	8	10	10	4	6	6.5	—	9	6	6	5
$C_{18:2}$	41	32	41	32	43	33	39	—	32	17	12	8
$C_{18:3}$	41	35	40	37	33	29	50	—	26	22	16	9

[a] Abbreviations: I.M., inner mitochondrial membrane; O.M., outer mitochondrial membrane; PC, phosphatidylcholine; PE, phosphatidylethanolamine; DPG, diphosphatidylglycerol; PI, phosphatidylinositol; and PG, phosphatidylglycerol. [From Bligny and Douce (1980).]

(monogalactosyldiacylglycerol, digalactosyldiacylglycerol, sulfolipid, and phosphatidylglycerol) revealed that C_{16} fatty acids (hexadecatrienoic acid, *trans*-3-hexadecenoic acid, and palmitic acid) are preferentially esterified to the 2-position. Interestingly, the major molecular species contained linolenate at the 1-position (Harwood, 1980; Douce and Joyard, 1980). Consequently, this positional specificity of fatty acids in galactolipids, sulfolipid, and phosphatidylglycerol in plastids contrasts with their distribution in the phospholipids of mitochondria which preferentially carry C16 : 0 acids at the C-1 position.

The fatty acids of plant mitochondrial membrane phospholipids are much more unsaturated that their mammalian counterparts. It is also worth noting that mitochondrial membranes from the same plant will contain higher amounts of linolenate when isolated from leaf tissue rather than the root (Oursel *et al.*, 1973; Mackender and Leech, 1974).

The exact function of the phospholipids of plant mitochondria is uncertain. It is clear that phosphatidylcholine and phosphatidylethanolamine are the principal constituents of the fluid lipid bilayer that acts both as a permeability barrier to polar molecules and as a flexible framework capable of accomodating a variety of proteins. There are two kinds of membrane proteins—peripheral and integral (Singer, 1974). The former are operationally defined as ones easily removable by salts and the latter by the need of detergents for removal, because they are hydrophobically bound. Some are confined to one side of the bilayer and some partially or completely penetrate it. Consequently, it is very likely that the proportion of the mitochondrial inner membrane surface area consisting of lipids is small and that lipids occupy only the gap between the protein molecules. It is also conceivable that there may exist regional differences in lipid composition within a single membrane. Such differences could play an important role in determining the insertion sites for proteins if their lipid binding sites are specific for certain types of lipids. In addition, several lines of investigation indicate that the composition of each plant mitochondrial membrane is assymetric with respect to the outer and inner layers (T. S. Moore, 1982), in terms of both lipid and protein composition. Exactly how this is achieved is unknown. The exact nature of phospholipid interactions with plant mitochondrial membrane proteins is not clearly understood. Specific associations of cardiolipin with cytochrome oxidase (Awasthi *et al.*, 1971) and Complex III (Shimomura and Ozawa, 1981) within the inner mitochondrial membrane in plants has not yet been conclusively demonstrated (see, for example, Matsuoka and Asahi, 1982). Likewise, it is not known whether all plant mitochondrial integral membrane proteins are surrounded by a boundary layer of relatively immobilized lipids (annulus), distinct from the phospholipid molecules in the bilayer outside of the annulus. However, the dependence of the reactions of the plant respiratory chain upon the general presence of mitochondrial phospholipids is now well established. In early studies, treatment by various phospholipases (phospholipases A, C, and D) indicated the phospholipid-dependence of mitochondrial oxidations, although products of enzymatic degradation, such as fatty acids, lysophospholipids, diacylglycerol, and phosphatidic acid, were not always excluded as inhibitory agents (Douce and Lance, 1972). Recently, more definitive investigations have been conducted using better procedures for the removal of phospholipids (extraction with 90% acetone). These studies have conclusively established the dependence of electron transport systems upon the presence of membrane phospholipids. NADH-cytochrome c oxidoreductase, succinate-cytochrome c oxidoreductase, and cytochrome oxidase were almost totally inactivated in the phospholipid-depleted mitochondria. A positive correlation was observed between the quantities of phospholipids extracted and the decrease of these enzymatic activities. Restoration of NADH-cytochrome c oxidoreductase activity was affected by the addition of micellar suspensions of mitochondrial phospholipids to the phospholipid-depleted membranes. The reactivation of the electron transport chain parallelled the amount of phospholipid added, until a saturating value was reached (Jolliot *et al.*, 1978). This activation effect is a general phenomenon and probably results from a physical effect on the state of aggregation of the respiratory enzymes.

According to Lyons (1973) and Raison and Chapman (1976), mitochondrial membranes exhibit a melting transition over a particular temperature range which depends critically on the nature of the fatty acid residues of their polar lipids; it is low for unsaturated polar lipids and much higher for saturated ones. Below the melting temperature the hydrocarbon chains are rigid, whereas above it they are free to move. Consequently, a very large number of reports have associated thermal transitions in the membrane lipids with changes in the activities of membrane-bound enzymes (Raison, 1973). One of the complicating factors is the presence in mitochondrial membranes of a range of molecular species of lipids that together with the membrane proteins tend to modify the phase-transition behavior of the lipids. A change in both mitochondrial membrane lipid composition and fluidity has been observed in some plants during acclimation to variation in growth temperature (Miller *et al.*, 1974), and an increase in the freezing tolerance of Jerusalem artichoke tubers during winter dormancy correlated with a decrease in the transition temperature of mitochondrial lipids (E. Chapman *et al.*, 1979). Likewise, an increase in membrane lipid fluidity has been directly correlated with an increase in the Arrhenius activation energy of membrane-associated enzyme systems of the mitochondria of Jerusalem artichokes and mung bean hypocotyls (Raison, 1980). More recently, however, Pomeroy and Raison (1981) have demonstrated that neither changes in membrane fluidity nor transition temperature are a necessary feature of cold acclimation in wheat. Furthermore, Rebeillé *et al.* (1984c), working with sycamore cells in suspension cultures, have shown that major changes in the degree of unsaturation of the fatty acids in mitochondrial membranes, which are controlled by the concentration of O_2 in the culture medium, could occur without any change in the Arrhenius-type plots of the rate of mitochondrial respiration. These results cast doubt on the importance of membrane lipid fluidity in the Arrhenius activation energy of membrane-associated enzyme systems of plant mitochondria. In fact, as suggested by different authors (see Raison, 1980; Quinn and Williams, 1983), some of the Arrhenius discontinuities observed at the level of membrane enzymes could be the consequence of intrinsic thermotropic changes in protein arrangement independent of lipid fluidity. The involvement of more subtle lipid–protein interactions cannot, of course, be ruled out. In other words, the origin of these discontinuities may be associated with the properties of the individual proteins rather than with the properties of the polar lipid matrix (Quinn and Williams, 1983). Obviously, the concept of membrane lipid fluidity must be refined and quantitated and the relationship between the orientational order and the rates of motion better understood. In particular, the nature of the boundary lipids surrounding the integral membrane proteins will require further study.

C.   Protein and Electron Carrier Composition

The plant outer mitochondrial membrane comprises only 6.8% of the total organelle protein, while the inner membrane accounts for 30% of the total, the rest being matrix protein (Bligny and Douce, 1980; Table 2.4). The distribution of some classical marker enzymes between the inner and outer membranes is similar to that reported for mammalian mitochondria (Hanson and Day, 1980). Consideration of the composition of various biological membranes shows that membranes with a high complexity of functions, such as the inner mitochondrial membrane, have a high protein content (over 70% on a dry weight basis) and a large range of polypeptide types. Thus, the ratio of phospholipid–protein is much higher in the outer membrane, which has fewer enzymatic functions than the inner membrane (see Table 2.3).

The mitochondrial inner membrane is one of the most complex of all the biological membranes. It is a highly specialized system for oxidative phosphorylation and energy-linked ion translocation. Energy capture, transduction, and utilization are achieved in a number of reactions that are energetically interlocked and involve at least several hundred polypeptides. Using a two-dimensional procedure for gel electrophoresis that combined two different gradient gel electrophoresis methods [a linear concentration gradient of acrylamide (7.5–15%) in lithium dodecylsulfate at 4°C and a 12–18% acrylamide gradient gel (with 8 M urea) in sodium dodecyl sulfate at room temperature], we have resolved up to 200 polypeptides, many of which are not extracted by NaOH, indicating that they are integral membrane polypeptides. The relatively high protein content of the inner mitochondrial membrane implies that proteins are not free to diffuse rapidly and randomly in the plane of such membranes. The polypeptide pattern of the inner membrane is very different from that of the matrix or of the outer membrane; very few polypeptides from the three fractions have identical mobility in the different electrophoretic systems used. However, careful comparison of the polypeptide patterns of the inner membrane and the matrix reveal that the inner membrane fraction contains some matrix protein. The mitochondrial inner membrane is negatively charged and, in the presence of cations, some matrix enzymes may be bound nonspecifically. Matrix proteins also may be trapped in inner membrane vesicles in high-salt buffer. Sonication of the inner membrane in the presence of 0.4 M NaCl, however, removed only a part of the malate dehydrogenase, suggesting that association of the malate dehydrogenase with the inner membrane may be of physiological significance. In fact, matrix enzymes differed only in the tightness of their binding to the inner membrane.

The membrane proteins that are exposed to the outer surface of the inner membrane have not been identified, probably because several difficulties complicate these types of experiments. In the first place, the outer mitochondrial membrane limits the access of macromolecular reagents (lactoperoxidase involved in the lactoperoxidase–iodide–hydrogen peroxide iodination system, specific proteases such as thermolysin, etc.) to the inner membrane. Secondly, both mitochondria and inner membrane–matrix vesicles (mitoplasts) are fragile, and one must be sure that these membranes are initially intact and do not become disrupted during the labeling procedure. The plant inner mitochondrial membrane contains several membrane-bound redox components (flavoproteins, iron–sulfur proteins, ubiquinone-10, and cytochromes) that catalyze the reduction of molecular oxygen by the reduced coenzymes (NADH, FADH$_2$) generated intramitochondrially (i.e., in the matrix space). The energy released during the transfer of electrons along the respiratory chain is used to drive energy-requiring reactions such as phosphorylation of ADP (mitochondrial ATPase). Substrate availability to the mitochondrial ATPase and to specific enzymes of the tricarboxylic acid (TCA) cycle is strongly regulated by the permeability of the inner mitochondrial membrane to particular intermediates, and transport of metabolic anions across the plant mitochondrial inner membrane occurs by carrier-mediated one to one exchange mechanisms that are specific for individual anions. Finally, we need to know a great deal more about almost every aspect of the structure, organization, and dynamics of the many proteins that comprise the bulk of this membrane and perform most of its specific functions.

1. Mitochondrial Cytochromes

Cytochromes are defined as hemoproteins whose principal biological function is electron transport by virtue of a reversible valency change of their heme iron. The plant mitochondria examined so far contain at least three major cytochromes: (a) cytochromes in which the heme prosthetic group contains a formyl side chain; (b) cytochromes with protoheme as their prosthetic group; and (c) cytochromes with covalent linkages between the heme and the protein, that are spectrally and functionally similar to those of animal mitochondria (Hartree, 1957; Hackett, 1963; Baker and Lieberman, 1962; Bonner, 1961). Each of these cytochromes exhibit intense light absorbtion in the visible region when reduced. Recording of the spectra at low temperature gives much greater resolution of the absorbance peaks (Keilin and Hartree, 1949; Bonner, 1961; Bonner, 1965; Lance and Bonner, 1968). According to common usage, the peak with the longest wavelength is termed the α-peak, the next peak towards the blue is termed the β-peak, and the peak in the 400 nm region is termed the γ-peak. The α-peaks of the three major cytochromes are sufficiently different to allow separate identification of various cytochromes. In most instances, cytochromes are therefore measured in the α-region of the spectrum. Even though the plant mitochondrial cytochromes absorb light quite intensely, their direct estimation in mitochondrial membranes by absorption spectroscopy is difficult, owing to the extreme turbidity of the membrane suspension. It is therefore customary to determine plant mitochondrial cytochromes by difference spectroscopy. Following the recommendation of the IUB cytochrome nomenclature subcommittee, the cytochromes will be named according to the position of their α-peak in reduced minus oxidized difference spectra at room temperature. The instrument used to obtain difference spectra is the split beam spectrophotometer (Chance, 1972). In this pro-

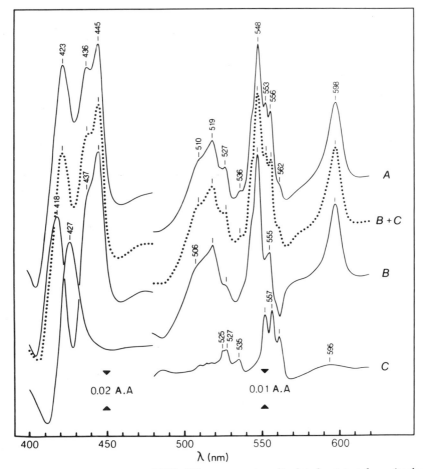

Fig. 2.4. Low temperature (77 K) difference spectra of isolated potato tuber mitochondria. Curve A: NADH (2 mM) reduced minus aerobic. Cytochrome aa_3 has a peak at 598 nm. The α-bands of cytochrome b appear at 562, 556, and 553 nm. In the soret region, cytochrome aa_3 shows a typical double peak at 445 and 436 nm and the b plus c type cytochromes give a common peak at 423 nm. Curve B: NADH (2 mM) reduced minus NADH (2 mM) aerobic with antimycin A (5 μg/ml). Under these conditions the contribution of b type cytochromes is virtually canceled out. Curve C: NADH (2 mM) aerobic with antimycin A (5 μg/ml) minus aerobic without antimycin A. Under these conditions, the contributions of cytochromes aa_3 and c are canceled out. Curve B plus C: spectrum B plus spectrum C reconstructed. This reconstructed spectrum compares favorably with spectrum A.

cedure, an aerobic mitochondrial suspension is distributed into two equal cuvettes (one for the sample to be measured, one for the reference sample). A "baseline" is recorded by measuring the difference in absorbance between the two cuvettes as a function of the wavelength between 500 and 650 nm. Subsequently, a suitable reducing agent (succinate until anaerobiosis, sodium dithionite, or ascorbate) is added to the sample cuvette, and the spectra is recorded again. Low-temperature difference spectra are usually taken with special cuvettes (light path 1–2 mm) made of Plexiglas. The mitochondria are suspended in buffer and divided into two equal portions; one portion is oxidized with O_2 and the other reduced with dithionite. The two samples are then placed in the cuvettes and quickly frozen in liquid nitrogen. This procedure intensifies absorption peaks by a factor of 4–10 and also results in a sharpening of the peaks so that cytochromes *b* and *c* can be readily distinguished. In addition, for most cytochromes the absorption maxima are shifted toward shorter wavelength by 3–4 nm (Bonner, 1965; Lance and Bonner, 1968) upon freezing. Reduced minus oxidized difference spectra of potato tuber mitochondria taken at liquid nitrogen temperature are shown in Fig. 2.4. A strong similarity of the cytochrome composition of the respiratory chain exists in all plant mitochondria, including those of photosynthetic tissues (Table 2.6).

TABLE 2.6

Cytochrome Components of the Respiratory Chain in Purified Plant Mitochrondria as Determined at Room Temperature

Material	Carrier concentration (nmoles/mg protein)[a]		
	cyt aa_3	cyt *b*	cyt *c*
Mung bean hypocotyls[b,c]	0.38	0.33	0.60
Potato tubers[b,c]	0.28	0.27	0.42
Sycamore cells[d]	0.37	0.35	0.55
Spinach leaves[e]	0.16	0.14	0.19
Spinach petiole[e]	0.34	0.33	0.40

[a] Abbreviation: cyt, cytochrome.
[b] Lance and Bonner, 1968.
[c] Douce *et al.*, 1972.
[d] Bligny and Douce, 1977.
[e] Gardeström *et al.*, 1983.

a. **Cytochrome Oxidase.**   Keilin and Hartree (1949) first demonstrated that mammalian cytochrome oxidase (cyt aa_3) in the reduced state exhibited a two-banded spectrum with maxima at 605 and 448 nm. They deduced that the main portion of the 605-nm band was contributed by component *a* and the main portion of the 448-nm band by component a_3. Cytochrome oxidase in plant mitochondria shows some optical differences compared with the mammalian enzyme. For example, the combined α-band of cytochrome *a* plus a_3 has an absorption maximum at 602–603 nm rather than at 605 nm (Bendall and Bonner, 1966). This is also true of a purified preparation of plant cytochrome oxidase (Ducet *et al.*, 1970). The low-temperature difference spectra in the Soret (γ-band) region of cytochrome oxidase in mung bean hypocotyl mitochondria shows a peak at 445(446) with a shoulder at about 437 nm (Fig. 2.4); the α-band is found at 599(598) nm (Bendall and Bonner, 1966 (Fig. 2.4). The spectra of cytochromes *a* and a_3 were individually studied in plant mitochondria by Yonetani's (1960) method for the optical sparation of these two components (Storey, 1970a; Bendall and Bonner, 1971). The reduced minus oxidized difference spectrum for cytochrome *a* was obtained from the difference spectrum for aerobic mitochondria in the presence of ascorbate plus tetramethyl-*p*-phenylenediamine (TMPD) and cyanide minus the fully oxidized state. The contribution of cytochrome a_3 was suppressed because the ferric a_3–cyanide compound in the sample cuvette and the ferric a_3 in the reference cuvette cancel each other. This spectrum shows a double Soret peak at 439 and 446 nm with an α-peak at 600 nm. The difference spectrum of cytochrome a_3 was obtained from the spectrum of anaerobic mitochondria reduced with ascorbate and TMPD minus the aerobic mitochondria in the presence of ascorbate plus TMPD and cyanide. The spectrum obtained (difference ferrous a_3 minus ferric a_3–cyanide) has little or no α absorption and a single Soret peak at 445 nm. Consequently, the α-peak in the difference spectrum (reduced minus oxidized) of plant cyt aa_3 is clearly derived from a normal cytochrome *a* spectrum differing from that of the mammalian component only in that the peak is shifted 2–3 nm toward the violet.

Central to the resolution of both electron transport and energy conservation in mitochondria is the problem of determining the midpoint redox potentials of the electron transport carriers. These midpoint potentials provide a means for determining the interrelationship between the electron transport carriers of the respiratory chain and the path of electron transport from substrate to O_2. A combined spectrophotometric–potentiometric technique developed by Dutton (1971) for measuring the midpoint potentials of membrane-bound carriers using redox mediators (Dutton, 1978), under strictly anaerobic conditions, has been used effectively by Dutton and Storey (1971) to determine the midpoint potentials of the plant cytochromes a and a_3 in mung bean hypocotyl mitochondria. They found that the redox potential of cytochrome a_3 is $+380$ mV in the absence of ATP while the potential of cytochrome a is as low as $+190$ mV. These are very close to the values obtained for the mammalian cytochromes (Dutton *et al.*, 1970).

b.   b-Cytochromes.   It is generally agreed that the mammalian electron transport chain contains at least two and possibly three b-type cytochromes. The mammalian mitochondrial inner membrane contains equimolecular amounts of cytochrome b_{562} (α-band at 560 nm and 77 K) and cytochrome b_{566} (α-band at 562.5 nm at 77 K) which may be distinguished from each other by their midpoint potential at pH 7.2 (Rieske, 1976; Nelson and Gellerfors, 1978). Cytochrome b_{562}, referred to as b_K (in honor of D. Keilin) by some authors, has a relatively high redox potential while b_{566} (or b_T) has a relatively low potential.

Knowledge of the b-type cytochromes in plant mitochondria is not yet as detailed as it is in animal mitochondria. Original studies on plant mitochondria showed the presence of three b-type cytochromes with α-bands at 557(556), 560, and 566 nm. The α-peaks of these cytochromes are shifted to slightly shorter wavelengths at low temperature (77 K) and appear at 553, 556(557), and 562(563), respectively (Lance and Bonner, 1968; Fig. 2.4). Lambowitz and Bonner (1974) have shown that plant mitochondria contain two extra b-type cytochromes. One of the new species is (α-peak, 25°, at 557–561 nm) reducible only by dithionite and may be a denatured form of the respiratory chain b-cytochromes. Redox titrations have shown that cytochrome b_{557} and cytochrome b_{560} have relatively high redox potentials of $+75$ and $+40$ mV, respectively, while cytochrome b_{566} has a relatively low potential of -75 mV (Dutton and Storey, 1971; Lambowitz and Bonner, 1974). The low-potential b-cytochrome (b_{566}) appears to be identical in terms of its spectra to the corresponding component of animal mitochondria (b_T). Both have the α-band in the reduced state at a longer wavelength compared to the other b-cytochromes, and both have midpoint potentials at around -80 mV. The analogy is strengthened by experiments with anaerobic succinate-reduced mitochondria showing enhanced reduction of the plant and the animal cytochrome b_{566} components in response to antimycin plus oxidant (O_2 or ferricyanide) or to energization (Lambowitz and Bonner, 1973, 1974; Storey, 1972). This extra reduction is rapidly relaxed upon addition of a redox mediator phenazine methosulfate (PMS), suggesting that the reduction is merely due to reversed electron transport (Lambowitz *et al.*, 1974; Wikström and Lambowitz, 1974). In fact, the essential difference between plant and animal systems is that the mediator is more effective in short-circuiting reversed electron transport in the former than in the latter (PMS brings the system closer to equilibrium by rapidly shunting electrons driven to cytochrome b_{566} back to cytochrome c).

Cytochrome b_{560} is readily reduced under anaerobic conditions by succinate, NADH, and ascorbate plus TMPD and rapidly reoxidized by O_2 with a half-time of 8 msec (Storey, 1969). This cytochrome appears to correspond to cytochrome b_{562} (b_K) in animal mitochondria, at least with respect to the fact that the α-peaks of both cytochromes shift slightly (by 2 nm) to the red in the presence of antimycin A (Lambowitz and Bonner, 1973). The time course of the transition from fully oxidized to fully reduced upon anaerobiosis was measured for cytochromes b_{560} and b_{566}. Under these conditions, the two b-type cytochromes were in redox equilibrium (Storey, 1973), and b_{560} transfers electrons directly to c-type cytochromes.

The second of the plant high-potential components, b_{557}, seems to have no counterpart at all in animal mitochondria. The oxidation kinetic studies of Storey (1969, 1970b, 1973) suggest that this component functions in respiration, but its specific role in the metabolism of plant mitochondria remains to be established. Cytochrome b_{557} is completely reduced by succinate under anaerobic conditions. If, however, azide is added to block the cytochrome oxidase, b_{557} remains oxidized even under anaerobic conditions; the full explanation of this strange phenomenon is unknown.

Recently, Ducet and Diano (1978) have shown that the three b-type cytochromes (b_{566}, b_{560}, and b_{557}) can be isolated in enzymatically active forms as part of a respiratory chain segment that catalyzes the electron transport from ubiquinone to cytochrome c. Attempts at further purification of the b-cytochromes, however, led to strong modifications. The low-temperature difference absorption spectrum of purified cytochrome b shows only one band in the α-region at 557 nm whereas three cytochrome b α-bands are seen in the low-temperature absorption spectrum of intact mitochondria or isolated respiratory chain segments. It is possible, therefore, that the different properties related to the b-type cytochromes *in situ* may result from differences in the environment of the membrane-bound protein (Yu *et al.*, 1979). In other words, it is not yet clear whether these spectroscopically defined entities are really different molecular species. Aroid spadices contain an additional cytochrome of the b-type, b_7 (Bendall and Hill, 1956). This particular cytochrome is probably not linked to the electron transport chain because neither malate nor NADH can reduce cytochrome b_7, even upon anaerobiosis. Finally, the plant outer mitochondrial membrane appears to be similar to its mammalian counterpart in that it has an antimycin A-insensitive NADH-cytochrome c reductase consisting of a flavoprotein and cytochrome b_{555} (Moreau and Lance, 1972; Douce *et al.*, 1973b; Fig. 2.5).

c.   *c-Cytochromes.*   The mammalian electron transport chain contains two c-type cytochromes, cytochrome c and cytochrome c_1. Cytochrome c is a positively charged (high isoelectric point) protein that has been structurally and functionally characterized on the molecular level. For example, the amino acid sequence of cytochrome c is known for more than 50 species. It differs from the other membrane electron carriers in that it is a hydrophilic molecule (Lemberg and Barrett, 1973). Cytochrome c possesses an absorption peak at 550 nm that shifts to 547(548) nm at 77 K and very often a satellite absorption band at around

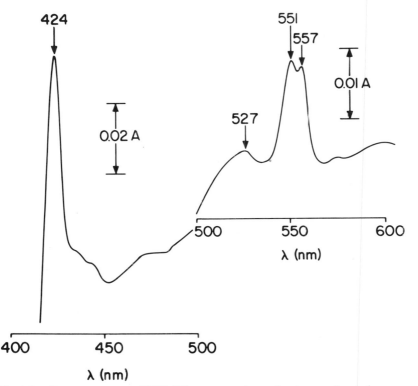

Fig. 2.5.   Low temperature (77 K) difference spectrum of outer membrane from mung bean hypocotyl mitochondria. Dithionite-reduced minus aerobic. Path length: 2 mm; 1.4 mg protein/ml. [From Douce *et al.* (1973b).]

544(545) nm is observed (Estabrook, 1966; Fig. 2.4). A half reduction potential of about +250 mV has been determined for cytochrome c, indicative of its localization in the respiratory chain between the b-type and a-type cytochromes. Cytochrome c_1 can be solubilized from the inner mitochondrial membrane with only bile salts, chaotropic agents, or detergents (Lemberg and Barrett, 1973). Cytochrome c_1 has been isolated from various sources and its absorbance characteristics are not grossly different from those observed in submitochondrial particles (Yu *et al.*, 1972). The reduced cytochrome c_1 shows peaks at 418 and 553(554) nm. At 77 K the absorption bands are sharpened and the α-peak is shifted to 552(553) nm. Consequently, cytochrome c_1 is notoriously difficult to measure by absorption spectroscopy of organelles since its absorption maximum is very close to that of cytochrome c and b. Consequently, there is always the possibility that a peak attributed to cytochrome c_1 may in fact represent another pigment (cytochrome b?) with a similar absorption maximum. The midpoint oxidation–reduction potential of purified yeast cytochrome c_1 measured at pH 7.0 is about +200 mV (Ross and Schatz, 1976).

The isolation and purification of soluble cytochrome c was first accomplished from wheat germ by Goddard (1944). Apparently, the types of cytochrome c found in all higher plants differ from those found in animals by having two trimethyllysine residues instead of lysine in positions 72 and 82, 111 or 112 total residues instead of 104, and an additional eight residue tail with an acetylalanine residue at the N-terminal end of the chain (Dayhoff, 1972; Lemberg and Barrett, 1973; Martinez *et al.*, 1974). Plant types of cytochrome c differ from one another in only 21 positions. For all the types of cytochrome c examined so far, however, 34 residues remain invariant, for example, Cys-14,-17 (heme binding), Met-80 (heme iron binding), and His-18 (heme iron binding). The heme is held in a hydrophobic crevice by the two thioether bridges formed by Cys-14 and Cys-17 (Dickerson, 1972). Therefore, certain units cannot be changed at all without destroying the protein's activity. From these data a phylogenetic tree of evolution has been constructed (Thompson *et al.*, 1971).

When mitochondria from Jerusalem artichokes are thoroughly washed for several hours in a phosphate buffer (pH 7.2, 0.1 M; Lance and Bonner, 1968) or 0.25% NaCl (w/v) (Bonner, 1965) with gentle stirring, cytochrome c is solubilized and can be separated from the bulk of the mitochondria by high speed centrifugation. It appears that the spectrum of the plant soluble cytochrome c is indistinguishable from that of mammalian and yeast cytochrome c, both at room temperature and at 77 K. The cytochrome c thus extracted has a sharp asymetrical α-band at

547(548) nm and a Soret peak at 415 nm in an ascorbate-reduced minus a ferricyanide-oxidized difference spectrum (77 K; Bonner, 1965; Lance and Bonner, 1968). When cytochrome c has been removed from mitochondria by extensive washing, an ascorbate-reducible cytochrome c, analogous to the c_1 component of animal and yeast mitochondria, remains strongly bound to the mitochondria (Bonner, 1965; Lance and Bonner, 1968; Storey, 1969). This cytochrome c_1 has been extracted from potato tuber mitochondria by the use of surface active agents (Diano and Ducet, 1971; Ducet and Diano, 1978). At 77 K the reduced cytochrome c_1 shows peaks at 419, 517, and 549 nm (552 nm at room temperature), and its absorbance characteristics are not grossly different from those observed in submitochondrial particles (Ducet and Diano, 1978a). Because of the close proximity of reduced absorption bands, the two c-type cytochromes appear in the difference spectra of intact mitochondria as a single component and the b_{553} (77 K) component should not be confused with cytochrome c_1 on the basis of its chemical and spectral similarity to the mammalian and the yeast cytochrome c_1 (Fig. 2.4). The c-type cytochromes of plant mitochondria cannot be distinguished by different midpoint potentials. The E_m values of membrane-bound cytochrome c_1 and cytochrome c in plant mitochondria (235 mV) are similar to those found in animal mitochondria (Dutton and Storey, 1971; Dutton and Wilson, 1974).

2. Mitochondrial Ubiquinones

Ubiquinones (coenzyme-Q) are derivatives of 2,3-dimethoxy-5-methyl-1,4-benzoquinone substituted at C-6 with an isoprenoid side chain comprising 6–10 isoprene units (Crane, 1965; Fig. 2.6). They are designated by the number of isoprenoid units on the side chain as Q-6, Q-7, and so on, up to Q-10. Unless otherwise stated, in the present book the term ubiquinone represents the homologue with the C-50 side chain (Q-10), which is the most abundant member of the series. The most characteristic marker for the ubiquinone structure is its absorption peak at 275 nm. The absorption of the compound decreases at 275 nm during a reduction reaction that will convert the quinone (ubiquinone) to the hydroquinone (ubiquinol) (Crane *et al.*, 1959). The ease of extraction of ubiquinone from membranes with hydrocarbon solvents has been taken as evidence that the quinone is not tightly bound to protein. Furthermore, ubiquinone in the inner membrane has the same 275-nm absorbance maximum as ubiquinone in ethanol. Therefore, no spectral shift indicative of binding to protein is observed (Szarkowska and Klingenberg, 1963). The existence of semiquinone anions (ubisemiquinone; Fig. 2.6), however, has been reported in the inner mitochondrial membrane

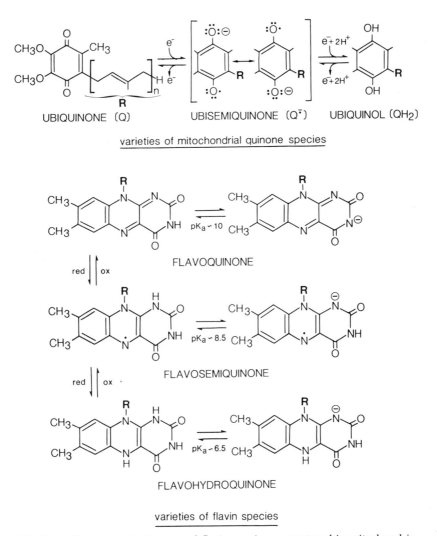

Fig. 2.6.   Structure of quinone and flavin species encountered in mitochondria.

(Trumpower, 1981a,b). The fact that the stability constant of ubisemiquinone measured in a lipophilic environment is extremely low (10^{-10};
i.e., the instability of semiquinone in solution is considerable) suggests
that protein-bound ubisemiquinones do exist and participate as redox
components in the mitochondrial respiratory chain (Yu and Yu, 1981;
Gutman, 1980). It is not clear whether the quinones, which are well in
excess of the other components of the electron transport chain, are freely
mobile to form a homogeneous pool or whether there are several pro-

tein-bound forms that may be considered distinct species and that may structurally comprise a "solid-state" electron transport chain (Rich, 1981, 1984). The trans double bonds in the molecule (Fig. 2.6) introduce a periodic rigidity into the isoprenoid side chain, and an examination of space-filling models indicates that the vicinal methyl groups further constrain intramolecular rotation around single bonds by steric hindrance (Trumpower, 1981a). It seems likely, therefore, that the side chain of ubiquinone has been selected for some function (i.e., unique positioning of these quinones in the hydrophobic core of the lipid bilayer) other than "flexibility" or "flip-flop" motions. Several lines of evidence (for review see Rich, 1984), however, have pointed to the free quinone itself as being the mobile agent that electronically connects the multiprotein donors and acceptors.

Again, knowledge of the ubiquinones in plant mitochondria is not yet as detailed as that in mammalian mitochondria, probably because plants contain several other quinones in relatively large amounts which interfere with assays for ubiquinone. Assays on green plants are especially complicated in view of the presence of plastoquinone, tocopherolquinone, and phylloquinone in the plastids. However, in the few cases that have been investigated, ubiquinone has been found in all tissues of higher plants (Crane, 1959), which would seem to place ubiquinone in the mitochondria (Pumphrey and Redfearn, 1960; Dilley and Crane, 1963). Its functional role in the plant mitochondrial electron transport chain has been established by Beyer *et al.* (1968). Bonner and Rich (1978) and Huq and Palmer (1978) have extracted the ubiquinone from *Sauromatum guttatum*, *Arum maculatum* spadices, and mung bean hypocotyl mitochondria. The extracts were examined by thin-layer chromatography, and the results of these investigations showed that the quinones of plant mitochondria were undoubtedly ubiquinone Q-10 (see, however, Schindler *et al.*, 1984). On a protein basis, the ubiquinone Q-10 content of plant mitochondria (7 nmoles/mg protein) is similar to that of other mitochondria (Ikeda *et al.*, 1980). The stopped-flow oxygen pulse experiments of Storey and Bahr (1972) and the application of the ordering theorem (Higgins, 1963) to the oxidation and reduction kinetics of ubiquinone identify this component as a link between the dehydrogenases and the cytochrome *b* of the respiratory chain. Strong support for the existence of a discrete pool of ubisemiquinone in plant mitochondria was provided by the observation that in a variety of plant mitochondria, as in animal mitochondria, there is an electron paramagnetic resonance (EPR) detectable paramagnetic component that is due to ubisemiquinone [it is seen as a rapidly relaxing split around $g = 2.00$ signal (Rich *et al.*, 1977a,b)]. In addition, the split signal behaved in a number of steady-state conditions as a single component; it could be reduced by NADH, succinate, and malate, and its level of reduction in state 4 was greater than in state 3 or the uncoupled state. Hence, it appears that the component(s) (ubisemiquinones?) giving rise to the EPR spectrum is indeed part of, or in rapid equilibrium with, the main respiratory chain.

3. Mitochondrial Iron–Sulfur Proteins

Iron–sulfur proteins contain iron and usually noncysteinyl sulfur in a center (or cluster) capable of sustaining reversible, one-electron transfer between two active-site oxidation levels. Centers composed of Fe_2-S_2 or Fe_4-S_4 are currently recognized in mammalian mitochondria (Beinert, 1977; Fig. 2.7), although the recently identified trinuclear structure has not been completely excluded (Beinert and Thomson, 1983).* The fact that iron–sulfur proteins were discovered only recently is due partly to their lack of unique optical features. Electron paramagnetic resonance spectroscopy remains the method of choice for their detection and assessment of function. The phenomenon and theory of EPR have been amply described elsewhere (see, for example, Jardetsky and Jardetsky, 1962). Measurements in aqueous solutions are hampered by the severe dielectric losses due to water. Ice, on the other hand, does not have this drawback. Therefore, EPR spectra, usually those of transition metals, are often sharpened by lowering the temperature (to 13 K); this is due to a decrease in the efficiency of certain relaxation processes. According to Holm and Ibers (1977), iron–sulfur structures may occur in several oxidation states of which three have thus far been observed in biological systems; these are oxidized high-potential iron–sulfur protein [H_iPIP, $(Fe_n-S_n)^-$], reduced H_iPIP, which is equivalent to the state of oxidized ferredoxins $(Fe_n-S_n)^{2-}$, and reduced ferredoxins $(Fe_n-S_n)^{3-}$. Iron–sulfur clusters in the +1 $(Fe_n-S_n)^-$ and +3 $(Fe_n-S_n)^{3-}$ oxidation states exhibit EPR resonance at low temperatures (detectable below 35 K), while an $(nFe-nS)^{2-}$ cluster is silent in EPR. Up to the time that biochemists started practicing EPR spectroscopy at temperatures in the range of 77 K, the number of known mitochondrial iron–sulfur centers was three: those of the NADH (complex I) and succinate (complex II) dehydrogenases and that of the cytochrome $b-c_1$ complex (complex III). The increased spectroscopic resolution realized at temperatures below 25 K has made it clear that each dehydrogenase contains a number of distinct iron–sulfur centers. However, EPR studies have not yet yielded an unambiguous picture of the number and type(s) of iron–sulfur centers present in mitochondria.

*A three volume series, "Iron–Sulfur Proteins," has been published (Lovenberg, 1977).

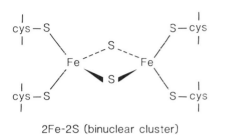

2Fe-2S (binuclear cluster)

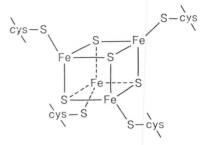

4Fe-4S (tetranuclear cluster)

IRON-SULFUR PROTEIN

Fig. 2.7.   Structure of iron–sulfur proteins encountered in mitochondria.

Little is known concerning the role of iron–sulfur proteins in plant mitochondria. Their presence was first deduced from the initial observation of Schonbaum *et al.* (1971), showing an EPR spectrum of a submitochondrial particle from a skunk cabbage spadix at 77 K. This spectrum showed a clear signal at **g** = 1.94 and another at 2.0, which are characteristic of iron–sulfur proteins (Fig. 2.8). In addition, careful analysis of plant mitochondria made by Dizengremel *et al.* (1973) showed that there is a 1 : 1 stoichiometry between the nonheme iron and the acid labile sulfur. More recently, EPR signals comparable to those of the iron–sulfur proteins of mammalian mitochondria complex I (Cammack and Palmer, 1973, 1977; J. M. Palmer, 1976), complex II (J. M. Palmer, 1976; Cammack and Palmer, 1977; Rich and Bonner, 1978a,b), and complex III (Prince *et al.*, 1981; Bonner and Prince, 1984) have been observed in plant mitochondria. These iron–sulfur proteins thus appear to be widely occurring in mitochondria (Table 2.7).

4.   Mitochondrial Flavoproteins

Riboflavin 5'-phosphate [flavin mononucleotide, (FMN), E_{m7} = −208 mV] and the phosphoanhydride between FMN and adenosine 5'-phosphate [flavin adenine dinucleotide (FAD), E_{m7} = −219 mV] are the flavin compounds found in complex flavoprotein dehydrogenases (NADH dehydrogenase, succinate dehydrogenase, lipoate dehydrogenase, etc.) of mammalian mitochondria. The functional distinction between flavoenzyme dehydrogenases and oxidases is that the flavin dehydrogenase holoenzyme prevents the intrinsicly high reactivity of reduced flavins with O_2 (Walsh, 1978). Instead, the physiological electron acceptors for flavoprotein dehydrogenases are molecules such as quinones or non-

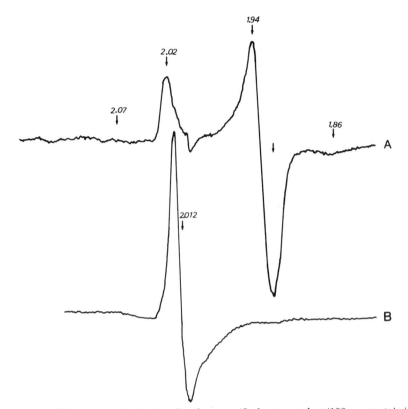

Fig. 2.8.   EPR spectra of mitochondria from purified potato tuber (100 mg protein/ml) reduced with dithionite (A) and (B) oxidized with O_2. The ordinate is the first derivative of microwave absorption (dX"/dH). The gain settings were adjusted to compensate for protein concentration. Spectra were measured at 8 K with the following instrument settings: microwave power 0.2 mW, frequency 9.25 GHz; modulation amplitude 1.25 mT, and frequency 100 kHz. Arrows indicate **g** values. (Courtesy of Dr J. Gaillard.)

heme iron–sulfur proteins. Consequently, these flavoprotein dehydrogenases may be defined as electron-transferring enzymes that contain a bound flavin prosthetic group (NADH dehydrogenase contains FMN whereas succinate dehydrogenase contains FAD). They catalyze redox reactions in which the flavin moiety accepts one or two electrons (Fig. 2.6). The flavoproteins of the mitochondrial respiratory chain play a prominent role in the redox reactions of electron transport and energy conservation which occur during oxidation of substrates. Three types of procedures are now in general use for analysis of mixtures containing FAD and FMN; these include (a) enzymatic analysis, (b) spectrophotometric analysis, and (c) fluorometric analysis. Unlike the cyto-

TABLE 2.7

Iron–Sulphur Centers Detected by EPR Spectroscopy of *Arum Maculatum* Submitochondrial Particles[a]

State	G Values	Temperature (K)	Midpoint potential (mV)	Assignment
Reduced	2.02; 1.932; 1.922	77	−7	Succinate dehydrogenase center S-1
Reduced	2.02; 1.93	18	−240	Succinate dehydrogenase center S-2
Reduced	(2.02); 1.933; 1.928	65	~−20	Alternative NADH oxidation pathway?
Reduced	(2.03); 1.93	77	−240	NADH-ubiquinone reductase center N-1b?
Reduced	2.05; (1.92)	18	−110	NADH-ubiquinone reductase center N-2
Reduced	1.87	12	−275	NADH-ubiquinone reductase centers N-3 and N-4?
Oxidized	2.02; 2.00	12	85	Succinate dehydrogenase center S-3
Oxidized	2.02; 2.00	18	−100	Oxidized $Fe_4–S_4$ centers?
Partially oxidized	2.038; 1.984	12	n.d.	Spin–spin interaction between QH· and QH· or succinate dehydrogenase center S-3

[a] Midpoint potentials are at pH 7.4, relative to the standard hydrogen electrode, and values are ± 15 mV. The nomenclature of Ohnishi (1976) is followed in the assignment of iron–sulphur centers. [From Commack and Palmer (1977) and reprinted by permission from *Biochem. J.* **166**, pp. 347–355. Copyright © 1977. The Biochemical Society, London.]
[b] Abbreviations: QH·, ubisemiquinone radical; n.d., not determined.

chromes, flavoproteins all have the same absorbance maximum around 460 nm due to the flavin prosthetic group in the oxidized form, which disappears on reduction (Chance and Williams, 1956). Consequently, all the components have essentially the same difference spectrum and individual components cannot be clearly resolved. Spectral changes observed in mitochondrial preparations in the region 450–520 nm cannot be specifically assigned to flavoproteins as was formerly thought, however, because the spectral characteristics of nonheme iron proteins overlap with those of flavoproteins. Chance *et al.* (1967) demonstrated that certain of these flavoproteins fluoresce to varying degrees in the oxidized state and that this fluorescence is lost on reduction. The different ratios of fluorescence–absorbance changes, designated FA ratios, which are observed in the presence of selected substrates and inhibitors, can be used to differentiate between the various mitochondrial flavoproteins. A total of six flavoproteins could be distinguished in animal mitochondria by this method (Chance *et al.*, 1967). The redox potential of the flavin in flavoproteins covers a very wide span because the free radical of flavin is rather stable and this stability is enhanced in many flavoproteins; this permits shuttling between flavohydroquinone and the radical, flavohydroquinone and flavoquinone, or the radical and flavoquinone (Fig. 2.6), each redox couple having a different potential. The stability of the flavin radical (flavosemiquinone) is also responsible for the ability of flavoproteins to accept and donate either one electron or two. This probably permits bridging between the oxidation of two-electron donor substrates, such as NADH and succinate, and the reduction of obligate one-electron acceptors, such as iron–sulfur proteins.

The problem of differentiating the various flavoproteins of plant mitochondria is technically exceptionally difficult because mitochondria from plant sources do not respond to many of the respiratory chain inhibitors (e.g., rotenone; Ikuma and Bonner, 1967b) which have proved so useful in differentiating the flavoproteins of animal mitochondria. In addition, plant mitochondria contain several additional flavoproteins absent from mammalian mitochondria (Storey, 1980). We believe that some of these flavoproteins derive from the extramitochondrial membranes present in all the plant mitochondria isolated so far. Nonetheless, the oxidation–reduction potentials of the flavoproteins of various plant mitochondria have been measured under anaerobic conditions by means of a combined spectrophotometric and fluorimetric–potentiometric method. At least five components were resolved with midpoint potentials (+170– −155 mV) considerably higher than those characteristic of the flavoproteins from yeast mitochondria, which vary between +50 mV and −320 mV (Storey, 1970c, 1971a), and animal mitochondria, which vary between −45 and −220 mV (Erecinska *et al.*, 1970). It has proved difficult to relate the flavoprotein species identified in plant mitochondria with those identified in mammalian mitochondria. The low-potential fluorescent component (−155 mV) is most probably the flavoprotein of lipoate dehydrogenase involved in the oxidation of the endogenous α-ketoacids of the TCA cycle in plant mitochondria (Storey, 1970c). It is clear, therefore, that there is much more to be learned about the plant mitochondrial flavoproteins. None of those that mediate redox reactions in plant mitochondria have been characterized as an isolated, fully active enzyme. There are, however, two recent reports in the literature on the isolation of succinate dehydrogenase from pea (Nakayama and Asahi, 1981) and sweet potato mitochondria (Hattori and Asahi, 1982). It appears that the plant succinate dehydrogenase containing FAD resembles that of yeast and mammalian mitochondria. Plant mitochondria do not appear to contain the flavoproteins associated with the oxidation of fatty acids (Cooper and Beevers, 1969; Hutton and Stumpf, 1969).

II. FUNCTIONS OF PLANT MITOCHONDRIAL MEMBRANES

The fundamental structure of the electron transport chain and phosphorylation system of higher plant mitochondria is remarkably similar to that found in mammalian systems. It is clear that this basic system, developed at an early stage of evolution, has been highly conserved throughout the development and divergence of animal and plant species (A. L. Moore and Rich, 1980). It is becoming increasingly apparent, however, that mitochondria isolated from higher plants have a considerably more complex respiratory chain, which may enable them to fulfil a more complex metabolic function in the cell. Likewise, in addition to the basic metabolite transport system found in mammalian mitochondria, specific carriers or transporters have been found in mitochondria isolated from plant tissues.

A. Mechanisms of Electron Transport

The immense complexity of the respiratory chain becomes apparent when it is realized that a single respiratory chain unit (mol. wt. $\sim 1.52 \times 10^6$) contains as many as 40 redox centers and 50 polypeptides together with significant amounts of phospholipids.

For many years, the organization of the respiratory chain of plant mitochondria was thought to be very similar to mitochondria from more extensively studied animal sources, such as rat liver or beef heart. In fact, the plant mitochondrial cyanide-sensitive electron pathway (i.e., the sequence of electron carriers that mediate the flow of electrons from respiratory substrates to O_2 via cytochrome oxidase) appears similar to that found in mitochondria from animal tissues. It is now recognized, however, that there are a number of distinct differences between plant and animal mitochondria; these include the cyanide-resistant electron pathway, which is also encountered in the mitochondria of microorganisms (Lloyd, 1974), and the rotenone-resistant electron pathway. At one time, these differences were not felt to be real but rather to be artifacts due to the difficulties associated with isolating mitochondria from plant tissues. Fortunately, this view is no longer widely held.

1. Cytochrome Oxidase Pathway

The plant mitochondrial respiratory chain is remarkably compact and is responsible for electron transfer from endogenous NADH and succinate to O_2. Both NADH and succinate are formed in the mitochondrial matrix and their initial oxidation by the electron transfer chain takes place on the inner surface of the inner membrane. The plant respiratory chain, similar to its more extensively studied counterpart in animal mitochondria, consists of only five protein complexes: complex I, complex II, complex III (usually called the cytochrome $b–c_1$ complex), cytochrome c, and complex IV or cytochrome c oxidase. Except for cytochrome c, these complexes are very hydrophobic (the fractionation procedure involves the use of deoxycholate and cholate) and are soluble in the "fluid" lipid bilayer medium of the mitochondrial cristal membrane, more generally known as the coupling membrane (Mitchell, 1980).

*a. **Mitochondrial Complex I.***   Complex I is the segment of the respiratory chain responsible for electron transfer from NADH to ubiquinone. Consequently, complex I is the entry point for the redox equivalents of NADH, produced in the matrix space during the course of substrate (malate, α-ketoglutarate, pyruvate, citrate) oxidation, as given in the following reaction schema:

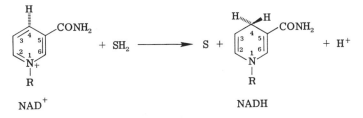

Enzymatic reduction of NAD^+, which is stereospecific, occurs at the 4-position of the nicotinamide ring that is attached to ribose in a β-glycosidic linkage. Complex I contains a noncovalently bound flavin mononucleotide (FMN), several iron–sulfur centers, and probably two molecules of ubiquinone. In this complex, labile sulfur and nonheme iron are present in equal amounts (16–18 g-atoms of non heme iron/mole of flavin). Upon reduction with NADH, four iron–sulfur clusters of the low-potential ferredoxin type (centers N-1–N-4) became detectable by EPR at 12 K, each at an approximately equal concentration with the flavin (Beinert, 1977). Both kinetic analysis and determination of the midpoint redox potentials of these components suggest that the sequence of reaction in complex I may be in the order of centers 1–4–3–2

(Singer and Gutman, 1974). In addition, potentiometric titrations of mitochondria with dithionite in the presence of mediator dyes (Ohnishi, 1975) revealed an apparent splitting of center 1, the lowest potential center, into two components (1_a and 1_b) with $E_{m7.2}$ values of -380 and -240 mV, respectively (the potential of the former being well below that of the FMN/FMNH$_2$ couple). Because complex I contains four, or at the most five, EPR-detectable iron–sulfur centers and 16 g-atoms of nonheme iron, it appears that one or more of the clusters are EPR silent. Depending on the type of iron–sulfur center present, one would expect to find four clusters if they were Fe_4–S_4 centers or twice as many if they were of the Fe_2–S_2 type. The structural nature of the iron–sulfur clusters of NADH dehydrogenase from animal mitochondria has been studied by the cluster extrusion technique (Paech *et al.*, 1981). Whenever extrusion was nearly complete, both binuclear and tetranuclear clusters were found at a mole ratio of 2 : 1. It is clear, therefore, that EPR studies have not yet yielded an unambiguous picture of the number and type(s) of iron–sulfur centers present. The preparation of complex I isolated by Galante and Hatefi (1979) was reported to contain 1.4–1.5 nmoles FMN/mg protein; from this content, a minimum molecular weight of 670,000 can be calculated for complex I. In contrast to the values of the other respiratory chain complexes, an unusually high number of polypeptides (16–18) has been reported for complex I (Hatefi *et al.*, 1979). According to Ragan *et al.* (Smith and Ragan, 1980; Early and Ragan, 1980), the flavoprotein fragment (probably the subunits of NADH dehydrogenase) forms the core of the enzyme, buried in a shell of other more hydrophobic subunits. The oxidized minus NADH-reduced spectrum of complex I shows a wide absorption spectrum between 400 and 500 nm, which is mainly due to the combined absorbances of the flavin and the iron–sulfur centers. More than 50% of the absorbance of the enzyme at 450 nm is bleached with NADH or dithionite. Destruction of the iron–sulfur centers with high concentrations of mersalyl, followed by reduction of the flavin with dithionite, showed that the enzyme so treated had no other absorption in the visible region of the spectrum.

A clean and active NADH dehydrogenase containing 13.5–14.5 nmoles FMN/mg protein can be prepared from complex I with an appropriate chaotropic salt (which weakens hydrophobic interactions), followed by fractionation with ammonium sulfate (Galante and Hatefi, 1978). In soluble flavoprotein, NADH dehydrogenase containing 5–6 g-atoms of nonheme iron/mole of flavin, N_1 type iron–sulfur EPR signals were detected, suggesting their close association with the flavoprotein subunit of complex I (Ohnishi *et al.*, 1981). The exact arrangement of the

FMN, iron–sulfur centers, and ubiquinones in complex I is still unknown.

Rotenone and the most powerful inhibitor piericidin A, an antibiotic isolated from *Streptomyces mobaraensis* (Salerno *et al.*, 1977), inhibit NADH-ubiquinone reductase activity of complex I (Fig. 2.9); both bind at specific sites and appear to act on the O_2 side of the iron–sulfur centers. Other redox reactions carried out by this enzyme, such as $K_3Fe(CN)_6$ or 5-hydroxy-1,4-naphtoquinone (juglone) reduction, are not affected by these inhibitors, suggesting that they are reduced at a site on the substrate side of iron–sulfur center 2. In the mitochondrial membrane, complex I functions as the "first coupling site," carrying reversible electron flux from NADH to quinone coupled to generation of $\Delta\mu H^+$ (see Section II,B). Thus, it has been shown that passage of a pair of electrons from NAD^+-linked substrates to ubiquinone in the mitochondrial respiratory chain results in the translocation of four H^+ ions from the matrix into the medium (for a review see Mitchell, 1980).

$$NADH + H^+ + Q + 4\,H^+_{in} \rightarrow NAD^+ + QH_2 + 4\,H^+_{out} \qquad (2.1)$$

The mechanism of the proton translocation at site 1 of energy conservation is not yet elucidated (FMN-mediated proton transfer? ubiquinone-mediated proton transfer?).

Biochemical characterization of complex I in plant mitochondria has not been undertaken so far. Cammack and Palmer (1973), however, have demonstrated unambiguously that complex I in Jerusalem artichoke mitochondria contains at least three distinguishable iron–sulfur centers. Two of these correspond closely to iron–sulfur centers N_1 ($g = 2.02$ and 1.93; $E_m = ^-300$ mV) and N_2 ($g = 2.05$ and 1.92; $E_m = -20$ mV) found in the NADH-ubiquinone reductase segment of the animal mitochondrial respiratory chain. A third signal at $g = 2.10$ and 1.87, seen at temperatures below 20 K, may be the counterpart of centers N-3 and N-4, although it was not possible to resolve the two centers as is possible with animal or yeast mitochondria. In complex I, apparently only the very low-potential center $N-1_a$ is absent in plants. Potentiometric titrations suggest that the sequence of iron–sulphur centers in complex I of plant mitochondria is $N-1 \rightarrow N-3 \rightarrow N-2$. (Cammack and Palmer, 1977).

b. Mitochondrial Complex II.   Complex II is the segment of the respiratory chain responsible for electron transfer from succinate to ubiquinone. The preparation of complex II isolated by Hatefi *et al.* (1976) was reported to contain 4.6–5.0 nmoles FAD/mg protein. The smallest possible unit (M_r 125,000) consists of only four protein subunits. Complex II can be dissociated by means of chaotropic agents into two sub-

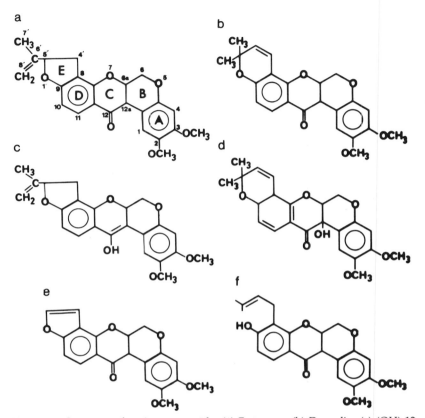

Fig. 2.9. Structure of various rotenoïds. (a) Rotenone; (b) Deguelin; (c) (OH)-12 rotenone; (d) tephrosin; (e) elliptone; and (f) rot-2′-enoic acid. [Adapted from Ravanel *et al.* (1984b).]

complexes (Davis and Hatefi, 1971). The first subcomplex, containing FAD and several nonheme iron centers, is water soluble and consists of two large polypeptides (M_r 70,000 and 27,000). This subcomplex (succinate dehydrogenase) cannot transfer electrons from succinate to ubiquinone but can catalyze redox reactions, such as N-methylphenazonium methyl sulfate reduction. Succinate dehydrogenase contains about 10 nmoles FAD/mg of protein (Davis and Hatefi, 1971) and constitutes about 50% of the protein of complex II; its interaction with other complex II components is mainly hydrophobic in nature. The other subcomplex is very hydrophobic and consists of two small polypeptides (M_r 7000 and 13,000). It has been suggested that one of these hydrophobic subunits (apo-Q protein) carries a ubiquinone-binding site (Salerno *et al.*, 1977). In addition, one of them appears to contain a b-type heme.

The flavin component of the enzyme is covalently attached to the peptide chain by way of a bond from the nitrogen (N-3) of a histidine residue to the 8α-CH_2 group of the FAD isoalloxazine ring system (W. H. Walker *et al.*, 1972). The color and fluorescence of the flavin moiety permits its ready localization in the M_r 70,000 subunit. Although the presence of nonheme iron and of labile sulfur in succinate dehydrogenase has been recognized for nearly 25 years (Beinert, 1977), the chemical nature of the iron–sulfur centers of the enzyme has remained unsettled despite extensive studies in many laboratories. Most investigators agree that, in relatively intact preparations, complex II contains 8 moles each of nonheme iron and acid-labile sulfur and one covalently bound flavin per mole. Two electron paramagnetic resonance–detectable iron–sulfur centers subject to reduction by the substrate have been recognized in particulate and soluble preparations of the enzyme in stoichiometric amounts with the flavin (Beinert *et al.*, 1977). One center, S-1, in the reduced state generates an EPR signal centered near $g = 1.94$, which is typical of binuclear (Fe_2–S_2) centers. The other center, S-3, when oxidized (H_iPIP-type cluster?) exhibits an EPR signal at $g = 2.01$ and has been assumed to be a tetranuclear (Fe_4–S_4) center. However, the increasing awareness that nearly isotropic $g = 2.01$ originals are a recognition mark of Fe_3 clusters raises doubt as to whether the present assignment of the signal in complex II to a (Fe_4–S_4) cluster is indeed correct (Beinert and Thompson, 1983). Reduction with dithionite but not with succinate elicits the appearance of an additional EPR signal with a different line shape at $g = 1.94$ and has been ascribed to a third center, S-2, also assumed to be binuclear (Beinert *et al.*, 1977; Ohnishi *et al.*, 1976). More complex signals have also been observed, in particular a split signal centered around $g = 2.00$ which has been attributed to an interaction of quinones in the partially reduced form (ubisemiquinone) and center S-3 of succinate dehydrogenase in its oxidized form (Ruzicka *et al.*, 1975). Values for the midpoint potentials measured in various preparations for centers 1, 2, and 3 are 0–30, −260, and +60–80 mV, respectively (Beinert, 1977). The sequence of the redox centers in the path of electron flow is only partially understood so far. Thenoyltrifluoroacetone is a selective inhibitor of succinate dehydrogenase in membrane preparations. This inhibitor blocks the reoxidation of all EPR detectable components of the enzyme, but not their reduction by the substrate (Ackrell *et al.*, 1977). Complex II is subject to a regulatory mechanism and can exist in at least two forms. The inactive form is stabilized by oxaloacetate, which binds tightly to the complex in a 1 : 1 stoichiometry (Gutman, 1976). The enzyme can be converted into the active form by incubation with an activating ligand, such as some dicarboxylic acids (Wojtczak *et al.*, 1969), ATP (Gutman *et al.*, 1971a), or ubiquinol (Gutman *et al.*, 1971b). One striking difference between the active and inactive forms is seen in the half-reduction potential of FAD. It changes from about 0 mV to less than 190 mV when the enzyme is converted from the active into the inactive form. The electron flow from succinate to ubiquinone is not connected with a vectorial translocation of protons from the matrix into the medium (for a review see Mitchell, 1980).

$$\text{Succinate} + Q \rightarrow \text{fumarate} + QH_2 \qquad (2.2)$$

According to Hattori and Asahi (1982), succinate dehydrogenase isolated from sweet potato root mitochondria contains two subunits with molecular weights of about 26,000 and 65,000. The fluorescence of the flavin moiety permits its ready localization in the M_r 65,000 subunit. Furthermore, Hattori and Asahi (1982) indicated that in plant mitochondria the two subunits of succinate dehydrogenase are weakly associated with additional hydrophobic polypeptides in the inner mitochondrial membrane. Very recently, Burke *et al.* (1982) have developed a useful procedure (utilization of Triton X-100 extraction, followed by ammonium sulfate precipitation) for the partial purification of succinate dehydrogenase from mung bean hypocotyl and soybean (*Glycine max*) cotyledon mitochondria. The final fraction was enriched in two polypeptides with approximate molecular weights of 67,000 and 30,000, contained a *b*-type cytochrome, and exhibited the characteristic ferredoxin-type and high-potential iron–sulfur protein-type electron paramagnetic resonance signals reported for the iron–sulfur centers of mammalian succinate dehydrogenase. J. M. Palmer (1976) has also identified two iron–sulfur centers in plant mitochondria which correspond closely to the iron–sulfur centers S-1 and S-2 associated with complex II in animal mitochondria. Redox titrations showed that the midpoint potentials for centers 1 and 2 are -10 and -230 mV, respectively. In addition to centers S-1 and S-2, J. M. Palmer (1976) and Rich and Bonner (1978b) have detected a high-potential iron–sulfur center (H_iPIP?) in plant mitochondria analogous to center S-3 of mammalian systems (signal centered at $g = 2.01$) in terms of its physical properties and its association with succinate dehydrogenase. The plant center S-3, which is also paramagnetic in the oxidized form, has been reported to have a midpoint potential of $+65 \pm 10$ mV which is fully pH independent between pH 7.0 and 8.5 (Rich and Bonner, 1978b). This result suggests that center S-3 is an electron carrier rather than a hydrogen atom carrier. Potentiometric titrations (Cammack and Palmer, 1977) and analysis of power saturation data (Rupp and Moore, 1979) indicate that electron transfer from succinate to ubiquinone probably occurs in the sequence FAD,S-1 → S-3 → ubiquinone. All of these results demonstrate the close analogy between the plant and the animal complex II.

Oestreicher *et al.* (1973) have demonstrated an oligomycin-insensitive activation of succinate dehydrogenase by ADP and ATP in mung bean hypocotyl mitochondria. Furthermore, Silva-Lima and Pinheiro (1975) working with *Vigna sinensis* mitochondria have shown that, after a short period of incubation with an uncoupler, succinate dehydrogenase was extensively deactivated unless the medium contained ATP. These results demonstrate, therefore, that complex II in plants is also subject to a regulatory mechanism.

c. Mitochondrial Complex III.

Complex III (cytochrome c reductase) is the segment of the respiratory chain responsible for electron transfer from ubiquinol (QH_2) to cytochrome c. The organization of complex III (or cytochrome b–c_1 complex) in the electron transport chain of mitochondria is partially known (Trumpower, 1981b). In addition to cytochromes b and c_1 and ubiquinones, an essential component of this complex is an iron–sulfur center with g values of 2.025, 1.89, and 1.81 or 1.78 and a midpoint potential of +300 mV. In a sense, this protein is, therefore, a high-potential iron–sulfur protein, although it only exhibits an EPR signal in its reduced state, typical of reduced ferredoxins. It was first isolated from beef heart mitochondria (Rieske *et al.*, 1964) and has been called the Rieske iron–sulfur protein thereafter. This protein has been purified in active form, reconstituting electron transport from succinate to cytochrome c in depleted preparations of the succinate–cytochrome c reductase complex from beef heart mitochondria (Trumpower, 1981a,b). According to the reported analyses for labile sulfur and iron, there is one Fe_2–S_2 center per cytochrome c_1 and per every two molecules of b-type cytochromes. Kinetically, this iron–sulfur center responds to reductants in a fashion very similar to that of its associated cytochromes (Beinert, 1977). Its reduction is inhibited by antimycin A but not its reoxidation (Rieske *et al.*, 1964). According to Trumpower (1981a,b), the Rieske iron–sulfur protein and not the reduced quinone is the direct donor for cytochrome c_1. The tenacious association between cytochrome c_1 and the iron–sulfur protein may reflect such a functional relationship (Trumpower, 1981a,b). The Rieske protein and cytochrome c_1 are exposed on the cytosolic side of the inner membrane (Beattie *et al.*, 1981). Animal complex III (minimal M_r 250,000) can be resolved into 8–9 polypeptides of M_r ranging from 50,000–8000 (Hatefi, 1978; Engel *et al.*, 1980). The two largest polypeptides (M_r 49,000 and 47,000) or "core proteins" do not carry redox centers. On the other hand, the three polypeptides of intermediate molecular weight containing the prosthetic groups have been characterized as cytochrome b (M_r 30,000), cytochrome c_1 (M_r 29,000), and the Rieske iron–sulfur protein (M_r 24,000). Interestingly, cytochrome b, containing two hemes per polypeptide, has been purified from complex III of yeast mitochondria, possibly providing an explanation for the presence of two spectral forms, b_{562} and b_{566}, in mitochondrial cytochrome b (see Widger *et al.*, 1984). The remaining two to three polypeptides (M_r 12,000–8000) contain no redox center (for a review see Trumpower, 1981a,b). Thirty percent of the enzyme mass is buried in the membrane lipid bilayer (Wingfield *et al.*, 1979). The function of the two larger polypeptides and of the smaller components is unknown, although they constitute half of the mass of the complex. According to Von Jagow and Engel (1980), when complex III is isolated in the presence of mild detergents and studied by gel chromatography and ultracentrifugation, it behaves as a protein of M_r 500,000 (2 × 250,000?). The mechanism of oxidation and reduction of ubiquinone in the mitochondrial respiratory chain at the level of complex III is not fully understood. This is partly because the kinetic studies have been largely limited to the optically detectable cytochromes c_1, b_{562}, and b_{566}. This problem is of special interest because ubiquinone possesses the intrinsic properties necessary to function as a redox-linked hydrogen carrier by a direct chemiosmotic mechanism (Mitchell, 1980) and because the ubiquinone pool is a crossroad between electron fluxes emerging from different dehydrogenases and the cytochromes through to O_2. In the mitochondrial membrane, complex III functions as the "second coupling site," carrying reversible electron flux from ubiquinol to cytochrome c coupled to the generation of $\Delta\mu H^+$. Thus passage of a pair of electrons from ubiquinol to cytochrome c in the mitochondrial respiratory chain results in the translocation of four H^+ ions from the matrix into the medium. The $H^+/2e^-$ ratio for "site II" was measured by utilizing succinate as the electron donor and ferricyanide as the electron acceptor from mitochondrial cytochrome c (Alexandre *et al.*, 1978).

$$QH_2 + 2 \text{ cytochrome } c^{3+} + 2 H^+{}_{in} \rightarrow Q + 2 \text{ cytochrome } c^{2+} + 4 H^+{}_{out} \quad (2.3)$$

Since mitochondria have a molar excess of ubiquinone, relative to the cytochrome b–c_1 complex (for a review see Hatefi, 1976), and since most, if not all, of the ubiquinone is needed for sustaining a normal electron flux, it was proposed that the total amount of redox-active ubiquinone was kinetically homogeneous, having sufficiently rapid lateral mobility so that the rates of reduction and oxidation of ubiquinone are determined by the turnover numbers of dehydrogenase and complex III, respectively (Kröger and Klingenberg, 1973). Further evidence for the mobility comes from experiments of Schneider *et al.* (1980) which showed that diffusion distance between components could be changed by lipid dilution. The existence of several forms of stable ubisemi-quinone in complexes II and III (and perhaps in complex I), however, strongly supports the view that bound ubisemiquinones (it is generally believed that the pK_a for QH· is about pH 5) participate as redox compo-nents in the mitochondrial respiratory chain (Trumpower, 1981a,b; Yu and Yu, 1981). In order to explain (a) the enigmatic oxidant-induced reduction of cytochrome *b,* (b) the interesting observation of Wikstrom and Berden (1972) showing that when the *b*-type cytochromes were titrated by the fumarate–succinate couple, in the presence of antimycin A but not in its absence, two electrons were required for each equivalent of cytochrome *b* reduced, and (c) the observation that the iron–sulfur protein of the cytochrome $b-c_1$ segment is required for reduction of cytochrome c_1 by succinate (in the presence of antimycin A it is required for reduction of both cytochrome *b* and c_1), Mitchell (1975) and Trum-power (1981b) formulated the proton-motive Q cycle (Fig. 2.10). In this cycle, it is speculated that the semiquinone forms be bound at least several orders of magnitude more tightly than the ubiquinone and ubiq-uinol forms. The last two forms might dissociate and diffuse, indicating that they are not tightly bound to an enzyme but behave as diffusible carriers. The stability constant of free ubisemiquinone in a hydrophobic milieu has been estimated to be 10^{-10} (Mitchell, 1976); consequently, some polypeptide (iron–sulfur protein? *b*-type cytochrome? or closely associated polypeptides?) must stabilize $Q^{\cdot-}$ by approximately four or-ders of magnitude. Recent reviews of this subject by Gutman (1980), Trumpower (1981b), and Rich (1984) should be consulted for more de-tailed discussions. The interactions between complex I (or II) and com-plex III are not clearly understood. Complex III is probably associated with complex II via a specific Q-binding protein (King, 1978), which may bind a pair of interacting Q molecules near to the iron–sulphur center 3 of succinate dehydrogenase. Complexes I and III apparently interact in a 1 : 1 molar ratio to give an NADH-cytochrome *c* oxidoreductase that contains equimolar FMN and cytochrome c_1. The two complexes appear to be linked by a mobile pool of ubiquinone molecules (two or three molecules of ubiquinone-10 in a special environment), and phos-pholipids are required to maintain high rates of electron transport (Her-on *et al.*, 1978). Consequently, a picture is emerging that electron trans-fer from complexes I and II to the cytochrome *b–c* complex is brought about by long-range diffusion within the membrane of reduced ubiqui-none.

The structure and function of complex III in higher plant mitochondria are unknown. An EPR signal comparable to that of the iron–sulfur pro-

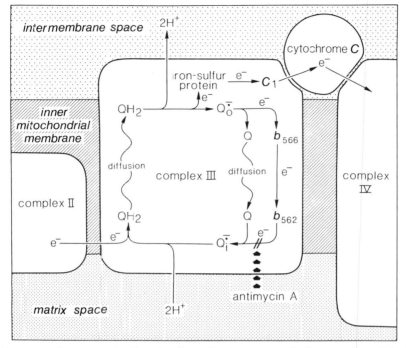

Fig. 2.10. Proton-motive Q cycle scheme for electron transfer from succinate dehydrogenase (complex II) through the cytochrome $b-c_1$ complex (complex III). The scheme depicts electron transfer reactions in the forward direction. On the inner side of the membrane Q_i^- (it is very likely that ubisemiquinone stability is enhanced by suitable protein ligands) is reduced to QH_2 by an iron–sulfur center of the dehydrogenase (complex II), resulting in the uptake of two protons from the matrix. The QH_2 thus formed "diffuses" in the membrane and is then oxidized at the cytoplasmic side of the membrane by a reaction in which the Rieske protein (iron–sulfur protein) of complex III transfers one electron from QH_2 to cytochrome c_1 and a second electron from the Q_o^- thus formed to cytochrome b_{566}, forming Q and releasing two H^+ into the external medium. The Q diffuses in the membrane and is reduced by cytochrome b_{562} at the matrix side, because the electron from reduced cytochrome b_{566} is transferred through the membrane to cytochrome b_{562} via an antimycin-sensitive reaction. The cytochrome b polypeptide of complex III has been found to be transmembranous. The semiquinones are depicted in the anionic form on the basis of potentiometric studies of ubisemiquinone in the isolated reductase complex. Whether the iron–sulfur protein is the direct oxidant for ubiquinol and, if so, the manner in which the resulting ubisemiquinone might participate in reduction and/or oxidation of cytochrome b is currently the subject of extensive investigation and discussion.

tein of mammalian complex III, however, has been observed in plant mitochondria (Prince *et al.*, 1981; Bonner and Prince, 1984), and Rich and Moore (1976) have suggested that the ubiquinone cycle also operates in plant mitochondria. The data obtained by Cottingham and Moore (1983) are good evidence in favor of the idea that plant mitochondria, like their mammalian counterparts, use a mobile quinone pool as a carrier of reducing equivalents in the respiratory chain. Interestingly, it seems that complex III in plant mitochondria has an absolute preference for electrons generated from complex I (Day and Wiskich, 1977a; Dry *et al.*, 1983a; Bergman *et al.*, 1981). This last result strongly suggests that diffusion distance between complexes I and III is shorter than that between complexes II and III.

d. Mitochondrial Complex IV.   Complex IV (cytochrome *c* oxidase) is the terminal complex of the electron transport chain. It has been isolated with detergents as a multipeptide aggregate containing two cytochromes and two atoms of copper (Cu_A and Cu_B) (for a review see Malmström, 1979). The two cytochromes in the cytochrome *c* oxidase differ only in the nature of their axial ligands. The first one, cytochrome *a*, reacts with cytochrome *c*, the other one, cytochrome a_3, reacts with O_2 and is accessible to exogenous ligands (e.g., CN^-, N_3^-, and CO; Lemberg and Barrett, 1973). Furthermore, complex IV spans the membrane asymmetrically, cytochrome *a* is on the side facing the outer membrane whereas cytochrome a_3 is situated on the matrix side. Cytochrome *c* oxidase is either totally immobilized in the membrane or it carries out only limited rotational diffusion around a single axis coinciding with the symmetry axis of heme a_3 (Junge and DeVault, 1975). The precise reaction catalyzed by cytochrome *c* oxidase is the four-electron reduction of O_2 shown below:

$$4 \text{ cytochrome } c^{2+} + 4 \text{ H}^+ + O_2 \rightarrow 4 \text{ cytochrome } c^{3+} + 2 \text{ H}_2O \qquad (2.4)$$

The electrons are provided via the sequential oxidation of four molecules of ferrocytochrome *c*, whereas the protons are drawn from the aqueous environment. In its purified state (molecular weight 240,000), cytochrome *c* oxidase is composed of at least seven subunits plus a phospholipid. The subunits are labeled I, II, III, IV, V, VI, and VII. All four redox centers are associated with subunit I and II. Ferricytochrome *c* and the cupric atom Cu_A are paramagnetic and hence are detectable by EPR spectroscopy; in contrast, Cu_B is antiferromagnetically coupled to cytochrome a_3 at the active site of the enzyme and is therefore "invisible". Low-temperature kinetic studies have indicated that cytochrome *c* oxidase participates in a complex reaction cycle that involves a series of well-defined redox intermediates (Fig. 2.11) (Clore *et al.*, 1980). Starting with the oxidase in its fully oxidized state ($Cu_A^{2+} \cdot a^{3+} \cdot Cu_B^{2+} \cdot a_3^{3+}$), the first part of the cycle involves the formation of the reduced species $Cu_A^+ \cdot a^{2+}$ at the expense of two molecules of ferrocytochrome *c*, followed by the further transfer of these two electrons to the active site of the oxidase

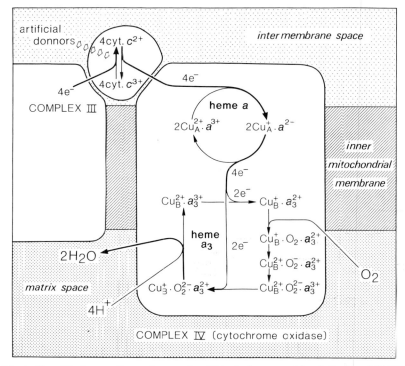

Fig. 2.11. Electron pathway in complex IV (cytochrome oxidase). Reduced TMPD (N, N, N', N'-tetramethyl-p-phenylenediamine) is an efficient artificial electron donor for complex IV via cytochrome c. Reduction of molecular oxygen by cytochrome c oxidase is the only irreversible step in oxidative phosphorylation.

to form $Cu_B^+ \cdot a_3^{2+}$. The latter then binds a molecule of O_2 to produce an oxygenated bridge complex that undergoes successive internal electron transfers to form a bound peroxide of very high redox potential ($Cu_B^{2+} \cdot O_2^{2-} \cdot a_3^{3+}$; $E_m = +700$ mV). This complex subsequently receives two more electrons from ferrocytochrome c, via $Cu_A^+ \cdot a^{2+}$ and finally reacts with four H^+ ions to yield two molecules of water and to regenerate the oxidase in its fully oxidized state (Fig. 2.11). Consequently, the oxidase passes to and from the fully oxidized state via a family of oxygenated and nonoxygenated, mixed-valence states. During the course of this reaction cycle, cytochrome c seems to shuttle very rapidly between cytochrome c_1 and cytochrome c oxidase (Fig. 2.11). Cationic residues, mostly lysines, are responsible for binding cytochrome c in an ordered manner to its neighbors, cytochrome c_1 and cytochrome oxidase. Thus according to Salemme (1977), the ionic interaction formed between the cytochrome c molecule and the membrane is relatively weak, giving the cytochrome c two-dimensional mobility such that it may reversibly and alternately interact with its physiological oxido-reductases (i.e., cytochrome a and cytochrome c_1).

Denis and Clore (1981) have investigated the reaction of mixed valence state cytochrome oxidase ($Cu_A^{2+} \cdot a^{3+} \cdot Cu_B^+ \cdot a_3^{2+}$; prepared by partial oxidation with ferricyanide of the CO complex of cytochrome oxidase) with O_2 in purified potato mitochondria by low-temperature flash photolysis (Chance et al., 1979) (the photodissociation of the cytochrome a_3–CO compounds occurred very rapidly, within 10^{-12} sec of absorption of the photon) and rapid scan spectrometry in the visible region (reaction sample minus unliganded mixed valence state cytochrome oxidase; Denis and Ducet, 1975), in order to compare and contrast the spectral and kinetic properties of the plant and mammalian mitochondrial systems. Two distinct optical species were resolved in time and wavelength. The first species (I_M) was characterized by a peak at 591 nm with a shoulder at 584 nm, and the second species (II_M), which behaved as a real endpoint for the reaction, was characterized by a split α-band with a large peak at 607 nm and a side peak at 594 nm. The split α-bands of the intermediate compounds observed in potato mitochondria suggested the possibility of different environments for heme a_3 (Denis and Bonner, 1978). A comparison of the kinetics and spectral features of the reaction of mixed valence state cytochrome oxidase with O_2 in the potato tuber and beef heart mitochondrial systems revealed the two following major features; (a) only two species (I_M, II_M) were seen in the intact potato mitochondrial system, in contrast with the beef heart mitochondrial system in which three species (I_M, II_M, and III_M) have been resolved (Denis, 1981) and (b) the reaction was much slower in the potato mitochondrial system than in the beef heart mitochondrial system at 173 K (by a factor ranging from 3–6). In addition, kinetic studies of CO rebinding to the mammalian reduced heme a_3 revealed an essentially monophasic process (De Fonseka and Chance, 1982). On the other hand, Denis and Richaud (1982) have presented an investigation on CO rebinding to reduced cytochrome c oxidase in potato mitochondria after flash photolysis in the low-temperature range 160–200 K. Four steps have been resolved in the rebinding process, which was found to be CO concentration–independent. Copper (B) must play an important role in this unexpected mechanism (Richaud and Denis, 1984), however, the observed differences do not seem to affect the general process: the redox state assigned to the final stage of the reaction is identical with the one attributed to the mammalian enzyme i.e., O_2^- forms a bridging ligand between Cu_B^{2+} and cytochrome a_3^{2+} (Denis and Clore, 1981).

Sweet potato cytochrome *c* oxidase was purified with a high yield by solubilization of the enzyme from submitochondrial particles with deoxycholate, followed by diethylaminoethyl (DEAE)-cellulose column chromatography and fractionation with ammonium sulfate (Maeshima and Asahi, 1978). The reduced form of sweet potato cytochrome *c* oxidase exhibited an α-band at 601 nm and a γ-band at 438 nm. The difference spectrum between the reduced and oxidized forms shows maxima at 443 and 601 nm. There were similarities in the wavelengths of the bands between cytochrome *c* oxidases from sweet and white potato (Ducet *et al.*, 1970). Five polypeptides were present with the following molecular weights: 39,000; 33,500; 26,000; 20,000; and 5700. In addition, the purified enzyme contained approximately 12 nmoles of heme *a*/mg of protein and 2.5% phospholipid. Pea cytochrome *c* oxidase resembled the sweet potato enzyme with respect to immunological properties and absorption spectra as well as the subunit composition (Matsuoka *et al.*, 1981). The interesting data of Maeshima and Asahi (1978) suggest that plant cytochrome *c* oxidase consists of two sets of at least five different subunits, in contrast to the animal situation. The reason for this discrepancy is unknown at present. It is clear, therefore, that a considerable number of additional experiments are required to establish the subunit composition of and arrangement in plant cytochrome *c* oxidase. Studies by Bligny and Douce (1977) have shown that copper deficiency in suspension-cultured sycamore cells inhibits cytochrome aa_3 appearance. Thus, in the case of copper-deficient mitochondria, they observed a striking reduction of the cytochrome *c* oxidase peaks (603 and 445 nm) (Fig. 2.12). More important is the observation that the low amount of cytochrome *c* oxidase present in copper-deficient mitochondria does not limit electron flow at the level of the inner mitochondrial membrane. This is in good agreement with early suggestions that the cytochrome *c* oxidase is present in very large excess in plant mitochondria (Chance and Hackett, 1959).

Recent experiments carried out by Krab and Wikström (1978), in contrast with a previous suggestion (Moyle and Mitchell, 1978), demonstrate that cytochrome *c* oxidase is a proton pump that conserves redox energy by converting it into an electrochemical proton gradient through electrogenic translocation of H^+. In other words, the development of a membrane potential and a pH gradient during cytochrome *c* oxidase functioning is not exclusively attributable to consumption of the protons required in water formation from the matrix phase. More recent results of Papa *et al.* (1983) and Mitchell and Moyle (1983), however, suggest that cytochrome *c* oxidase *in situ*, in ox heart and rat liver mitochondria, does not display any proton-pumping activity during the transition from

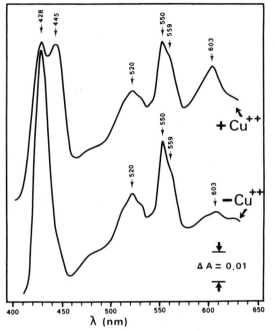

Fig. 2.12.   Difference spectra of normal and copper-deficient sycamore cell mitochondria at room temperature (25°C). Succinate (20 mM) reduced minus aerobic. Optical path:
10 mm (quartz cuvettes). Mitochondrial protein: with copper, 2.2 mg/ml; without copper,
2.8 mg/ml. Note that in the case of copper-deficient mitochondria, there is a striking
reduction in the cytochrome aa_3 peaks (α-band at 603 nm and γ-band at 445 nm). [From
Bligny and Douce (1977).]

the anaerobic reduced state to the active pulsed state. Whatever the
mechanism, the gap of nearly 200 mV between the midpoint potentials
of cytochromes a and a_3 indicates that a site of energy conservation of
the respiratory chain is to be found between cytochromes a and a_3 (see,
for example, Bonner and Plesnicar, 1967).

2.   Cyanide-Resistant Electron Pathway

a.   Characterizaton.   Practically all of the plant tissues examined so
far show a residual respiration in the presence of CN^-, CO, or N_3^-,
even when these inhibitors are present at concentrations sufficient to
completely inhibit cytochrome c oxidase (Hackett, 1959; Ducet and
Rosenberg, 1962; Day *et al.*, 1979); it occurs naturally or can be induced
under appropriate experimental conditions (Dizengremel and Lance,
1982). Among animals, tissue respiration is usually extremely sensitive
to cyanide, with rates being reduced to 1% or less of those found in its

absence. In addition to higher plants, the cyanide-resistant electron pathway has been reported in a wide range of fungi, including a large number of yeasts and *Neurospora*, and various microorganisms (Lloyd, 1974). The extent of the inhibitor-resistant respiration (alternative pathway) observed in higher plants ranges from a small proportion of the total respiration in some tissues to those cases in which hardly any sensitivity is observed. Tissues such as the spadices of the Aracae are capable of very high rates of O_2 uptake, most of which are cyanide resistant; consequently, much of the early work on cyanide-resistant respiration focused on these spadices in species such as the tropical *Sauromatum guttatum* and the European *Arum italicum* and *Arum maculatum* (cuckoo pint), during the course of the respiratory crisis associated with pollination (for a review see Meeuse, 1975). During this crisis, the spadix becomes thermogenic for a limited period and the heat releases odiferous amines that attract insects involved in the pollination process (Smith and Meeuse, 1966). Henry and Nyns (1975) made a compilation of species in which cyanide-resistant respiration had been reported and numerous additional examples have appeared since (Solomos, 1977; Lambers, 1980; Lance, 1981). Cyanide-insensitive tissues generally yield mitochondria that are also cyanide insensitive (Table 2.8). Outstanding in this respect are mitochondria isolated from the spadices of aroids which show little, if any, sensitivity to cyanide inhibition. However, this is not true with potato tubers. Oxygen uptake by mitochondria isolated from fresh tubers shows no appreciable cyanide resistance (Dizengremel and Lance, 1976), whereas respiration in the intact tuber is not only resistant to but actually stimulated five- to sixfold upon addition of cyanide (Solomos and Laties, 1975). In this case, the cyanide-resistant O_2 uptake may not have been originally associated with the mitochondria. If potato tuber slices are first allowed to "age" for several hours in aerated medium, however, mitochondria isolated from the resulting tissues exhibit a strong cyanide-resistant respiration (Thimann *et al.*, 1954; Hackett *et al.*, 1960a,b). According to Laties (1978, 1982), the loss of mitochondrial cyanide resistance in potato tubers could be attributable to a massive mitochondrial membrane phospholipid breakdown caused by the release of lipolytic acyl hydrolase upon slicing or mincing. This assumption should be reevaluated (Lance, 1981), however, because the phospholipid composition of mitochondria isolated from fresh and aged potato tissue slices are very similar on a quantitative and a qualitative basis (Dizengremel and Kader, 1980). Furthermore, cyanide-resistant mitochondria can be isolated from fresh tuber tissue if the intact tubers are first held for several hours at room temperature in the presence of ethylene (Day *et al.*, 1978a; Rychter *et al.*, 1979). Ob-

TABLE 2.8

Comparison of Percentage of Cyanide Resistance of
Oxygen Uptake in Intact Roots and Leaves and in
Mitochondria Isolated Therefrom[a]

Species	Tissue	Cyanide-resistance (%)	
		Whole tissue	Mitochondria
Cotton	Roots	36	22
Bean	Roots	61	41
Spinach	Roots	40	34
Wheat	Roots	38	35
Maize	Roots	47	32
Pea	Leaves	39	30
Spinach	Leaves	40	27

[a] Mitochondrial substrates were 10 mM malate plus 10
mM succinate; KCN (0.2 mM) was added in the presence
of ADP. The cyanide-resistant O_2 uptake was inhibited
by SHAM and disulfiram. Intact tissue and mitochondrial measurements were made on the same batch of plants.
Mitochondrial respiration rates ranged from 50–100
nmoles/mg protein/min but varied between species and
batches. [From Lambers *et al.* (1983).]

viously, the fate of the cyanide-resistant O_2 uptake on isolation of the
mitochondria from untreated tubers is entirely unknown.

Bendall and Bonner (1971) established that cyanide- and antimycin A-
insensitive respiration was mediated by an additional electron transport
pathway consisting of the same set of dehydrogenases as the normal
respiratory chain, but entirely bypassing the cytochromes via a second
oxidase. In isolated mitochondria, the K_m for O_2 of the alternative pathway is found to be somewhat higher than that of cytochrome oxidase
but nonetheless quite low (1–2 µM; Bendall and Bonner, 1971). Several
studies on tissue slices have reported high K_m values for O_2 (10–25 µM
or higher) on the alternative pathway, but these values are usually associated with diffusion-limited steps (Solomos, 1977). The two oxidases in
higher plant mitochondria represent the end of two branches with a
common origin rather than the terminal components of two parallel
chains. In support of this suggestion, S. B. Wilson (1978) has demonstrated that purified mung bean hypocotyl mitochondria utilize O_2 with
an ascorbate–TMPD mixture as an electron donor in the presence of
CN^-. This electron flow, which is controlled by ATP, occurs by re-

versed electron transport through complex III and then to O_2 through the cyanide-insensitive pathway. It should be noted that, besides cytochrome oxidase and the alternative oxidase, O_2 reduction can occur at the level of the flavoprotein region of complex I and at the level of the ubiquinone-cytochrome b region of complex III. These latter two processes produce superoxide anions and could represent a nonphysiological "short-circuiting" of the electron transport chain (Rich and Bonner, 1978a; Huq and Palmer, 1978d).

Several workers have concluded that energy conservation is not associated with the cyanide-insensitive oxidase and, therefore, ascribe a thermogenic role to this independent "noncytochrome" oxidase (Hackett *et al.*, 1960b; Storey, 1971a; Passam and Palmer, 1972). Rather surprisingly, S. B. Wilson (1980) has suggested that the cyanide-insensitive oxidase is able to support a variety of energy-linked functions; namely phosphorylation, reversed electron transport, and the maintenance of a membrane potential detected by the dye probes 8-anilino-naphthalene-1-sulphonate and safranine. Obviously these conclusions are contrary to many of those reported in the literature. In fact, convincing measurements of the magnitude of the proton-motive force in the presence of antimycin A indicate that the alternative pathway of plant mitochondria is nonelectrogenic and, consequently, not coupled to ATP synthesis (A. L. Moore and Bonner, 1976). In addition, the recent experiments of A. L. Moore and Bonner (1982) on *Sauromatum guttatum* mitochondria quite categorically indicate that there is no energy conservation by the alternative oxidase, there being no membrane potential generated during cyanide-insensitive respiration.

Schonbaum *et al.* (1971) reported that substituted benzohydroxamic acids (XC_6H_4-CONHOH) inhibited the cyanide-insensitive respiratory pathway in isolated plant mitochondria. The substituted hydroxamic acids, of which salicylhydroxamic acid (SHAM) has been the most widely used, have no discernible effect on the cyanide-sensitive pathway at concentrations that completely inhibit the alternative pathway. Likewise, Siedow and Girvin (1980) found that the antioxidant *n*-propyl gallate (3,4,5-trihydroxybenzoic acid propyl ester) specifically inhibited the alternative pathway in isolated mung bean hypocotyl mitochondria, with a K_i value five- to tenfold lower than that obtained using benzohydroxamic acids. Siedow and Bickett (1981) also demonstrated that the dihydroxyl function on propyl gallate, not the ester component, was the substituent responsible for inhibition. Specifically, pyrogallol, which lacks the ester moiety, was still a relatively good inhibitor. Kinetic analysis indicated that propyl gallate inhibition of the alternative pathway occurred at, or very near, the site of inhibition of the alternative pathway by salicylhydroxamic acid (Siedow and Girvin, 1980), probably because there is a geometric similarity between the benzohydroxamic acids and propyl gallate. The well-characterized ability of hydroxamic acids to chelate ferric ions (Anderegg et al., 1963) has led some workers to invoke the involvment of an iron–sulfur protein in the alternative pathway (Bendall and Bonner, 1971; Lance et al., 1978). This hypothesis is strengthened by the fact that propyl gallate also has a marked ability to chelate ferric ions (Soloway and Rosen, 1955). This concept is now generally thought not to be correct (Bonner and Rich, 1978), however, because all of the mitochondrial iron–sulfur proteins that could be considered as candidates for the alternative oxidase are unaffected by hydroxamates (Rich and Bonner, 1978a). Disulfiram (tetraethylthiuram disulfide) was also found by Grover and Laties (1981) to be a potent and selective inhibitor of the alternative respiratory pathway of plant mitochondria ($K_i = 7$ μM). Although both the hydroxamic acids and disulfiram are capable of complexing metal ions, the evidence for separate binding sites and the fact that metal ions fail to reverse disulfiram inhibition both suggest that the two inhibitors have distinct mechanisms of action. In fact, the facile reversal of disulfiram inhibition by thiols, such as mercaptoethanol, strongly suggests that the inhibitory effect of disulfiram involves the formation of mixed disulfides with one or more sulfhydryl groups in the alternative pathway.

Storey (1976) by means of an O_2 pulse on CO-saturated skunk cabbage mitochondria clearly identified ubiquinone as the carrier common to both the cytochrome and alternative oxidase pathways. Ubiquinone is rapidly oxidized under these conditions, and this rapid rate of oxidation is strongly inhibited by m-chlorobenzhydroxamic acid. Consequently, according to Storey (1976), ubiquinone is the component of the plant mitochondrial respiratory chain which best fits the criteria for the point of interaction of the alternative pathway with the main pathway (Fig. 2.13). Rich et al. (1978) reported that several soluble redox enzymes, such as peroxidase and tyrosinase, are strongly inhibited by substituted hydroxamic acids. Investigations on the mode of hydroxamate inhibition on these two enzymes have shown that the inhibition is competitive with respect to the enzyme's phenolic substrate and is not competitive with respect to O_2 concentration. By extrapolating the above results to considerations of the alternative oxidase, which is also not competitive with respect to O_2 and for which the K_i apparent is dependent on electron flux, Rich et al. (1978) have suggested that inhibition of the alternative oxidase by hydroxamic acids might be due to competition with the reducing substrate of the alternative oxidase, probably reduced ubiquinone. Inhibition of cyanide-resistant respiration by chloroquine has

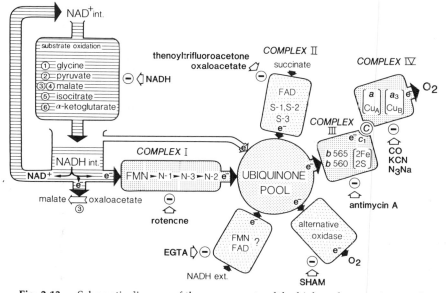

Fig. 2.13. Schematic diagram of the components of the higher plant respiratory chain arranged as a continuous sequence from NADH (low potential), generated by various internal NAD^+-linked dehydrogenases, to oxygen (high potential) via either complex IV (cytochrome oxidase) or the alternate oxidase. The four complexes and their sequential arrangement in the cytochrome oxidase pathway (complex I, II, III, and IV), as deduced from chemical fractionation and reconstitution experiments in Green's laboratory (Green, 1966). Plant mitochondria oxidize endogenous NADH ($NADH_{int}$) by three separate pathways. The first pathway, which is sensitive to rotenone, involves complex I and is coupled to three energy-transducing sites. The second pathway, which is insensitive to rotenone, presumably involves another NADH dehydrogenase located on the matrix side of the inner membrane and is coupled to only two energy-transducing sites. This dehydrogenase, in contrast with complex I, exhibits a low affinity for NADH. The third pathway, which "short-circuits" the electron chains, involves the reduction of oxaloacetate (if present) by malate dehydrogenase. Finally, external NADH is oxidized by a distinct NADH dehydrogenase located on the outer surface of the inner membrane and bypassing complex I and the "first site" of ATP synthesis. Symbols: (1) glycine decarboxylase and serine hydroxymethyltransferase; (2) pyruvate dehydrogenase; (3) malate dehydrogenase; (4) NAD-dependent malic enzyme; (5) isocitrate dehydrogenase; and (6) α-ketoglutarate dehydrogenase.

been reported (James and Spencer, 1982). The structural similarity of chloroquine to ubiquinone supports the involvement of ubiquinone in cyanide-resistant respiration. Collectively, these results suggest that the cyanide-resistant electron transport system consists of a branch from the conventional electron transport system, beginning with ubiquinone and terminating with an "alternative oxidase" (see, however, Beconi *et al.*,

1983). Rich and Moore (1976) have proposed that the "alternative oxidase" of higher plants is closely associated with part of the Q cycle, to explain how electrons might branch from the main pathway. They speculate that the specific location of the branch point of electron flow into the alternative pathway occurs at the reversal of the step of complex I or II reduction of QH· to QH_2 on the matrix side of the inner membrane, possibly close to succinate dehydrogenase. In support of this suggestion, Solomos (1977) has pointed out the well-known observation that, in mitochondria from nonthermogenic tissues, the reducing equivalents from exogenous NADH are preferentially channeled toward the cyanide-sensitive oxidase while reducing equivalents from endogenous NADH and succinate are available to the cyanide-resistant oxidase. Furthermore, the results of steady-state experiments carried out by Rich and Bonner (1978a) indicate that the operation of the alternative respiratory oxidase, in the presence of succinate and cyanide, tends to keep center S-3 oxidized. This suggests that the alternative oxidation site is on the electron-donating side of center S-3. Huq and Palmer (1978a) found that partial extraction of ubiquinone from *Arum maculatum* mitochondria by pentane treatment resulted in a preferential loss of the alternative pathway, relative to the main pathway. A similar result was obtained with "partial extraction" of quinones in aged potato tuber mitochondria by Triton X-100 treatment (Dizengremel, 1983). One possible explanation for these observations is that the affinity of complex III for ubiquinol is higher than that of the alternative oxidase. Thus, removing part of the ubiquinone complement will affect the rate of electron flow through the alternative pathway more than that through the cytochrome chain. Such an explanation requires that ubiquinol must bind to complex III and the alternative oxidase in order to donate electrons. Different affinities of complex III and the alternative oxidase for ubiquinone may also explain why the alternative pathway operates only when electron flow through the cytochrome chain is either restricted or very rapid (i.e., approaching saturation; see later). Obviously, and as pointed out judiciously by Siedow (1982), a better understanding of how the quinone pool and the recently discovered protein-bound ubiquinone species interact is needed before we can fully understand how electrons are shunted between the main and alternative pathways.

Since the alternative oxidase is indistinct both in its electron paramagnetic resonance (Rich *et al.*, 1977) and in its spectrophotometric paramaters, the nature of the alternative oxidase has remained elusive. Additional problems encountered in attempts at characterization of the alternative oxidase are the lack of an artificial donor that may be used for the direct assay of the alternative oxidase and the apparent lability of the oxidase (Lance *et al.*, 1978). The suggestion that the alternative oxidase might be a quinol oxidase led Huq and Palmer (1978b) and Rich (1978) to use quinols (reduced quinone) as artificial electron donors to assay for the alternative oxidase. Menaquinol, ubiquinol-1, and duroquinol have been shown to be oxidized in *Arum*, sweet potato, and cassava mitochondria via the cyanide-resistant SHAM-sensitive respiratory pathway. It is interesting to note that, in the absence of KCN, the oxidation of duroquinol by the mitochondria from aged potato slices was rapid and tightly coupled to phosphorylation (Dizengremel, 1983). These quinols seem to donate electrons to a point that is at or very close to the alternative O_2 consuming step (Rich and Bonner, 1978a,b; Huq and Palmer, 1978a) and offer an invaluable tool for the further investigation of the oxidase itself. Huq and Palmer (1978c) and Rich (1978) have liberated a quinol oxidase by detergent solubilization from *Arum* spadix mitochondria. Huq and Palmer (1978c) have partially purified and characterized this quinol oxidase using a series of DEAE-cellulose steps and find that it contains a fluorescent compound, which is probably a flavoprotein, and significant quantities of copper. Care must be taken in interpreting results obtained utilizing reduced quinone as an artificial electron donor to assay for the alternative oxidase, however, since quinol and especially the short-chain ubiquinols, can directly interact with O_2. Furthermore, the autooxidation of menaquinol, duroquinol, and the short-chain ubiquinols is inhibited by salicylhydroxamic acid and stimulated by various nonprotein compounds, such as purine nucleotides and metals, especially copper (Vanderleyden *et al.*, 1980). In a comprehensive review on the nature of the cyanide-resistant pathway in plant mitochondria, Siedow (1982) has pointed out that the alternative oxidase catalyzes the same reaction as that of a laccase-type polyphenol oxidase i.e., a *p*-diphenol–O_2 oxidoreductase. Laccase is a copper enzyme that has been described in fungi but scarcely mentioned in plants (for a review see Mayer and Harel, 1979). Paramagnetic copper of types 1 and 3 appears in laccase; consequently, the failure to observe EPR signals arising from some of the copper species in plant mitochondria strongly suggests that copper is not associated with the alternative oxidase.

Goldstein *et al.* (1980) indicated that the cyanide-insensitive O_2 uptake measured with unpurified mitochondria from wheat seedlings was due to the addition of molecular O_2 across a double bond in polyunsaturated fatty acids, catalyzed by a nonheme iron-containing dioxygenase (lipoxygenase). *In vitro* experiments have shown that these fatty acids are converted to hydroperoxides. Furthermore, lipoxygenase is inhibited by most known inhibitors of alternative respiration (Parrish and Leopold, 1978; Miller and Obendorf, 1981). Taken together, these observations might suggest that cyanide-resistant respiration was due to lipoxygenase activity adhering to, or present in, the mitochondria (Dupont, 1981; Kelly, 1982). We consider this to be a trivial explanation for the alternative oxidase, however, because purification of mitochondria by discontinuous sucrose or Percoll density gradient centrifugation leads to a marked reduction in lipoxygenase levels (Siedow and Girvin, 1980; Neuburger *et al.*, 1982), but, with the exception of wheat (Goldstein *et al.*, 1980), no drastic reduction in the level of the alternative pathway has been reported (Siedow, 1982). In addition, the structural features of propyl gallate necessary for inhibition of lipoxygenase were found to differ from those required for inhibition of the plant mitochondrial alternative pathway (Peterman and Siedow, 1983). Finally, the differential effects of disulfiram on cyanide-resistant respiration and on lipoxygenase strongly suggest that, although lipoxygenase is present in most of the plant mitochondria isolated so far, it is not responsible for cyanide-resistant respiration (Miller and Obendorf, 1981). Recently Rustin *et al.* (1983a,b) have made the provocative suggestion that the cyanide-insensitive alternative pathway in plant mitochondria depends on the occurrence in the membranes of peroxidative activity that requires the presence of peroxidizable unsaturated fatty acids. [The generation of superoxides at the levels of ubiquinone or the flavoprotein dehydrogenases (Forman and Boveris, 1982) could easily initiate the peroxidation of fatty acids.] According to this scheme, ubiquinol (QH_2)—or a pool of it—can be reoxidized in the presence of free unsaturated fatty acid peroxyl radicals (ROO·) acting as electron acceptors and present in trace amounts in mitochondrial membranes. In support of this hypothesis, they have suggested that reduced quinones (duroquinol) and fatty acid peroxyl radicals (ROO·) can interact together in a way that is insensitive to cyanide but is very sensitive to the inhibitors of the alternative pathway (Rustin *et al.*, 1983b). In this scheme, the most probable reduction product of ROO· is ROH, and its reoxidation is cyanide resistant and yields H_2O not H_2O_2, as a terminal product. In support of this suggestion, all of the inhibitors of the alternative pathway, such as SHAM, propyl gallate, and disulfiram, seem to behave as radical scavengers (R· and ROO·). Since their usual concentrations (0.1– 1mM) in reaction media by far exceed the concentrations of R· and ROO· in mitochondrial membranes, they could readily stop the cyanide-resistant O_2 uptake by trapping the R· and ROO· radicals (Rustin *et al.*, 1983b). However, this attractive hypothesis does not explain the reason potato tuber mitochondria, which contain large amounts of polyunsaturated fatty acids and lipoxygenase potentially capable of generating unsaturated ROO·, are strongly cyanide sensitive. Nor does it explain the thermolability (40°C during 30 min) of the alternative electron transport pathway observed by Chauveau *et al.* (1978) in higher plant mitochondria. Furthermore, the appearance of the alternative pathway in *Vigna mungo* cotyledons during germination of the axis seems to depend on cytoplasmic protein synthesis (Morohashi and Matsushima, 1983), and the same is true of *Neurospora* cells (Edwards *et al.*, 1974). Another convincing feature indicative of a protein is the isolation of a *p*-quinol: O_2 oxidoreductase from *Arum maculatum* mitochondria (Bonner *et al.*, 1985). This heat-labile quinol oxidase oxidizes *p*-quinols in a cyanide-insensitive hydroxamate-sensitive manner. Finally, the major question arising from such a model is the problem of its regulation.

b. The Physiological Role of the Alternative Pathway. The physiological role of the alternative pathway is still uncertain. With the discovery of the aromatic hydroxamic acids as inhibitors of the alternative pathways, it has become possible to estimate the distribution of electrons between the two pathways. According to Bahr and Bonner (1973a,b), the cyanide insensitive pathway seems to be regulated by the activity of the normal cytochrome pathway. Electrons from TCA cycle substrates are diverted to the alternative pathway only when the cytochrome pathway approaches saturation, either by inhibition (including state 4) or by flooding with electrons. In other words, as long as the capacity of the cytochrome pathway is sufficient to handle the electron flux entering ubiquinone, the alternative pathway will not be engaged. This hypothesis is supported by work with slices from storage tissues in which it has been shown that, the alternative pathway is inoperative unless the cytochrome pathway is saturated with electrons, by the addition of an uncoupler (Theologis and Laties, 1978). Likewise in the leaves of many plant species, the degree to which the alternative pathway participates in normal leaf respiration is related directly to the level of sugars in the leaf (Azcón-Bieto *et al.*, 1983a,b). However, if the availability of ADP is the main factor which restricts cytochrome oxidase *in vivo*, how then does site 1 of oxidative phosphorylation function, in preference to the other two sites, allowing the operation of the alternate oxidase? Since it is very unlikely that preferential disengagement of oxidative phosphorylation at site 1 is involved, the electrons from the oxidation of the NAD^+-linked substrates must bypass this step. In all of the plant mitochondria isolated so far, there exists a rotenone-insensitive NADH dehydrogenase located on the inner surface of the inner membrane. This pathway allows site 1 of energy transduction to be bypassed (Fig. 2.13), thereby allowing the operation of the alternate oxidase. Thus, if TCA cycle intermediates are urgently needed elsewhere in the cell, the alternative pathway could be used as a well-adapted regulatory mechanism whenever traffic through the cytochrome pathway is constrained by the energy charge (Bomsel and Pradet, 1968).

It is possible that low temperatures could play an important role in the engagement of the cyanide-insensitive pathway (Yoshida and Tagawa, 1979; McCaig and Hill, 1977). In other words, the thermogenic nature of the alternative pathway has led to suggestions that it plays a role in frost resistance (Bonner, 1973; Knutson, 1974; Chauveau and Lance, 1982). In a few cases, such as in *Arum lilies,* inhibitor-resistant electron transport is apparently directly related to thermogenic metabolism (Meeuse, 1975). On the day of flowering, the alternative pathway becomes operational, and starch that has been accumulated up to this stage is burned up in a few hours, due to prodigiously high rates of glycolysis (ap Rees *et al.,* 1977) and respiration (Lance, 1972). The site of this crisis is generally the sterile part of the inflorescence, and the excess of energy that is not used for ATP synthesis is liberated as heat. At climax, this phenomenon can lead to an increase in temperature to 15°C above ambient in *Arum maculatum* and even more in *Alocasia pubera.* In many tissues, however, there is no such correlation, and the existence of a cyanide-insensitive pathway appears to be a wasteful energy process. Lambers (1980) has speculated that the main purpose of the alternative pathway is to drain away so-called "excess reducing power," such as malate and citrate stored in the large vacuole reservoir. If indeed the alternative pathway participates in an "energy overflow" metabolism, then it will have a substantial impact on the interpretation of experiments carried out on carbon metabolism and energy requirements. For example, it will need to be taken into consideration in the determination of growth and maintenance respiration (Szaniawski, 1981). Lambers *et al.* (1983) have demonstrated that the alternative pathway activity *in vivo* can be reliably estimated by the judicious use of high SHAM concentrations for short periods. They also demonstrated the widespread presence of the alternative pathway in plants and that the extent to which this pathway is engaged varies substantially between species. According to Azcón-Bieto *et al.* (1983a,b) and Lambers *et al.* (1983), the extent to which the cyanide-resistant alternative pathway participates in the respiration of whole tissue appears to be governed by mitochondrial substrate supply. For example, when leaf sugar levels are substantial (i.e., after a period of several hours in the light), the alternative pathway becomes engaged on top of a restricted cytochrome chain. When leaf sugar levels are low (at the end of the night), respiration is limited by substrate supply to the mitochondria and the alternative pathway is not expressed.

Obviously, much remains to be done in order to understand how the control of the alternative pathway operates under physiological situations. The mechanism whereby the alternative pathway is engaged and the extent to which it operates are of overriding importance in the physiological role of the cyanide-resistant respiration.

3. Rotenone-Resistant Electron Pathway

Lehninger (1964) found that NADH added to liver mitochondria was not oxidized. If the mitochondria were gently disrupted by hypotonic swelling, however, oxidation of NADH was considerably enhanced via the universally distributed, rotenone-sensitive, respiratory chain-linked NADH dehydrogenase (complex I). Indeed, isotopic studies clearly showed that NADH readily penetrated the outer mitochondrial membrane but exchanged at an insignificant rate with NADH in the matrix compartment (Von Jagow and Klingenberg, 1970). It is now generally accepted in animal mitochondria that the inner membrane is totally impermeable to NADH.

a. Oxidation of Exogenous NADH and NADPH by Plant Mitochondria. In contrast to animal mitochondria, the mitochondria of higher plants and microorganisms, such as *Neurospora crassa* and yeast, catalyze a rapid oxidation of exogenous NADH in the absence of added cytochrome *c* (Bonner, 1967; Lloyd, 1974). This observation has been interpreted by many workers to result from the rapid permeation of NADH across the inner membrane and its subsequent oxidation by complex I. This apparent "leakiness" was thought to be due to mitochondrial damage induced by the rigorous grinding used in isolating the organelles (Lieberman and Baker, 1965). With the advent of more rigorous criteria for mitochondrial integrity, however, it became clear that this was not the case (Douce *et al.*, 1972). In fact, it is now generally agreed that exogenous NADH can be oxidized by two different external NADH dehydrogenases. The first dehydrogenase, located on the outer surface of the inner membrane, is specific for the β-hydrogen of NADH and feeds electrons directly to complex III (Douce *et al.*, 1973b), bypassing complex I and the first site of H^+ translocation (Douce *et al.*, 1972; J. M. Palmer, 1979). This pathway, which is inhibited by antimycin A, does not seem to be connected with the alternate oxidase (except in the case of mitochondria from *Arum lilies*). Consequently, NADH oxidation by this external NADH dehydrogenase, which does not require NADH translocase, is insensitive to rotenone and has an ADP/O ratio similar to that of succinate (Fig. 2.14). According to Cowley and Palmer (1980) and Dry *et al.* (1983a), plant mitochondria preferentially oxidize endogenous

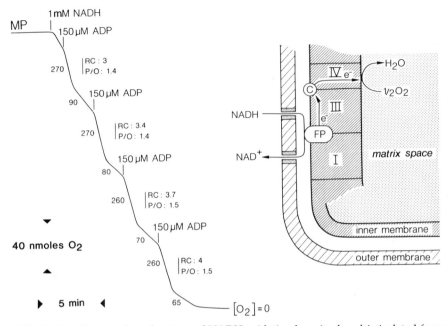

Fig. 2.14.   Rates and mechanisms of NADH oxidation by mitochondria isolated from higher plants. The reaction medium contained: 0.3 M mannitol; 5 mM MgCl$_2$; 10 mM KCl; 10 mM phosphate buffer, pH 7.2; 0.1% (W/V) defatted bovine serum albumin, and 0.6 m eq of Ca^{2+} (Ca^{2+} very often stimulates NADH oxidation; see Møller *et al.*, 1981a). Note that external NADH is directly oxidized by a specific dehydrogenase (FP) located on the outer surface of the inner membrane (Douce *et al.*, 1973b), one which does not require NADH translocase. This dehydrogenase, which is not inhibited by rotenone, feeds electrons directly to complex III.

NADH when confronted with a mixture of NADH and TCA cycle substrates. The mechanism for this interaction remains unresolved. Taking into account quinone mobility, it is possible that diffusion distances between complex III and either external NADH dehydrogenase or complex I are not identical. The calcium ion (Ca^{2+}) plays a specific role in the oxidation of external NADH (Miller *et al.*, 1970; Coleman and Palmer, 1971; Møller *et al.*, 1981a; Moore and Åkerman, 1982), because the rate of oxidation of exogenous NADH is strongly inhibited by the addition of EGTA and this inhibition can be reversed by adding Ca^{2+}. According to J. M. Palmer and Møller (1982), Ca^{2+} could alter the conformation of the enzyme such that it interacts with the substrate more efficiently. In support of this suggestion, it appears that mitochondria actively oxidizing NADH bind Ca^{2+} more tightly than resting organelles. However, the suggestion that interaction occurs because of spatial alterations between the proteins and the membrane is very difficult to justify experimentally. In addition, cations unspecifically enhanced the oxidation of exogenous NADH by plant mitochondria (Johnston *et al.*, 1979). The mechanism by which this cation stimulation takes place is interesting. It is not dependent on the chemical identity of the cation, only on its valency, and the order of efficiency is X^{3+}, X^{2+}, X^+ (Møller *et al.*, 1981c). This enhancement is closely associated with the ability of the cations to screen the fixed charges associated with the surface of mitochondrial membranes (ionized polar groups of membrane proteins and phospholipids form the basis of the surface charges). Indeed, this screening will decrease the repulsion of the negatively charged substrate NADH and thus cause an increase in the effective substrate concentration near the active site of the membrane-bound NADH dehydrogenase and, therefore, an apparent increase in affinity between the enzyme and the substrate (Wojtczak and Nalecz, 1979; Møller *et al.*, 1981; Møller and Palmer, 1981b; Table 2.9). Parenthetically, plant mitochondria also oxidize exogenous NADPH, apparently via a Ca^{2+}-dependent dehydrogenase located on the outer surface of the inner membrane (Koeppe and Miller, 1972; Arron and Edwards, 1979, 1980a,b; Arron *et al.*, 1979b). There are similarities between the characteristics of the oxidations of external NADH and NADPH; these include ADP/O ratios below 2, rotenone-insensitivity, and sensitivity to cyanide and antimycin A (Koeppe and Miller, 1972; Arron and Edwards, 1979). Møller and Palmer (1981a) showed conclusively, however, that the pH optimum for

TABLE 2.9

Effect of Cations on the Kinetic Parameters of NADH Oxidation[a]

Source of mitochondria	Parameter	Low-cation conditions		High-cation conditions		Effect of cations	
Jerusalem artichoke	$K_m{}^b$	91	(1)	54	(1)	−41%	
	$V_{max}{}^c$	395	(1)	555	(1)	+41%	
Arum maculatum	$K_m{}^b$	31 ± 4	(2)	19 ± 1	(2)	−36 ± 9%	(2)
	$V_{max}{}^c$	1260 ± 50	(2)	1590 ± 160	(2)	+29 ± 19%	(2)

[a] The oxidation of NADH was measured at 340 nm in a spectrophotometer. To obtain high-cation conditions, 2 mM decamethylene-bis-(trimethylammonium) bromide [(DM)Br$_2$] was added. Data are given as x ± S.E. with the number of preparations in parentheses. [From Møller and Palmer (1981a) and reprinted by permission of *Biochem. J.* **195**, pp. 583–588. Copyright © 1981. The Biochemical Society, London.]

[b] Units: μM.

[c] Units: nmoles/min/mg protein.

NADPH oxidation is lower than that for NADH oxidation. In addition, the responses of external NADPH and NADH oxidations to both chelators and mersalyl are quite different. The NADPH oxidation is inhibited when no effect is observed on the NADH oxidation. This would seem to support the idea of two separate dehydrogenases, one specific for NADH and the other for NADPH (Koeppe and Miller, 1972). Neither a phosphatase, converting NADPH to NADH, nor a nicotinamide nucleotide transhydrogenase was involved in the oxidation of NADPH by plant mitochondria (Møller and Palmer, 1981c). The rates of O_2 uptake with purified pea leaf mitochondria show that their potential for NADPH oxidation *in vitro* is quite high and suggest that *in vivo* oxidation of cytoplasmic NADPH may occur at reasonably fast rates (Nash and Wiskich, 1983). This contrasts with the results obtained using potato tuber mitochondria, for which NADPH oxidation was about 40% as fast as NADH oxidation (Arron and Edwards, 1980a) and with Jerusalem artichoke mitochondria, for which the rate of NADPH oxidation at pH 7.2 was less than 5% of that of NADH oxidation (Møller and Palmer, 1981c; Table 2.10). Finally, Day and Wiskich (1974a,b) presented evidence that mitochondria isolated from red beet (*Beta vulgaris* L.) tissue do not oxidize extramitochondrial NADH via the inner membrane electron transport chain as do other plant mitochondria. However, mitochondria isolated from red beet slices that have been aged for 24 hr or more develop a capacity to oxidize external NADH, with oxygen as the electron acceptor (Day *et al.*, 1976). This development seems to be dependent on protein synthesis (Rayner and Wiskich, 1983). The nonionic detergent lauryl dimethylamine *N*-oxide in conjunction with affinity chromatography on 5'-ADP-sepharose 4B have been used to isolate the external NADH dehydrogenase. The rotenone-insensitive NADH dehydrogenase of *Arum maculatum* mitochondria appears to be a flavoprotein (the enzyme probably contains FAD) and does not contain any electron spin resonance–detectable iron-sulfur centers (Cook and Cammack, 1984; see, however, Klein and Burke, 1984; Cottingham and Moore, 1984).

The second dehydrogenase, located on the outer membrane, contains a flavoprotein that is associated with cytochrome b_{555}, is specific for the α-4 hydrogen atom of NADH (Douce *et al.*, 1973b), and is insensitive to antimycin A (Douce *et al.*, 1972, 1973b; Moreau and Lance, 1972; Day and Wiskich, 1975). In addition, this dehydrogenase readily interacts with ferricyanide, in contrast with the external NADH dehydrogenase located on the outer surface of the inner membrane (unless mitochondria are hypotonically swollen) (Douce *et al.*, 1973b). The function of the NADH dehydrogenase–cytochrome b_{555} system of the outer mitochon-

TABLE 2.10

Oxidation of Malate, NADH, and NADPH by a Variety of Washed Plant Mitochondria[a,b]

Tissues and sources	Substrate	Respiration rate (nmoles O_2/min/ mg protein)		Respiratory control ratio	ADP/O ratio
		State 3	State 4		
Leaves					
Sedum praealtum	Malate	26.2	6.9	3.8	2.1
	NADH	28.8	6.3	4.6	1.6
	NADPH	28.3	6.0	4.7	1.5
Sunflower	Malate	104	35	3.0	2.5
	NADH	200	100	2.0	1.4
	NADPH	168	79	2.1	1.4
Spinach	Malate	100	28	3.6	2.3
	NADH	124	45	2.7	1.3
	NADPH	42	36	1.2	n.d.
Seedlings					
Corn	Malate	114	24	4.8	2.6
	NADH	202	50	4.1	1.5
	NADPH	96	40	2.4	1.4
Mung bean	Malate	217	48	4.5	2.4
	NADH	201	67	3.0	1.4
	NADPH	185	47	4.0	1.5
Stem tuber					
White potato	Malate	109	32	3.4	2.6
	NADH	146	69	2.1	1.6
	NADPH	102	53	1.9	1.5
Root					
Beetroot	Malate	62	15	4.2	2.5
	NADH	44	27	1.7	1.3
	NADPH	7.4	7.4	1.0	n.d.

[a] Final concentrations were 10 mM malate, 1 mM NADH, and 1 mM NADPH. With malate as substrate, 10 mM glutamate was included in the reaction medium to prevent oxaloacetate inhibition. ADP was added in pulses of 150 μmoles in the case of malate and 100 μmoles in the case of NADH and NADPH. The respiratory rate data given are the averages from the second and third ADP additions. The ADP/O ratios are the averages from the second and third ADP additions also.

[b] Abbreviation: n.d., not determined. [From Arron and Edwards (1979).]

drial membrane is not known. Evidence for electron transfer at a slow rate from the outer to the inner membrane in the presence of added cytochrome *c* has been obtained in animal (Bernardi and Azzone, 1981) and plant mitochondria (R. H. Wilson and Hanson, 1969; Dizengremel, 1977; Moreau, 1976). Under these conditions, aerobic oxidation of ex-

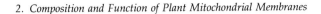

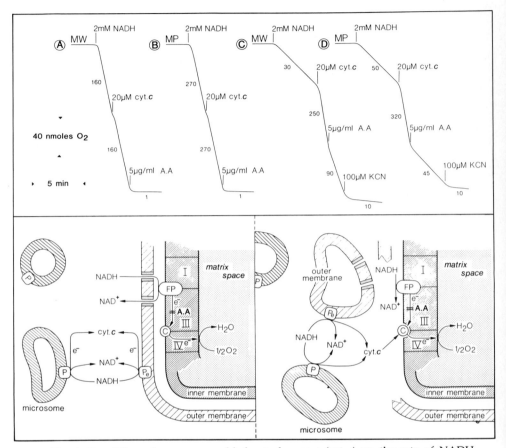

Fig. 2.15. Effect of exogenous added cytochrome c (cyt c) on the rate of NADH oxidation by unpurified (MW) and purified (MP) mitochondria isolated from higher plant mitochondria. (A) and (B) Intact mitochondria; (C) and (D) Swollen mitochondria (mitochondria were added to distilled water and were stirred for 3 sec before adding a double-strength isotonic medium to restore isotonic conditions following the osmotic shock). When the mitochondria are fully intact [(A) and (B), percentage intactness of 99–100%], cyt c does not stimulate the rate of NADH oxidation which is fully inhibited by antimycin A (A.A). In this case, added cytochrome c reduced by the microsomal (P) and/or the outer mitochondrial membrane (P_e) antimycin A-insensitive NADH-cytochrome c oxidoreductase is denied access to the inner membrane. Consequently, the rate of antimycin A-insensitive cytochrome c-dependent O_2 uptake is negligible unless broken mitochondria are present. On the contrary, when the mitochondria are damaged [(C) and (D), osmotically shocked mitochondria], the rate of NADH oxidation is strongly stimulated by exogenous added cytochrome c and shows appreciable antimycin A resistance. In the case of purified mitochondria [(D) i.e., without microsomal vesicles (P)], the antimycin A-insensitive cytochrome c-dependent O_2 uptake rate is very slow and is equal to that of cytochrome c reduction by the antimycin A-insensitive NADH-cytochrome c reductase, which is located near the external face of the outer membrane (Douce *et al.*, 1972). In the case of unpurified mitochondria [(C) i.e., with variable amounts of microsomal vesicles

ogenous NADH involving the following pathway is insensitive to anti-mycin A and has an ADP/O ratio below 1.

$$\text{NADH} \rightarrow \text{NADH-cytochrome } b_{555} \text{ reductase} \rightarrow \text{cytochrome } b_{555} \rightarrow$$
$$\text{intermembrane cytochrome } c \rightarrow \text{cytochrome oxidase} \rightarrow O_2 \qquad (2.5)$$

However, the demonstration that cytochrome c transfers electrons from the outer to the inner membrane implies that the mitochondrial prepara-tions contain almost exclusively intact mitochondria. Unfortunately, metabolically active preparations containing large numbers of intact mi-tochondria always contain a variable proportion of damaged or outer membrane–free mitochondria. This mitochondrial population has respi-ratory chains that can interact with exogenous cytochrome c (Douce *et al.*, 1972). The fact that both types of mitochondria occur in the same preparation has not always received the attention it deserves and has occasionally led to erroneous conclusions concerning the permeability of the outer mitochondrial membrane to cytochrome c. Consequently, the cytochrome c-dependent, antimycin A-insensitive, NADH oxidation ob-served by the previous authors could be entirely, or at least partially, attributable to broken mitochondria present in their preparations. In our laboratory, we have observed a direct relationship between the rate of antimycin A-insensitive O_2 consumption, involving the NADH-cytochrome c reductase of the outer membrane, and the percentage of broken mitochondria present in the mitochondrial preparation. This rate is negligible when the mitochondria are fully intact and maximal when all of the mitochondria exhibit damaged outer membranes (Fig. 2.15). Finally, we believe that the communication between the outer and inner membranes mediated by cytochrome c for exogenous NADH oxidation is of doubtful significance, especially in plant cells, because the external dehydrogenase located on the outer surface of the inner membrane is extremely active in oxidizing exogenous NADH.

The metabolic significance of the respiratory-linked, inner membranal external-dehydrogenase capable of oxidizing cytosolic NADH very rapidly and present in all of the plant mitochondria isolated so far, is unknown. It is clear that this dehydrogenase will favor the conversion of glyceraldehyde 3-P to glycerate 3-P and, therefore, the forward direction of glycolysis. Furthermore, photosynthetically generated ATP and

---

(P)], the antimycin A-insensitive cytochrome c-dependent O_2 uptake rate is much higher. Depending on the extent of microsomal membranes present in the mitochondrial suspen-sion, the antimycin A-resistant O_2 uptake ranges from a small proportion of the total O_2 uptake in some mitochondrial preparations to those cases in which hardly any sensitivity is observed (i.e., reduced cytochrome c supply to the outer surface of the inner membrane matches the capacity of complex IV). Abbreviation: FP, specific dehydrogenase involved in external NADH oxidation (see Fig. 2.14).

NADPH are not directly available for extrachloroplastic reactions, due to the quasi-impermeability of the chloroplastic inner membrane to these compounds (Douce and Joyard, 1979). A shuttle system involving glycerate 3-P and dihydroxyacetone phosphate, both of which move rapidly across the plastid envelope, has been reported and would allow the indirect transfer of ATP and NAD[P]H from a plastid to the cytoplasm (for review see G. E. Edwards and Walker, 1983). Depending on the redox potential of pyridine nucleotides inside and outside of the chloroplasts, the shuttle can operate in both directions. This shuttle system requires the oxidation of dihydroxyacetone phosphate to glycerate 3-P in the cytoplasm by an NADP-linked, nonreversible, glyceraldehyde 3-P dehydrogenase (Kelly and Gibbs, 1973) or an NAD-linked, reversible, glyceraldehyde 3-P dehydrogenase and a glycerate 3-P kinase. Consequently, the two mitochondrial external-dehydrogenases could play an important role in the indirect transport of ATP outside of the plastids by continuously reoxidizing cytosolic NADH or NADPH. It seems reasonable to expect, therefore, that the electron flux through these mitochondrial external-dehydrogenases is strongly regulated. Since these dehydrogenases are Ca^{2+}-dependent, *in vivo* control by modulation of the cytosolic free-Ca^{2+} levels is possible. Unfortunately, the importance of external NAD[P]H oxidation by Ca^{2+} is difficult to gauge, since the concentration of free-Ca^{2+} in the cytosol of plant cells is difficult and perhaps impossible to measure.

   b.   Oxidation of Endogenous NADH.   The oxidation of endogenous NADH in plant mitochondria appears to be more complex than its counterpart in mammalian mitochondria. The most obvious indication of this is that inhibitors such as piericidin A or rotenone, which inhibit the oxidation of endogenous NADH in animal mitochondria by interacting with the iron–sulphur centers associated with complex I, only cause a partial and sometimes an imperceptible inhibition in the plant mitochondrial system. By measuring the ATP formation, it is clear that the rotenone-resistant pathway is not coupled to the first site of ATP synthesis whereas the rotenone-sensitive pathway is (Table 2.11). It seems that plant mitochondria, in contrast with animal mitochondria, possess two internal NADH dehydrogenases on the inner surface of the inner membrane (Fig. 2.14). One of these internal-dehydrogenases readily oxidizes endogenous NADH in a rotenone-sensitive manner. This dehydrogenase is, therefore, coupled to the synthesis of three molecules of ATP and is probably similar to the complex I characterized in mammalian mitochondria. Current evidence suggests that this dehydrogenase has an apparent K_m for NADH of 8 μM (Møller and Palmer, 1982). We also

TABLE 2.11

Effect of Rotenone (22 μM) on Phosphorylation Efficiency by Broad Bean Mitochondria[a]

Substrate	ADP/O ratios	
	Without rotenone	With rotenone
Citrate	2.40 ± 0.15 (3)	1.45 ± 0.15 (3)
α-Ketoglutarate	3.10 ± 0.2 (4)	2.04 ± 0.15 (4)
Malate	2.30 ± 0.25 (5)	1.25 ± 0.20 (5)
NADH	1.60 ± 0.15 (3)	1.55 ± 0.1 (2)
Succinate	1.85 ± 0.1 (2)	1.75 ± 0.2 (2)
Ascorbate	0.9 (1)	1.0 (1)

[a] The data were calculated from an enzymatic assay of glucose-6-phosphate after incubation in the presence of glucose and hexokinase; in parentheses are the numbers of experiments. [From Marx and Brinkmann (1979).]

believe that complex I, which operates in close relationship with all of the NAD$^+$-linked TCA cycle dehydrogenases, utilizes a common pool of NAD$^+$ present in the matrix space (Neuburger and Douce, 1983; Fig. 2.13). The second dehydrogenase connected to the respiratory chain via the ubiquinone pool (Fig. 2.13) is insensitive to inhibition by rotenone and is coupled to the synthesis of only two moles of ATP (J. M. Palmer, 1976; Marx and Brinkmann, 1979; Table 2.11). This dehydrogenase, in contrast with complex I, exhibits a low affinity for internal NADH and differs from the rotenone-resistant NADH dehydrogenase associated with the outer face of the inner membrane inasmuch as it is not sensitive to EGTA or Ca^{2+} (Møller and Palmer, 1982). Furthermore, the specific inhibition of the external NADH dehydrogenase by the analogue N-4-azido-2-nitrophenyl-4-aminobutyryl-NADH, does not suppress the rotenone-resistant endogenous NADH oxidation (M. Neuburger and R. Douce, unpublished data). Interestingly, Ragan (1978) has demonstrated that exogenous quinones are reduced by the mitochondrial NADH dehydrogenase (complex I) by two pathways, one rotenone-sensitive and the other rotenone-insensitive. A minimal scheme to explain rotenone-resistance thus requires simply that these two mechanisms, or pathways of electron transfer, reflect two different sites of interaction for the quinones with complex I. Rotenone sensitivity is higher for the more lipophilic ubiquinone than for the more hydrophilic ubiquinone. In addition, correlation between rotenone sensitivity, H$^+$ extrusion, and $\Delta\Psi$ rise indicates that only the rotenone-sensitive site is related to the activation of the H^+ pump, as shown by Di Virgilio and Azzone (1982) in rat liver mitochondria. Consequently, it is possible that in the case of plant mitochondria complex I, in conjunction with a specific pool of ubiquinone, is perhaps associated with the external NADH dehydrogenase and may also be associated with the rotenone-resistant pathway. That is, there may not be two distinct internal dehydrogenases.

The physiological significance of the rotenone-resistant, internal NADH dehydrogenase is not understood. It has already been suggested that complex I may be associated with the cyanide-sensitive oxidase, whereas the nonphosphorylating internal dehydrogenase is associated with the cyanide-resistant oxidase, providing a totally nonphosphorylating pathway for the oxidation of endogenous NADH when the energy charge is high. Such an idea is very attractive, especially in the case of mitochondria from thermogenic tissues. There is no evidence to support this speculation, however, since potato tuber mitochondria have no cyanide-insensitive alternate oxidase, but they have two pathways for the oxidation of endogenous NADH (Marx and Brinkmann, 1979). Again it is clear that the mechanism whereby the rotenone-insensitive pathway is engaged and the extent to which it operates are of the utmost importance in the physiological role of this pathway. The concentration of NADH in the matrix space seems to play an important role in the regulation of the pathways responsible for endogenous NADH oxidation, because the affinity of the rotenone-sensitive NADH dehydrogenase for NADH is greater than the affinity of the internal, rotenone-resistant NADH dehydrogenase. This observation could explain why the rotenone-insensitive pathway is often more susceptible to mitochondrial aging than the rotenone-sensitive pathway. As the level of NADH progressively falls in the mitochondria, it becomes too low to act as a substrate for rotenone-resistant respiration but is still sufficient to engage the rotenone-sensitive pathway (Neuburger and Douce, 1983). Likewise, under state 3 conditions, the level of NADH falls and therefore becomes too low to act as a substrate for rotenone-resistant oxidation (Douce and Bonner, 1972). Conversely, under state 4 conditions or in the presence of rotenone, the level of NADH rises and the rotenone-resistant pathway is engaged (Tobin et al., 1980). Thus, the rotenone-resistant dehydrogenase may play a role when the phosphate potential restricts electron flow through the normal respiratory chain. Nonetheless, the mechanism of the rotenone-insensitive pathway in plant mitochondria remains obscure. It may only represent a "short-circuit" within the iron–sulfur centers of site-1 phosphorylation (i.e., in the presence of rotenone) and have little physiological significance.

B. Energetics of Electron Transport and Oxidative Phosphorylation

The transfer of electrons from substrate to O_2 via the cytochrome pathway is coupled to an electrogenic translocation of protons across the inner mitochondrial membrane (Fig. 2.16). For example, when a small quantity of dissolved O_2 is added to a lightly buffered anaerobic suspen-

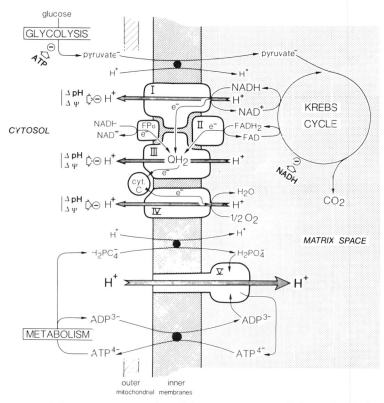

Fig. 2.16.   Schematic representation of the arrangement and electrochemical connection of the four electron-transferring complexes (I, II, III, and IV) within the inner membrane of a higher plant mitochondrion. Electrogenic proton movement established by complexes I (rotenone-sensitive NADH dehydrogenase), III (ubiquinol-cytochrome *c* oxidoreductase), and IV (ferrocytochrome *c*-oxygen reductase) during the transfer of two electrons from internal NADH or complex II (succinate-ubiquinone oxidoreductase) to O_2, and the utilization of such proton equivalents in ATP synthesis (complex V or ATPase complex) and anion transport (pyruvate, $H_2PO_4^-$, and ADP^{3-}) are indicated. Another dehydrogenase (FPo) located on the outer surface of the inner membrane, which feeds electrons directly to complex III and is involved in the oxidation of external NADH, is also indicated.

sion of intact mitochondria in a closed thermostatically controlled glass vessel, the surrounding medium undergoes a rapid acidification that is followed by a slow realkalinization as the ejected protons reequilibrate across the inner mitochondrial membrane (respiratory pulse method; Mitchell, 1980). The opposite occurs in submitochondrial particles. This reaction is clearly diminished by inhibiting the respiration with antimycin or by damaging the inner membrane with detergents. Protons have two different properties, acidity and electric charge, consequently, electrogenic proton translocation without cotransport of anions generates both a proton gradient (ΔpH) and a membrane potential [$\Delta\Psi$ negatively charged side (matrix side); positively charged side (cytoplasmic side)]. In this way, electron transport creates an electrochemical gradient of protons across the membrane. Both the lower proton concentration and any net negative charge on the inside tend to drive protons in again. The two parameters are additive, contributing to a "proton-motive force" differential across the membrane (according to the following equation).

$$\Delta\bar{\mu}\,H^+/F = \Delta p = \Delta\Psi - 2.303\,RT/F \times \Delta pH \qquad (2.6)$$

Experimental determinations of Δp ($\Delta\bar{\mu}\,H^+$)/F via the separate measurement of $\Delta\Psi$ (quantitated by measuring the distribution of a nonphysiological lipophilic cation across the membrane) and ΔpH (quantitated by measuring the distribution of a permeative weak acid) have yielded an average value of approximately 126–250 mV for state 4 respiration in plant mitochondria (A. L. Moore and Bonner, 1982; Mandolino *et al.,* 1983; Ducet *et al.,* 1983; Diolez and Moreau 1985; Table 2.12). This is a considerable value if one considers that an energy-transducing membrane with a membrane potential of 200 mV has an electrical field in excess of 300,000 V/cm across its hydrophobic core.

The number of protons translocated to the suspending medium by mitochondria is limited by two factors. In the first place, the observations relate to a steady-state condition. Protons are pumped out by electron transport but diffuse in again through various leaks or ports (i.e., by passive diffusion across the lipid bilayer and/or noncoupled leakage via the ATPase complex). When the mitochondrial membranes are badly injured during the course of their preparation, the leaks become considerable. Adding uncouplers such as dinitrophenol (Plantefol, 1922, 1932) or *p*-trifluoromethoxycarbonylcyanide phenylhydrazone (FCCP) increases the rate of leakage and thereby equilibrates the electrochemical H^+ gradient across the mitochondrial inner membrane. Potent uncouplers (protonophoric uncouplers) are all moderately weak acids, with an acid dissociation constant (pK_a) in the range 5–7, and are

TABLE 2.12

Typical Values for the Proton-Motive Force in Intact Plant Mitochondria[a]

Material	Conditions	$\Delta \psi$ (mV)	-60Δ pH (mV)	Δ p (mV)	Reference
Potato tuber	State 4	126	36	162	A. L. Moore *et al.*, 1978
Potato tuber	State 4	250	12	262	Ducet *et al.*, 1983
Mung bean hypocotyl	State 4	126	36	162	Moore *et al.*, 1978
Arum spadix	State 4	132	26	158	Moore *et al.*, 1978
Jerusalem artichoke tuber	State 4	220	n.d.	n.d.	Mandolino *et al.*, 1983

[a] Abbreviation: n.d., not determined.

known to increase H^+ transfer across model phospholipid membranes and energy-transducing membranes. Uncouplers release respiratory control completely and suppress all energy-linked reactions, such as ATP synthesis, active transport of ions, and reverse electron flow. Nigericin ($+K^+$), in permitting entry of H^+ for K^+ exchange (K^+, H^+ antiport) across the mitochondrial inner membrane, causes ΔpH to collapse, whereas the mobile carrier ionophore valinomycin ($+K^+$), by permitting electrogenic diffusion of K^+, dissipates the membrane potential (Fig. 2.17). [It is interesting to note, that in the absence of added K^+, valinomycin has little effect on respiring mitochondria, confirming that it does not increase H^+ conductance. Subsequent addition of K^+ increases the electrophoretic movement of K^+ that is mediated by valinomycin.] Nigericin alone ($+K^+$), in contrast with valinomycin ($+K^+$), does not uncouple the mitochondria. Secondly, if protons leave without any counter-ion flux, the membrane potential must rise almost immediately to a point that would halt further H^+ translocation and electron transport. The membrane potential directly influences the apparent redox potential of any component of the electron transport chain which "senses" the potential (Ducet, 1980; Mitchell, 1980). For example, $\Delta\Psi$ exerts a constraint on forward electron flow from cytochrome b_{566} to cytochrome b_{562}. This effect is explained as a consequence of the anisotropic location of cytochromes b_{566} and b_{562} in the membrane (Papa *et al.*, 1981; Ducet, 1980). Because the following reaction:

$$QH_2 + \text{Rieske protein}_{ox} \rightarrow Q^- + 2H^+ + \text{Rieske protein}_{red} \qquad (2.7)$$

(see Fig. 2.10) is highly pH-dependent ($\Delta E_h = 120$ mV/pH unit) and may be equally responsive to the membrane potential, it is possible that the iron–sulfur protein (Rieske protein) may be the site of respiratory control in this region of the respiratory chain. Likewise, in an interesting

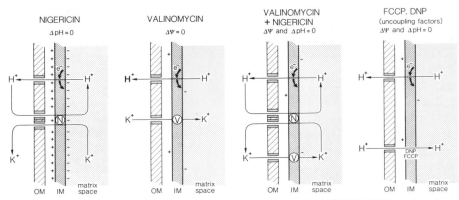

Fig. 2.17.   Effect of nigericin, valinomycin, and uncouplers (FCCP, DNP) on the electrochemical gradient of protons across the membrane generated by the electron chain (e^-). Nigericin, in permitting entry of H^+ for K^+ exchange (K^+, H^+ antiport) across the mitochondrial inner membrane, collapses only ΔpH whereas the mobile carrier ionophore valinomycin ($+K^+$), in permitting electrogenic diffusion of K^+ (K^+ uniport), dissipates the existing membrane potential ($\Delta\Psi$). Finally, the protonophoric uncouplers, such as 2,4-dinitrophenol (DNP) and phenylhydrazone (FCCP), in permitting diffusion of H^+, collapse the proton-motive force (ΔpH, $\Delta\Psi$) and release respiratory control completely (Terada, 1981).

study of the redox behavior of potato tuber mitochondria oxidizing succinate, Ducet (1980) has observed, in good agreement with Oshino *et al.* (1974), that only cytochrome *a* is reduced at anaerobiosis while cytochrome a_3 stays oxidized. It is only when the membrane potential is collapsed by an uncoupler that cytochrome a_3 becomes reduced in thermodynamic equilibrium with the succinate–fumarate couple. According to Ducet (1981), cytochromes *a* and a_3 are not buried at the same place in the inner membrane: in this way cytochrome a_3 senses the membrane potential but cytochrome *a* does not sense it. They appear equally reduced only when the membrane potential is collapsed. The number of protons that can leave, therefore, is strictly limited by the number of counter-ions that can be redistributed to neutralize the negative matrix. The internal volume of isolated plant mitochondria, measured with [^{14}C]sucrose, is estimated at approximately 3–4 μl/mg protein. [Potato tuber mitochondria purified by isopycnic centrifugation in density gradients of Percoll exhibited an intramitochondrial volume (3.5 μl/mg of mitochondrial protein) higher than that of mitochondria purified in density gradients of sucrose (1 μl/mg of mitochondrial protein; Neuburger and Douce, 1983).] Under these circumstances, it is obvious that proton extrusion cannot occur beyond the extent of available internal buffering groups, unless counter-ion movement occurs.

According to the chemiosmotic hypothesis (Mitchell, 1980), the respiration-linked proton pump provides a link between oxido-reduction and ATP synthesis. Indeed, the proton-motive force built up during the course of substrate oxidation has sufficient driving power (FΔp) to permit ATP synthesis. However, the precise mechanism by which the proton flow drives ATP synthesis is not yet known. There is now considerable evidence that ATP synthesis in mitochondria is accompanied by proton reentry via a membrane-localized, reversible, vectorial ATPase (ATP hydrolysis is characterized by proton ejection; Fig. 2.16). It is vectorial most importantly with respect to the protons involved in the dehydration of ADP plus P_i, leading to ATP synthesis as follows:

$$ADP^{3-} + H_2PO_4^- + nH^+_{out} \rightleftarrows nH^+_{in} + ATP^{4-} + H_2O \qquad (2.8)$$

Since it was calculated that 6 or 12 protons were translocated per oxygen atom consumed in the presence of NAD$^+$-linked substrates (H$^+$/O = 6 or 12) and 4 or 8 protons in the presence of succinate (H$^+$/O = 4 or 8), it was inferred that 2 or 4 H$^+$ ions were extruded per coupling site (i.e., at complexes I, III, and IV) (Wikström *et al.*, 1981; Alexandre *et al.*, 1980; Mitchell, 1976; Mitchell and Moore, 1984). Several authors have measured the electrochemical proton gradient ($\Delta\bar{\mu}$ H$^+$) and the Gibbs free energy of ATP synthesis in mitochondria. [$\Delta G_p = \Delta G'_o + RT$ log([ATP]/[ADP][P$_i$]), in which $\Delta G'_o$ is the standard free energy of hydrolysis.] A comparison of $\Delta\bar{\mu}$ H$^+$ with ΔG_p (ΔG_p = H$^+$/ATP $\times$ $\Delta\bar{\mu}$ H$^+$) indicates that the apparent H$^+$/ATP ratio (i.e., the number of H$^+$ ions ejected per pair of electrons per ATP-generating site) in mung bean hypocotyl mitochondria is 3.4 $\pm$ 0.2, while in phosphorylating submitochondrial particles it is 2.2 $\pm$ 0.1 (A. L. Moore and Bonner, 1981). Submitochondrial particles that are inside out with respect to intact mitochondria are prepared by sonication, followed by differential centrifugation to remove unbroken mitochondria (Passam and Palmer, 1971; Rich *et al.*, 1977a; Grubmeyer *et al.*, 1979; Møller *et al.*, 1981b). Since the classical techniques used for the preparation of submitochondrial particles result in a mixed population of right side out and inside out vesicles, submitochondrial particles must be extensively purified by a two-phase system (Møller *et al.*, 1981b). The lack of stimulation of respiration by the addition of exogenous cytochrome *c* can be used as a measure of polarity (Harmon *et al.*, 1974). In fact, ATP synthesis in mitochondria (in contrast with the situation observed in submitochondrial particles) takes place on the inside surface of the inner membrane, and consequently, part of the total energy available is used to move substrates and products against concentration gradients between the external medium (or cytosol) and the matrix space. In other words, some of the energy that would otherwise be available for ATP synthesis has been diverted for the translocation of adenine nucleotides and phosphate.

Several reviews have appeared recently dealing with the structure of the mitochondrial ATPase complex (Fillingame, 1980; Baird and Hammes, 1979; Futai and Kanazawa, 1980; Vignais and Satre, 1984) and the mechanism of ATP formation (Cross, 1981; Boyer *et al.*, 1977). The ATPase complex (mol.wt. 450,000–500,000) or complex V of yeast mito-chondria is analogous to the ATPase complex of mammalian mitochon-dria and consists of two empirically defined components F_1 and F_0. When bound to the inner mitochondrial membrane, the proton–ATPase com-plex catalyzes the dehydration of ADP plus P_i, leading to ATP synthesis. When physically separated from the membrane using potassium cholate, this complex is only capable of catalyzing the ATP–P_i exchange and the net hydrolysis of ATP. Both the exchange and the hydrolytic reactions are inhibited by oligomycins. (This family of antibiotics was discovered in the culture filtrate of *Streptomyces diastatochromogenes*.) In addition, un-couplers stimulate oligomycin-sensitive ATPase activity. The F_1 compo-nent (mol. wt. 350,000) can be detached from the membrane as a water soluble complex of five distinct subunits in the approximate ratio $\alpha_3\beta_3\gamma\delta\epsilon$. The molecular weights of the subunits determined by electrophoresis in sodium dodecyl sulfate and by sedimentation equilibrium in dissociating agents fall in general within the following ranges: α from 53,000–62,500; β from 50,000–57,000; γ from 28,000–43,000; δ from 12,000–21,000; and ϵ from 7500–14,000 (see Amzel, 1981). The α- and β-subunits are probably responsible for the catalytic activity of F_1, and the binding of nucleotides. The minor subunits have regulatory and/or structural functions. A sixth subunit (IF_1) that is clearly distinct from the ϵ subunit of F_1 and that may or may not remain attached to F_1 depending on the purification pro-cedure employed, is a potent inhibitor of ATPase activity. This ATPase inhibitor is a low-molecular-weight, heat stable, trypsin-sensitive protein (Satre *et al.*, 1979). It is possible that the inhibitor acts as a directional regulator of ATP synthesis by preventing the back flow of energy from ATP to the electron transport system and thereby preventing ATP hydro-lysis (see, for example, Ducet, 1981). In addition, strong evidence has been provided that the energy state of mitochondrial particles modulates the binding affinity of IF_1 for F_1 and that the proton-motive force devel-oped by mitochondrial respiration results in a release of IF_1 (Klein *et al.*, 1981). The second component of the ATPase complex named F_0 is an integral membrane complex of four or five distinct proteins that can be extracted from the membrane by the use of detergents. These are the uncoupler binding protein (UBP), the oligomycin sensitivity–conferring protein (OSCP), the dicyclohexylcarbodiimide (DCCD) binding protein (DCCD-BP), and minor polypeptides. One component of the complex, the oligomycin sensitivity–conferring protein (OSCP), binds the F_1 sector of the ATPase complex to the F_0 membrane sector of the complex. It is not the site of binding of oligomycin, but by its action of binding F_1 to the membrane, OSCP confers oligomycin sensitivity on the soluble F_1. One may speculate that OSCP participates in the proton translocation activity of the ATPase, perhaps by transmitting conformational changes that mediate the uptake and release of protons. The function of the DCCD-binding protein, which probably aggregates into a hexamer *in vivo*, is to form an H^+-translocating channel through F_0 and thus allowing the transmembrane movement of H^+ to and from the active sites of F_1. The uncoupler binding protein probably regulates transmembrane proton conduction (Galante *et al.*, 1979). Extensive studies are required, however, to establish if binding to the "uncoupler binding protein" is essential for uncoupling (Terada, 1981). The major question that remains unresolved is the chemical mechanism of Δp utilization.

Relatively few measurements of proton fluxes have been made with plant mitochondria. Likewise, our knowledge concerning the plant mitochondrial ATPase complex is fragmentary. Seldom have plant mitochondria been used by biochemists as primary objects for investigation of oxidative phosphorylation. It had been originally suggested that the values of the proton fluxes are relatively small in plant mitochondria compared with those obtained with animal mitochondria (Johnson and Wilson, 1972; Kirk and Hanson, 1973; Chen and Lehninger, 1973), and it has been suggested that this may be due to an inherent "leakiness" of the plant inner mitochondrial membrane with respect to protons, possibly reflecting a greater role of the electrical potential as opposed to the chemical potential in plant mitochondrial energy conservation (R. H. Wilson and Graesser, 1976). For example, when valinomycin is added to bean mitochondria incubated in KCl, a rapid ejection of H^+ into the external medium occurs but is followed by an immediate and spontaneous decline in the H^+ level. In more recent publications (A. L. Moore and Wilson, 1977, 1978), the proton permeability of purified plant mitochondria was indirectly estimated from the rate of swelling of nonrespiring mitochondria suspended in salt medium or directly estimated using an acid-pulse technique. These results clearly demonstrate the presence of a permeability barrier to protons within the inner membrane of plant mitochondria and markedly contrast with earlier experiments. For example, the effective rate of H^+ translocation under a known proton-motive force in turnip mitochondria (A. L. Moore and Wilson, 1978) is somewhat similar to that obtained by Mitchell (1980) in rat liver preparations. Parenthetically, Ducet (1979) provides good evidence that bovine serum albumin decreases the proton conductance of potato tuber mitochondria. Similarly, determinations of the total proton-motive force in plant mitochondria (A. L. Moore et al., 1978) also argue against a proton permeable inner membrane in plant mitochondria. In fact, the apparent leakiness of the plant mitochondrial inner membrane in early studies is almost entirely attributable to mitochondrial damage, contaminations, and fatty acids released and acting as uncouplers. A. L. Moore and Bonner (1982), following the work of Åkerman and Wikström (1976), demonstrated that the positively charged dye safranine can be used to monitor membrane potential in intact plant mitochondria, because the spectral shift in safranine absorbance (using the wavelength pair 511 and 533 nm) was linearly related to the developed membrane potential and could be calibrated with reference to a K^+ diffusion potential. Under these conditions, A. L. Moore and Bonner (1982) indicated that both respiration and ATP hydrolysis gave rise to a membrane potential of approximately 135 mV. Likewise, the membrane potential of potato tuber mitochondria has been investigated under various conditions, using a tetraphenylphosphonium electrode during the oxidation of substrates (Ducet et al., 1983). The results indicate that in plant mitochondria the proton-motive force arises mainly from the membrane potential (250 mV, negative inside, in state 4, as shown in Table 2.12), the proton gradient (measured with nigericin) of approximately 0.2 pH units was in good agreement with a previous result (Neuburger and Douce, 1980). In the absence of O_2 or reducing equivalents, the membrane potential slowly depolarizes (Ducet et al., 1983), presumably via the natural conductance (due to H^+ leak and ancillary ion movements i.e., K^+/H^+ antiporters; see Jung and Brierley, 1979; Hensley and Hanson, 1975; A. L. Moore and Wilson, 1977) of the membrane. Interestingly, A. L. Moore and Bonner (1982) observed that the membrane potential is higher following an initial ADP pulse, indicating that the membrane becomes less leaky to protons. This correlates with O_2 electrode studies that indicate an increase in respiratory control following a state 3–4 transition (Raison et al., 1973). According to a delocalized chemiosmotic model, the steady-state value of Δp should reflect a steady-state balance of the outward proton flux driven by the respiratory chain and the inward proton flux due to ATPase functioning and the passive proton leaks. In other words, in the delocalized chemiosmotic mechanism, a single-value correlation should exist between the extent of Δp and the kinetic parameters of respiration and ATP synthesis. Unfortunately, there are several observations that argue against this view. For example, in Jerusalem artichoke mitochondria when the rate of respiratory elec-

tron flow is decreased, a parallel inhibition of the rate of phosphorylation is observed while very limited effects can be detected on the extent
of Δp (Mandolino *et al.*, 1983). The safranine method to monitor membrane potentials in plant mitochondria demonstrated that, unlike the
addition of cyanide or antimycin A, rotenone has relatively little effect
upon the membrane potential. Because this inhibitor blocks proton
pumping in site 1, this experiment suggests that the potential is maintained by sites 2 and 3 (A. L. Moore and Bonner, 1982). Furthermore, the
stimulation of respiration upon addition of ADP in mung bean and
voodoo lilly (*Sauromatum guttatum*) mitochondria is accompanied by a
smaller decrease in Δp than is seen when respiration is increased to the
same extent with an uncoupler (A. L. Moore and Bonner, 1982). Such
observations are difficult to reconcile with Δp as the sole and obligate
intermediate between electron transfer and ATP synthesis, suggesting
that there may be either some intramembrane protonic coupling between individual redox chains and ATP synthetases or some activation
mechanism of the ATP synthetases by the redox reactions (R. J. P.
Williams, 1978; Zoratti *et al.*, 1983; Mandolino *et al.*, 1983). In fact, R. J. P.
Williams (1978) believes that there may be some degree of interaction
with the respiratory chain and ATPase. (See, however, Diolez and Moreau, 1985.) Furthermore, according to Deléage *et al.* (1983), efficient
protons appear to be localized in a "kinetic compartment." Consequently,
the localization of the efficient protons either in bulk phases or in specific circuits is still a matter of controversy. The delocalized chemiosmotic
hypothesis, although endorsed by most texts in cell biology, remains
unacceptable to some and is the subject of increasingly jesuitical debate.

The effects of oligomycin on the respiration of plant mitochondria
agree qualitatively with those on the respiration of animal mitochondria:
(a) Oligomycin blocks respiration coupled to phosphorylation but has no
effect on respiration in the presence of uncoupling agents (Ikuma and
Bonner, 1967b); (b) Oligomycin does not interfere with the generation of
an electrochemical potential or pH gradient across the mitochondrial
membrane nor with the utilization of such gradients to move ions
against a concentration gradient; (c) Oligomycin does not inhibit the
substrate-level phosphorylation of ADP catalyzed by succinic thiokinase
during the oxidation of α-ketoglutarate. Several workers have reported
that the ATPase activity of whole sonicated corn mitochondria is not
inhibited (Jung and Hanson, 1973; Sperk and Tuppy, 1977) or only partially inhibited (Grubmeyer and Spencer, 1978) by oligomycin, suggesting that mitochondria of corn may possess an unique energy transducing ATPase system. However, the mitochondrial ATPases in their
membrane-bound state (i.e., when associated with F_0) are noted for

their sensitivity to oligomycin, and this sensitivity, conferred upon AT-Pase F_1 by integral membrane components, is lost when ATPase F_1 (i.e., without F_0) is solubilized (Iwasaki and Asahi, 1983). Consequently, according to Grubmeyer and Spencer (1978), whole sonicated preparations can be misleading since they contain two distinct forms of the ATPase (soluble and membrane-bound). In addition, they may contain extra-mitochondrial ATPase and/or phosphatase. In fact, particulate preparations capable of substrate oxidation and oxidative phosphorylation (S. B. Wilson and Bonner, 1970a,b; Passam and Palmer, 1971; Grubmeyer *et al.*, 1979) showed an ATPase activity that was more than 95% inhibited by low levels of oligomycin (Grubmeyer and Spencer, 1978).

Tuppy and Sperk (1976), working with wheat seedling mitochondria, have failed to isolate an ATPase resembling that of yeast and animal mitochondria with regard to high molecular weight and cold lability. They suggested that this low-molecular-weight ATPase might correspond to one subunit (α or β?) of the ATPase complex found in other mitochondria. However, we are convinced that this low-molecular-weight ATPase isolated by Tuppy and Sperk (1976) represents an unspecific phosphatase or ATPase that contaminates unpurified plant mitochondrial preparations. Although reports are not extensive enough to permit detailed comparison with mammalian and yeast ATPase F_1, it seems that solubilized ATPase from plant mitochondria exhibits properties similar to the mammalian mitochondrial enzyme (Yoshida and Takeuchi, 1970; Grubmeyer *et al.*, 1977). More recently, the ATPase F_1 complex has been purified from maize mitochondria and shown to consist of five subunits with M_r values of 58,000 (α); 56,000 (β); 35,000 (γ); 22,000 (δ); and 8000 (ϵ) (Fig. 2.18). Further proof that the 58,000 and 56,000 molecular weight polypeptides were the α- and β-subunits was obtained by immunoprecipitation with antisera raised against the α- and β-subunits from yeast ATPase F_1 (Hack and Leaver, 1983). Interestingly, sweet potato ATPase F_1 is composed of six kinds of subunits $\alpha,\beta,\gamma,\delta,\delta'$, and ϵ while the enzyme from other sources consists of five subunits $\alpha,\beta,\gamma,\delta$, and ϵ (Iwasaki and Asahi, 1983). However, subunit δ' (or δ) may be a degradation product formed by the proteolysis of larger subunits. In addition, sweet potato ATPase F_1 may release some of its subunits (probably, subunits α and β, or both) during storage, giving a logical explanation to the results of Tuppy and Sperk (1976).

Trypsin treatment of sonicated potato mitochondria (Jung and Laties, 1976) increased Mg^{2+}-dependent, oligomycin-sensitive ATPase activity 10- to 15-fold, suggesting the presence of an ATPase inhibitor protein in plant mitochondria (Table 2.13). In potato tuber mitochondria, the inhibitor must be tightly associated with the ATPase since activation is

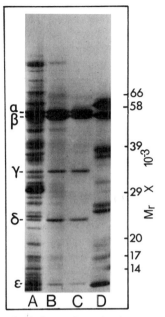

Fig. 2.18. One-dimensional gel electrophoresis of maize mitochondrial ATPase F_1 and chloroplast ATPase CF_1. Samples were electrophoresed on a 10–15% (w/v) SDS–acrylamide linear gradient gel and stained with Coomassie Brilliant Blue. (A) 50 μg total mitochondrial protein; (B) 15 μg ammonium sulphate fraction ATPase F_1; (C) 10 μg sucrose gradient purified ATPase F_1; and (D) chloroform-extracted ATPase CF_1. Gel electrophoresis provided by C. J. Leaver. [From Hack and Leaver (1983).]

achieved only after mitochondrial disruption and treatment with trypsin. The inhibitor in plant mitochondria may regulate the membrane-bound ATPase in the same manner as that found for beef heart mitochondria (Asami *et al.*, 1970), blocking ATP-driven but not respiratory chain–driven energy transfer reactions. Three different situations exist, however, with respect to ATPase in freshly isolated intact plant mitochondria. First, the situation in which uncouplers readily stimulate ATPase activity (dinitrophenol-stimulated, oligomycin-sensitive ATPase) as in corn mitochondria (Jung and Hanson, 1973). Second, that in which ATPase is activated by the disruption of the inner membrane. In this case uncoupler stimulation in intact mitochondria is observed only after incubation with a respiratory substrate (respiratory priming) or ATP (self-priming), as in cauliflower (Jung and Hanson, 1975) and sweet potato mitochondria (Carmelli and Biale, 1970). Third, that in which an inhibitor is firmly associated with ATPase, as in potato tuber mitochondria (Jung and Laties, 1976). Although respiration may cause dissociation of the inhibitor

TABLE 2.13

ATPase Activity of Sonicated Potato Tuber Mitochondria[a]

Additions	No treatment[b]	Trypsin-treated[b]
None	8.4	26.4
Mg^{2+}	13.2	158.8
Oligomycin	8.4	21.6
Mg^{2+}; oligomycin	14.4	43.2
Mg^{2+}; 2,4-dinitrophenol	14.4	154.6

[a] Sonicated mitochondrial protein used per determination was 0.24 mg. [From Jung and Laties (1976).]
[b] Units: nmoles P_i/min/mg protein.

during priming, it is most likely a secondary effect in ATPase enhancement. Since the inhibitor is largely dissociated in freshly isolated cauliflower mitochondria (disruption of membrane barriers alone enhances cauliflower ATPase up to 10-fold), we believe that the initiating factor in the priming is establishment of a proton gradient that drives net adenine nucleotide uptake, triggering in turn the transport–hydrolysis complex (Jung and Hanson, 1975). The transport of P_i (efflux) is essential for the continued hydrolysis of ATP. Consequently, the efflux of P_i occurring as a neutral exchange on the P_i/OH$^-$ transporter (Kaplan and Pedersen, 1983) cannot charge compensate for the ATP^{4-}/ADP^{3-} exchange (see mechanism of anion transport). It is the uncoupler that carries out the charge compensation by collapsing the proton gradient (Fig. 2.19). Furthermore, it has been demonstrated by Abou-Khalil and Hanson (1977, 1979a,b) that in most plant mitochondria examined so far, the total amount of mitochondrial nucleotide is low. They demonstrated also that if nucleotide-deficient mitochondria were incubated with ATP, the nucleotide pool size increased and that this net uptake was strongly accelerated if an electrochemical gradient of protons across the membrane was established. Conversely, this net uptake was strongly inhibited by uncouplers. In other words, very often the level of adenine nucleotides in plant mitochondria is too low for spontaneous ATP/ADP exchange and subsequent ATPase activity.

In conclusion, studies on the mitochondrial ATPase complex of higher plants are scant; a wider survey still awaits the solution of the formidable technical difficulties involved in obtaining large amounts of plant mitochondria free from extramitochondrial contaminations. Studies of plant respiration at the level of isolated mitochondria suggest that some of the processes of phosphorylation are somewhat similar to those oc-

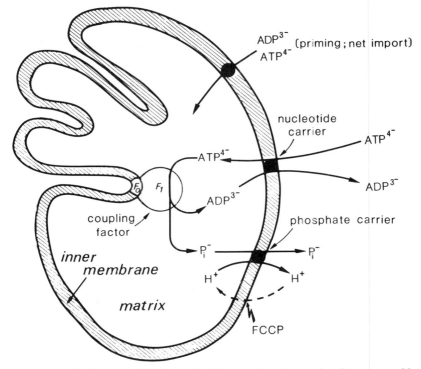

Fig. 2.19. Mechanism of ATP hydrolysis by intact plant mitochondria triggered by an uncoupler (FCCP). Mitochondria are first loaded with ATP in an energy manner (net import, priming) in order to trigger a rapid ATP/ADP exchange. Abbreviations: F_1 and F_0, components of the ATPase complex (coupling factor).

curring in mammalian or yeast cells and that more detailed investigations will prove rewarding.

C. Mechanisms of Anion Transport

The mitochondrion represents a distinct metabolic compartment within the cell. Its inner membrane is highly permeable to H_2O, O_2, and lipophilic molecules, and the available evidence indicates that CO_2 readily passes through the mitochondrial membrane whereas the bicarbonate anion does not (K. J. Chappell and Haarhoff, 1967; Neuburger and Douce, 1980). Although the mitochondria are the source of nearly all of the CO_2 produced in plant cells, its transport across the mitochondrial membrane has received little attention. During the transition from an isotonic to a hypotonic reaction mixture, the mitochondrial water content increases by 300% in a few seconds. Consequently, it is clear that the efflux of water, as a result of the oxidation of substrate by O_2 (1000–1400 nmoles/min/mg protein i.e., 10% of the intramitochondrial water must leave the matrix per min), is not limited by a diffusion barrier (Klingenberg, 1970). It is obvious that the problem is to decide whether this solute permeates by dissolution in the membrane fabric or by crossing the membrane through polar pathways, "aqueous pores" (Sha'afi, 1981), or lipophilic pathways. The inner membrane is also permeable to several monocarboxylic acids with relatively high pK values (such as acetate) which pass, after protonation, through the membrane as undissociated acids (K. J. Chappell and Haarhoff, 1967).

On the other hand, the inner mitochondrial membrane is in principle impermeable to glucose and larger uncharged molecules, H^+, mono-, di-, and trivalent cations, anions, amino acids, nucleotides, and inorganic phosphate. Indeed, the hydrophobic core possessed by lipid bilayers creates an effective barrier to the passage of charged species. The high activation energy required to insert an ion into a hydrophobic region is the reason for the extremely low ion permeability of bilayer regions. Interestingly, it has been shown that the inner membrane loses its selective permeability character when mitochondria are treated with thiol-alkylating agents. Consequently, it is possible that some sulfhydryl groups play a structural role and may control the integrity of the mitochondrial inner membrane (Lê-quôc and Lê-quôc, 1982). This overall unspecific impermeability of the inner membrane toward hydrophilic solutes is overcome by specific translocators, enabling the transport of pyruvate, dicarboxylates, tricarboxylates, ADP, ATP, and inorganic phosphate. In fact, flux rates of metabolites across the mitochondrial membranes must be considerable in order to sustain the full rate of respiration. Of interest is the question of whether or not the transport of metabolites across the mitochondrial membrane is a rate-influencing step in respiration. Finally, it must be realized that the kinetic properties of the translocators may vary from species to species and even from tissue to tissue in the same species.

1. Methodologies for Transport Studies

Transport in plant mitochondria is often faster than in animal mitochondria, and the measurement of transport rates is accordingly more difficult. Another experimental and general problem is the fact that anion transport is in most cases an exchange that requires the corresponding release of a specific metabolite necessary for the uptake of other components before anion uptake begins.

Methods for the measurement of mitochondrial permeability can be divided into two classes. The first relies on measurements of osmotically induced volume changes of mitochondria in iso-osmotic solutions of ammonium salts. The mechanism for this swelling in ammonium phosphate is represented in Fig. 2.20. Ammonia diffuses across the membrane, down a concentration gradient; inside the mitochondria it associates with a proton to form NH_4^+, leaving an excess of OH^- which then exchanges for phosphate on the phosphate carrier (see Section II,C,2). The increased osmotic pressure within the mitochondria induces swelling. This large mitochondrial swelling is made possible by the unfolding of the inner membrane. The resulting decrease in absorbance (e.g., measured at 520 nm) can be followed kinetically and, over a limited range, is proportional to the rate of anion penetration (K. J. Chappell and Haarhoff, 1967). The second method relies on measurements of exchange times of suitably labeled solutes in the absence of appreciable solvent flow. Measurements of metabolite distribution are made after the separation of the mitochondria from the incubation medium. There are basically two types of separation procedures: centrifugation and filtration. In the centrifugal sedimentation method, the mitochondria become anaerobic in the pellet, and thus the steady-state composition of the metabolite is altered. In the filtration with extremely permeative

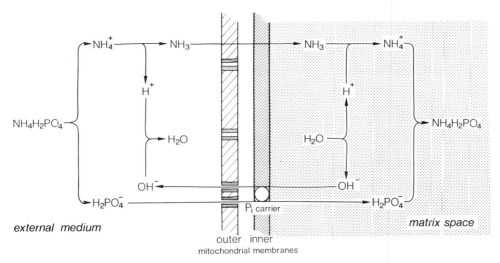

Fig. 2.20. Mechanism of swelling of intact plant mitochondria in iso-osmotic solution of ammonium phosphate ($NH_4H_2PO_4$). Ammonia diffuses across the membrane, down a concentration gradient. Inside the mitochondrion, NH_3 associates with a proton to form NH_4^+, leaving an excess of OH^- which then exchanges for phosphate ($H_2PO_4^-$) on the phosphate carrier. The increased osmotic pressure within the mitochondrion induces swelling.

compounds, the washing procedure employed to remove the external radioactivity also results in a rapid exodus of the internal solute. These problems are partially avoided by the centrifugal filtration method, which combines sedimentation of the mitochondria with filtration through a nonaqueous layer of silicon into an acid layer. The resulting immediate deproteinization prevents metabolism of the molecule transported and releases the soluble constituents of the mitochondria into the extract. Another advantage is that in silicon filtration, the less adherent external volume is carried over into the acid layer. Unfortunately, kinetic measurements of transport are limited by the time taken for all the mitochondria to pass through the oil layer. The highest resolution time is obtained with the inhibitor–stop method that is based on the use of a transport inhibitor (when available) to terminate the reaction (Palmieri and Klingenberg, 1979). The inhibitor–stop transport assay technique, however, makes the assumption that quenching of the transport reaction by inhibitor is rapid with respect to sampling time. Efflux studies can also be performed with this technique by first loading the mitochondria with an unlabeled anion, followed by addition of carrier-free isotopic anion to allow exchange. After washing and resuspending the mitochondria in an incubation medium, addition of a suitable counteranion will promote efflux of the isotopically labeled anion.

Accurate measurements of ^3H_2O, [^{14}C]sucrose, and [^{14}C]dextran ($M_r \simeq 60,000$) distributions in mitochondrial pellets (Fig. 2.21) have yielded the following information: The mitochondrial water space (total water space minus the space occupied by polydextran that does not penetrate the outer membrane) ranges from 1–5 μl/mg protein. Based upon the accessibility to sucrose, the mitochondrial space can be divided into two parts: (a) the space accessible to sucrose which represents the intermembrane space and (b) a space inaccessible to sucrose. The latter space comprises the matrix but varies considerably according to mitochondrial preparation and experimental conditions. Since isolated mitochondria behave as perfect osmometers, the matrix space fluctuates with changes in the osmolarity of the external solutes. It is clear, therefore, that by comparing the water space, the dextran space, and the sucrose-permeable space with that accessible to a given substance, it is possible to determine whether or not it permeates the mitochondrial membranes.

In mitochondria, carrier-mediated anion transport can be subdivided into three types. The three types are (a) electroneutral, (b) electroneutral proton compensated, and (c) electrogenic (Table 2.14). The first involves an electroneutral exchange of one anion with another of equal charge. The equivalent, tightly coupled process in which the transport of one ion is linked to the transport of another species in the opposite direction

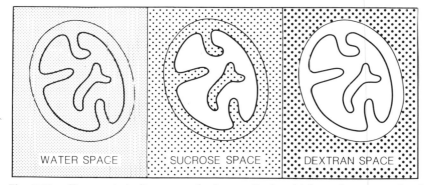

WATER SPACE   SUCROSE SPACE   DEXTRAN SPACE

Fig. 2.21. Transport studies across the inner mitochondrial membranes require the use of different molecules having different permeability properties toward the two mitochondrial membranes. Left: water penetrates very rapidly into all the different compartments of the mitochondria; center: sucrose penetrates the outer membrane but cannot pass through the inner membrane; and right: dextran cannot pass through any of the envelope membranes. Comparison of the distribution of a given molecule with that of water, sucrose, or dextran allows the study of its transport (or absence of transport) across the inner mitochondrial membrane.

is termed antiport or exchange diffusion. The second type involves exchange between anions of dissimilar charge in which electroneutrality is maintained by cotransport of a proton or direct cotransport of a monovalent anion with a proton. This latter exchange may also be regarded as exchange of A^- with OH^-. In fact, it is usually impossible to distinguish between the symport (i.e., a transport process involving the obligatory coupling of two ions in parallel) of a species with H^+ and the antiport of the species with OH^-. This type of transport is driven by a preestablished proton gradient. A ΔpH is maintained across the mitochondrial membrane by the electrogenic efflux of the protons associated with the oxidation–reduction reactions of electron transport as follows:

$$\log ([A^{n-}]mit/[A^{n-}] \text{ cyt}) = n \, \Delta pH \qquad (2.9)$$

in which $\Delta pH = pH_{mit} - pH_{cyt}$, n is the number of charges that the anion carries at the particular pH, pH_{mit} is the mitochondrial value, and pH_{cyt} is the cytoplasmic value. Consequently, proton cotransport provides a means by which the carriers can utilize the large electrochemical potential gradient of protons as an energy source and to give directionality to the transport processes. The third type is termed electrogenic ("creating a potential") or electrophoretic ("moving in response to a preexisting potential") transport because it involves an exchange between anions of dissimilar charge and, therefore, an interaction with the electrical potential established across the inner mitochondrial mem-

TABLE 2.14

Summary of Known Anion Transporters and Inhibitors in Plant Mitochondria[a]

Carrier	Type of exchange	Species exchanged	Inhibitors
Adenine nucleotide	*Electrogenic* ADP_{IN}/ATP_{OUT}	ADP, ATP	Atractyloside, carboxyatractyloside, bongrekic acid
Phosphate	*Electroneutral* P_i/hydroxyl P_i/P_i	P_i, arsenate (sulphite?)	Mersalyl, N-ethylmaleimide, p-hydroxymercuribenzoate
Monocarboxylate	*Electroneutral* Monocarboxylate–hydroxyl	Pyruvate, lactate	α-Cyano-4-hydroxycinnamic acid, mersalyl
Dicarboxylate	*Electroneutral* Dicarboxylate–P_i Dicrboxylate–dicarboxylate	Malate, malonate, succinate, sulphate	2-N-butylmalonate, pentylmalonate, benzylmalonate, mersalyl
Tricarboxylate	*Electroneutral* Tricarboxylate–dicarboxylate	Citrate, isocitrate, *cis*-Aconitate, phosphoenolpyruvate	Benzene-1,2,3-tricarboxylate, propane-1,2,3-tricarboxylate, phthalonate
Oxoglutarate	*Electroneutral* Oxoglutarate–dicarboxylate	Oxoglutarate	2-N-butylmalonate, phenylsuccinate
Glutamate	*Electroneutral* Glutamate–dicarboxylate	Glutamate, dicarboxylates	Phthalonate
Oxaloacetate	*Electroneutral* Oxaloacetate–hydroxyl	Oxaloacetate	Phthalonate

[a] [From Day and Wiskich (1984).]

brane. Since the membrane potential is normally negative inside, this serves as a source of energy to support a net efflux of negative charges or a net influx of positive charges. Transport of anions by this mechanism, therefore, will dissipate a preestablished proton electrochemical gradient, which will require expenditure of energy for its maintenance. Energy used for electrogenic transport is thus expected to lower the theoretical P/O ratios for each coupling site (Klingenberg and Rottenberg, 1977).

Transport of metabolites into plant mitochondria has recently been reviewed (Wiskich, 1977, 1980; Day and Wiskich, 1984). Probably the most important area for future research lies in the investigation of the role the carriers play in metabolic regulation. Unfortunately, an obvious drawback in assessing the role of transport in the control of metabolism is the lack of reproducible techniques for measuring metabolite concentrations in the cytosolic, plastidic, mitochondrial, and vacuolar compartments separately. Furthermore, in the case of the mitochondrial metabolic translocators, the kinetic characteristics of the translocator must be taken into account with respect to the substrate and product concentrations on either side of the mitochondrial membrane and to the characteristics of the interrelated enzymatic steps. Results presented in the literature show that inhomogeneity within the mitochondrial matrix, probably due to viscosity and folds in the cristae, makes it difficult to determine the activity of substrate seen by the carrier facing the matrix side of the membrane. For example, the concentration of the enzymes in the matrix of mitochondria is over 50% by weight, and they probably exist and behave as a multienzyme complex rather than as enzymes in solution (Srere, 1980).

2. Phosphate Transport

The phosphate transporter is responsible for the high flux of phosphate into the mitochondrial matrix during oxidative phosphorylation of ADP. Phosphate is taken up very rapidly by mitochondria in the exchange with OH^- (Table 2.14). This transport, which is electroneutral, is sensitive to thiol reagents, such as mersalyl or *N*-ethylmaleimide (see Fonyo, 1978). Plant mitochondria swell spontaneously when suspended in a solution of ammonium phosphate, as shown in Fig. 2.20. *N*-Ethylmaleimide inhibits this spontaneous swelling by more than 80% (Phillips and Williams, 1973). Furthermore, when the ΔpH is collapsed, rapid passive efflux of phosphate occurs and this is also inhibited by mersalyl (Hensley and Hanson, 1975; De Santis *et al.*, 1975). These results indicate that a phosphate carrier similar to that of animal mitochondria is present in plant mitochondria and that high matrix–phosphate content is maintained only so long as respiration maintains a ΔpH (Rebeillé *et al.*, 1984a). In studies performed by Wohlrab (1980), a mitochondrial phosphate transport protein with an M_r value of 34,000 has been extracted from animal mitochondria. The carrier has been incorporated in liposomes, and its properties in the reconstituted system closely resemble those observed in the intact mitochondria. A specific dependence on cardiolipin has been shown for the mitochondrial phosphate carrier (Kadenbach *et al.*, 1982).

The capacity of the translocator, as measured in isolated plant mitochondria, is higher than the maximum measured rate of oxidative phosphorylation. As the K_m value of phosphate for the translocator is about 0.2 mM (Hanson *et al.*, 1972) and the cytosolic phosphate content is approximately 5 mM (Rebeillé *et al.*, 1983), the phosphate translocator could operate at full capacity, and, thus, in physiological situations the capacity well exceeds the requirement.

3. Adenine Nucleotide Transport

The ADP/ATP transport is the most comprehensively investigated translocation system of mitochondria. Excellent review articles concerning the operation, energetics, and physiological significance of this transport system exist in the literature (Vignais, 1976; Klingenberg, 1980). The adenine nucleotide exchange is highly specific for ADP and ATP and is basically electrogenic (see, however, D. F. Wilson *et al.*, 1983). Only free ATP or ADP are translocated, the Mg^{2+} complexes being inactive (Table 2.14). This finding is highly significant in view of the probability that ATP in the mitochondria and in the cytosol are fully chelated with Mg^{2+}. Other nucleotides such as CTP, CDP, GTP, GDP, and AMP are connected to the ATP/ADP system via phosphate transfer reactions in both compartments. These nucleotides cannot directly communicate across the mitochondrial membrane, however, because no transport system is available to them. Since ADP carries three and ATP carries four negative charges at physiological pH, the ADP/ATP exchange causes a net movement of charges across the mitochondrial membrane. The transmembrane potential (negative inside) is the driving force for an electrophoretic, asymmetric nucleotide exchange; it drives ADP inside the mitochondria and ATP outside. Accordingly, at equilibrium the external ATP/ADP ratio is greater than that inside and the log of the ratios of the external ATP/ADP ratio to the internal ATP/ADP ratio is equal to $\Delta\Psi$. About 70% of cellular ATP is located in the cytosolic space whereas ADP is nearly equally distributed between the cytosol and the other compartments. It has been calculated that the energy required for the transport of adenine nucleotides and phosphate *in vitro* represents about 15% of the free energy necessary for the formation of ATP (Heldt *et al.*, 1972b).

Three specific and powerful inhibitors of the adenine nucleotide transport system are known: atractyloside and carboxyatractyloside (gummiferin), toxic principles of an Algerian thistle, and bongkrekic acid, an antibiotic formed by a mold in decaying coconut meals (Fig. 2.22). One of the peculiarities of the mitochondrial ADP/ATP carrier is its ability to bind atractyloside and bongkrekic acid and their derivatives, carboxyatractyloside and isobongkrekic acid, in an asymmetric manner. The atractylosides bind to the ADP/ATP carrier from the outside of the mitochondria and the bongkrekic acids from the inside (Vignais, 1976). The translocator has been isolated as a carboxyatractylate–protein complex (Lauquin *et al.*, 1978) that is probably a dimer, each subunit having an

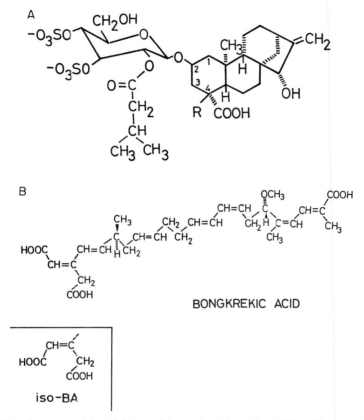

Fig. 2.22. Structure of the inhibitors of the nucleotide carrier. (A) R = H: atractyloside; R = COOH: carboxyatractyloside. (B) Bongkrekic acid (iso-BA = isobongkrekic acid). Figure provided by Dr. P. V. Vignais.

M_r value of 30,000. It seems that proteins that traverse membranes tend to have a dimeric (or tetrameric) structure in which the dimer is arranged asymmetrically across the membrane, with the axis of symmetry perpendicular to the membrane plane (Klingenberg, 1981). Finally, the ADP/ATP carrier is the most abundant integral protein in the inner membrane of mammalian mitochondria (Vignais, 1976).

Available evidence suggests that the nucleotide translocator in plant mitochondria carries out a very rapid ADP/ATP exchange, with low K_m values comparable to animal mitochondria (Earnshaw, 1977). However, the adenine nucleotide transport system in plant mitochondria differs from that of mammalian mitochondria by several features which include: (a) the small size of the internal pool of adenine nucleotides in plant mitochondria, which is 5 to 10 times less than that of mammalian mitochondria (Passam et al., 1973; Vignais et al., 1976; Abou-Khalil and Hanson, 1977; Silva-Lima and Denslow, 1979); (b) the low sensitivity to atractyloside, as compared to the high sensitivity of mammalian mitochondria to atractyloside (Jung and Hanson, 1973; Passam et al., 1973; Passam and Coleman, 1975). In short, the adenine nucleotide translocator in plant mitochondria is highly sensitive to bongkrekic acid and to carboxyatractyloside but much less sensitive to atractyloside; (c) the competitive and inhibitory effects of carboxyatractyloside on the ADP transport, as compared to the apparently noncompetitive effect of carboxyatractyloside in mammalian mitochondria, in spite of a similar K_d value for the binding of ^{35}S-labeled carboxyatractyloside to both types of mitochondria (Vignais et al., 1976); and (d) the binding properties of the adenine nucleotide translocator in plant and animal mitochondria are similar with respect to bongkrekic acid but somewhat different with respect to atractyloside and carboxyatractyloside.

The adenine nucleotides ADP, ATP, and AMP are present in high concentrations (11–13 nmoles/mg protein) in the matrix space of isolated animal mitochondria and rapid leakage across the inner mitochondrial membrane does not usually occur. Consequently, addition of ADP to the medium triggers the full rate of nucleotide exchange. In contrast, the total amount of adenine nucleotides present in fully intact mitochondria from plant tissues, such as potato (Vignais et al., 1976) or Jerusalem artichoke (Passam et al., 1973), is very low (1–2 nmoles/mg protein). Consequently, in plant mitochondria the kinetics of nucleotide transport are not easy to resolve. However, Jung and Hanson (1975) and Abou-Khalil and Hanson (1977) have clearly demonstrated the existence in plant mitochondria of an energy-linked mechanism for net uptake of ADP which is insensitive to carboxyatractyloside. Also, the K_m value for adenine nucleotide influx was several orders of magnitude higher than for the translocase. The collective characteristics and requirements of net adenine nucleotide uptake are different from those of the other known transport mechanisms, including the ADP/ATP exchanger. We can speculate that in vivo the matrix adenine nucleotide pool is retained against a concentration gradient by virtue of a one-way movement of adenine nucleotides. This net uptake can explain how the total adenine nucleotide content is established and maintained during mitochondrial proliferation. In support of this suggestion, it has been shown in rat liver that the large increase in the rate of ADP translocation immediately after birth appears to be related to the intramitochondrial adenine nucleotide content, which increases about fourfold in this period (Pollak and Sutton, 1980). In addition, such a system would allow regulation of the matrix adenine nucleotide pool size in vivo, either maintaining it at a constant level or allowing change in response to specific signals. In this way, metabolic pathways with adenine nucleotide–dependent enzymes localized in the matrix or the nucleotide translocase might be subject to modulation.

4. Pyruvate Transport

The availability or access of pyruvate to pyruvate dehydrogenase that is located in the mitochondrial matrix is of great metabolic importance for overall respiration. Pyruvate, although being a monocarboxylate, seems to require a carrier because of its relatively low pK. This evidence is obtained by saturation kinetics and the use of specific inhibitors (α-cyanocinnamates are specific inhibitors of this transport system; Halestrap and Denton, 1974). Since at equilibrium the ratio of pyruvate across the membrane is a direct function of ΔpH, and since the initial rate of pyruvate efflux is stimulated at alkaline pH and one proton disappears from the media for each pyruvate molecule accumulated in the matrix (LaNoue and Schoolwerth, 1979), pyruvate is probably transported across the mitochondrial membrane in an exchange with OH^- (Papa *et al.*, 1971; Table 2.14). Consequently, pyruvate transport is electroneutral. Studies on pyruvate transport have given conflicting results, however, because at high nonphysiological concentrations (above 2 mM), pyruvate is no longer only transported via the translocator but "passive diffusion" through the inner membrane also becomes important (Pande and Parvin, 1978).

Only in recent years has the pyruvate transport system in plant mitochondria been studied (Day and Hanson, 1977a). Pyruvate transport corresponds closely to that in animal mitochondria, with 2 μM α-cyano-4-hydroxycinnamic acid providing 50% inhibition (Day and Hanson, 1977). In addition "passive diffusion" also occurs at high pyruvate concentrations (D. A. Day, M. Neuburger, and R. Douce, unpublished data). An important question is whether or not mitochondrial pyruvate transport can regulate pyruvate oxidation. Inspection of the kinetic properties of the pyruvate translocator shows that limitation of the pyruvate metabolism by its transport in plant mitochondria is possible. Thus, for corn mitochondria the rate of pyruvate transport (estimated to be 20 nmoles/min/mg protein) is too low to support rapid respiration i.e., under conditions of high energy demand (Day and Hanson, 1977a). Since pyruvate can be generated intramitochondrially from malate via malic enzyme (see Section 3,II,A), Day and Hanson (1977a) believe that rapid malate transport may make up the difference, supplying pyruvate via malic enzyme and replenishing losses of tricarboxylic acid cycle intermediates. However, we have found higher rates of pyruvate transport in mung bean hypocotyl and potato tuber mitochondria (200 nmoles/min/mg protein).

5. Dicarboxylate and Tricarboxylate Transport

Only the phosphate, adenine nucleotide, and pyruvate translocators are directly involved in respiration and are common to all tissues, while the presence of the other translocators correlates with specialization of the tissue for a particular metabolic function.

The penetration of dicarboxylates into the mitochondrion was first noted through redox changes of intramitochondrial NAD^+, and the operation of a specific carrier was proposed on the evidence of osmotic swelling (K. J. Chappell and Hahaarhoff, 1967). Dicarboxylates can only accumulate in exchange for phosphate, and this is the main pathway that allows net uptake of TCA cycle substrates at the expense of the mitochondrial pH gradient (Table 2.14; Fig. 2.23). Competitive inhibition for uptake has been observed between malate, succinate, and malonate. Net exchange between various added and intramitochondrial dicarboxylates has been shown to have a 1 : 1 ratio (Meijer and Tager, 1969). The electroneutral dicarboxylate translocator is relatively insensitive to pH

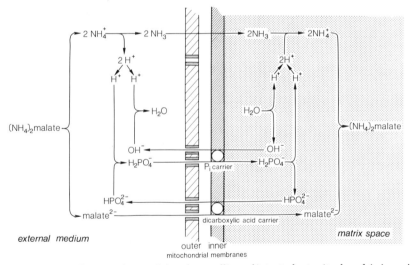

Fig. 2.23.   Mechanism that explains the swelling of intact plant mitochondria in an iso-osmotic solution of ammonium malate [$(NH_4)_2$ malate]. The swelling is triggered by the addition of a small amount of phosphate ($H_2PO_4^-$ / HPO_4^{2-}) to the incubation medium. In this process, $H_2PO_4^-$, which rapidly penetrates the inner membrane via the P_i/OH^- transporter (P_i carrier), exchanges for malate on the dicarboxylic acid carrier. Ammonia diffuses across the membrane, down a concentration gradient, and once inside the mitochondrion, associates with protons to form NH_4^+. The increased osmotic pressure within the mitochondrion induces swelling.

changes, since no ancilliary proton movement is involved. A number of dicarboxylic acids have been shown to inhibit the dicarboxylate transporting system of mammalian mitochondria; these include 2-butylmalonate, 2-pentylmalonate, 2-p-iodobenzylmalonate, and 2-phenylsuccinate.

The tricarboxylate exchanger mediates the exchange of citrate with malate. The tricarboxylate translocator also promotes an electroneutral exchange of anions but, since transport of a proton is involved to achieve electrical balance (at physiological pH the citrate anion carries three negative charges), this exchange is very sensitive to alterations of pH. Consequently, mitochondria can take up citrate via the tricarboxylate carrier by exchange diffusion with intramitochondrial di- or tricarboxylates in an energy-independent process (Table 2.14; Fig. 2.24). In

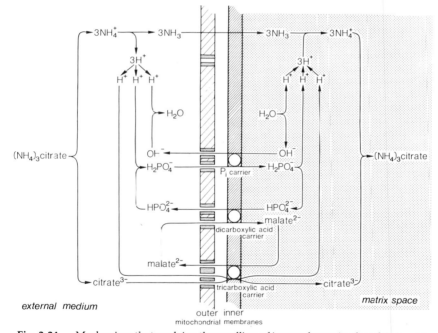

Fig. 2.24. Mechanism that explains the swelling of intact plant mitochondria in an iso-osmotic solution of ammonium citrate [$(NH_4)_3$ citrate]. The swelling is triggered by the addition of small amounts of phosphate ($H_2PO_4^- / HPO_4^{2-}$) and malate (malate^{2-}) to the incubation medium. In this process, $H_2PO_4^-$, which rapidly penetrates the inner membrane via the P_i/OH^- transporter (P_i carrier), exchanges for malate on the dicarboxylic acid carrier; then malate exchanges for citrate on the tricarboxylic carrier. Finally, ammonia diffuses across the membrane, down a concentration gradient, and once inside the mitochondrion, associates with protons to form NH_4^+. The increased osmotic pressure within the mitochondrion resulting from these cascade reactions induces swelling.

the presence of an energy source, net citrate accumulation can occur since the proton gradient generated by respiration can be used for P_i/OH^- exchange, followed by dicarboxylate–P_i exchange and citrate–dicarboxylate exchange, with net citrate accumulation (McGivan and Klingenberg, 1971). A number of tricarboxylic acid analogues of citrate, such as benzene-1,2,3-tricarboxylate, were shown to be inhibitory to the citrate transporting system (B. H. Robinson *et al.*, 1972; Table 2.14).

The entry of both α-ketoglutarate and citrate anions is activated by L-malate, however, the transporting systems do not appear to be the same (J. B. Chappell, 1968). In fact, it seems that the behavior of the α-ketoglutarate translocator, which participates very actively in the transfer of reducing equivalents from the cytosol to the mitochondrial respiratory chain in mammalian cells, has been shown to be very complex (Sluse *et al.*, 1980). In addition, this carrier is regulated by internal and external aspartate in opposite ways (Sluse *et al.*, 1980).

Although few in number, studies on the di- and tricarboxylate carriers in plant mitochondria have been in general agreement with the animal literature (Phillips and Williams, 1973; Wiskich, 1974; De Santis *et al.*, 1976; Table 2.14). For instance, potato tuber mitochondria swell in ammonium malate or succinate only after the addition of phosphate and in ammonium citrate only after the addition of both phosphate and a dicarboxylic acid. This swelling can be inhibited by N-ethylmaleimide (Phillips and Williams, 1973). On the other hand, internal phosphate is exchanged with malate, succinate, and sulfate, and these exchanges are inhibited by butylmalonate but are not affected by N-ethylmaleimide (De Santis *et al.*, 1976). Furthermore, oxidation of malate by plant mitochondria was shown to be inhibited by butylmalonate in mitochondria from Jerusalem artichoke (Coleman and Palmer, 1972) and cauliflower (Day and Wiskich, 1974a,b).

There is also an α-ketoglutarate–dicarboxylate carrier in bean mitochondria which differs from the dicarboxylate carrier in not utilizing phosphate and in being mersalyl-insensitive (De Santis *et al.*, 1976). Unlike that in liver mitochondria (Meijer *et al.*, 1976), however, α-ketoglutarate transport in plant mitochondria is not inhibited by phthalonate (Day and Wiskich, 1981a). This suggests that the two systems are fundamentally different.

There have been some anomalous results reported, especially with citrate uptake. Wiskich (1974) reported that beet root mitochondria would swell in ammonium citrate without the addition of malate and that castor bean and wheat coleoptile mitochondria show citrate swelling in the absence of both malate and phosphate. Jung and Laties (1979) reported that phosphate and malate stimulate respiration-driven

[^{14}C]citrate uptake by potato mitochondria but substantial accumulation was observed without these additions. Huber and Moreland (1979) found that mung bean mitochondria do not swell in ammonium malate or citrate, with or without catalytic additions of phosphate and malate. Finally, Birnberg *et al.* (1982) have demonstrated that, in corn mitochondria, a citrate$_{in}$–malate$_{out}$ exchange occurs, but at a rate too slow to account for observed citrate uptake. They concluded, therefore, that citrate can be rapidly accumulated by a mechanism other than by an exchange for dicarboxylates. The effect of uncouplers on respiration-driven [^{14}C]citrate accumulation and studies of passive swelling using ionophores and uncouplers, indicated that the major avenue of citrate uptake was by hydrogen ion–citrate cotransport. However, the low pH optimum of this hydrogen ion–citrate symport is not what would be expected for a transport enzyme functioning effectively *in vivo*. Day and Wiskich (1981a) have also shown that higher plant mitochondria can take up citrate by exchange with intramitochondrial isocitrate. In this experiment, added citrate entered the mitochondria and was converted to isocitrate by aconitase, and the efflux of isocitrate was monitored. Phthalonate was found to inhibit this exchange. It therefore appears that several different mechanisms exist in plant mitochondria for the uptake of citrate. Obviously more work is required to understand citrate transport in plant mitochondria, since potential utilization of citrate as a mitochondrial substrate from organic acid pools has been suggested (Hanson and Day, 1980).

Finally, Jung and Laties (1979) have clearly shown that once isolated, plant mitochondria, in contrast with mammalian mitochondria, appear to be depleted of exchangeable anions, resulting in a low capacity for uptake in the absence of respiration. Consequently, isolated plant mitochondria very often display depressed initial state 3 rates of respiration which rise thereafter to a maximum with several consecutive state 3–state 4 cycles. This development of state 3 respiration has been termed "conditioning" (Laties, 1973; Raison *et al.*, 1973). This is particularly conspicuous during the course of citrate and α-ketoglutarate oxidation (Journet and Douce, 1983). From the interesting data presented by Jung and Laties (1979), it appears that the depressed state 3 rates result largely from inadequate substrate uptake. In addition, they demonstrated the need for continued energy use to maintain anions in the matrix and suggested an inherent leakiness of the inner mitochondrial membrane. In fact, when a large proton-motive force exists due to respiration, high levels of anion accumulation can occur. It has to be pointed out that state 3 rates may also increase due to the effects of adenylates on respiratory and coupling enzyme systems (Laties, 1973; Oestreicher *et al.*, 1973) or to the increased contribution of reducing equivalents by subsequent oxidation steps, and we have shown recently that cofactor uptake (e.g., ADP) is an important part of the conditioning process.

6. Oxaloacetate Transport

Oxaloacetate is an intermediate in several physiologically important metabolic cycles and sequences, both catabolic and anabolic in nature. Furthermore, oxaloacetate may perform a catalytic function in the exchange of reducing equivalents between the mitochondrial and cytosolic compartments of the cell. In mammalian cells it has been assumed that the inner of the two membranes enclosing the mitochondrion may be considered to be impermeable to oxaloacetate under normal, physiological conditions, although reports have shown that it can be slowly transported via either the α-ketoglutarate translocator (Passarella *et al.*, 1977) or the dicarboxylate translocator (Gimpel *et al.*, 1973). With extramitochondrial concentrations of oxaloacetate close to physiological levels ($\cong 10$ μM), uptake rates of 3 nmoles/min/mg protein could be measured. This basal rate may be stimulated several fold in the presence of an energy source, but it remains relatively insignificant in comparison with the rates of those metabolic processes that require the transport of oxaloacetate out of the mitochondrion. In addition, since physiological levels of oxaloacetate are much lower than the K_m for transport, movement of oxaloacetate across mammalian mitochondrial membranes is not likely to occur to any significant extent *in vivo*.

In marked contrast, oxaloacetate has been found to rapidly traverse the membranes of all of the plant mitochondria isolated so far (Douce and Bonner, 1972). The validity of this finding is, of course, contingent on the integrity of the plant mitochondria isolated. It was demonstrated that the membrane was intact, since no membrane damage could be detected by electron microscopy and because exogenous cytochrome *c* did not have access to its site on the inner membrane. At an extramitochondrial oxaloacetate concentration of 100–200 μM, the influx of oxaloacetate was so severe that NAD$^+$-linked TCA cycle substrate–dependent O_2 consumption stopped, due to the competition for NADH by malate dehydrogenase (Fig. 2.25). Because of the very high activity of malate dehydrogenase in the mitochondrion (Douce *et al.*, 1973b), the reaction will be limited only by the transport of oxaloacetate. Alleviation of respiratory inhibition subsequently occurred as the oxaloacetate became reduced. While this "swamping" effect by excess oxaloacetate at best affords only a coarse type of control over electron transport, it amply demonstrates how a metabolic imbalance ensues when those enzymes normally limited by the supply of oxaloacetate are presented with a relative surfeit.

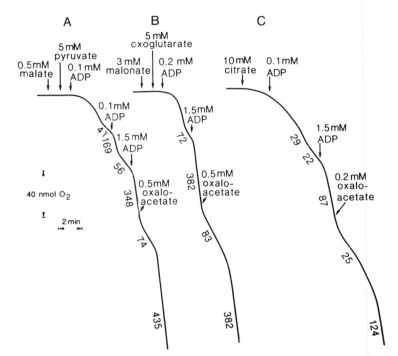

Fig. 2.25. Effects of limiting amounts of oxaloacetate on pyruvate (A); α-ketoglutarate (oxoglutarate) (B); and citrate (C) oxidations by isolated mitochondria from potato tubers. Numbers on the traces refer to nmoles O_2 consumed/min/mg protein. Note that, when oxaloacetate is added to state 3, a clear inhibition of the respiration rate occurs which is gradually reversed. Enzymatic analysis showed that, during the time of inhibition, oxaloacetate is converted to malate, a conversion that is dependent on the intramitochondrial NADH generated during the course of substrate oxidation.

In pea leaf mitochondria, malate transport is sensitive to butylmalonate while that of oxaloacetate apparently is not. Conversely, phthalonate has little effect on malate oxidation but severely restricts oxaloacetate transport (A. L. Moore *et al.*, 1979; Day and Wiskich, 1981b). A. L. Moore *et al.* (1979) suggested that the α-ketoglutarate carrier was involved, since phthalonic acid inhibits α-ketoglutarate transport in rat liver mitochondria. This is clearly not so, however, since phthalonate has virtually no effect on α-ketoglutarate oxidation by higher plant mitochondria (Chen and Heldt, 1983; Day and Wiskich, 1981b). Furthermore, a 1000-fold excess of malate and α-ketoglutarate or phosphate had little effect on the rate of oxaloacetate transport (Oliver and Walker, 1984). These results suggest, therefore, that malate efflux and oxaloacetate influx occur on separate carriers and that malate could exchange for external phosphate while oxaloacetate could enter at the expense of the transmembrane pH

gradient. Cauliflower bud mitochondria swell spontaneously when suspended in high (nonphysiological) concentrations of ammonium oxaloacetate, implying exchange for hydroxyl ions (Day and Wiskich,
1981a). It had previously been assumed that oxaloacetate could readily
penetrate plant mitochondria in exchange for malate on either the dicarboxylate (Woo *et al.*, 1980) or α-ketoglutarate (A. L. Moore *et al.*, 1979)
carriers. Indeed, such exchanges can be set up in bean mitochondria but
probably do not participate in rapid oxaloacetate transport because of
kinetic limitations similar to those in animal mitochondria (see earlier).
For the present, the details of oxaloacetate transport in plant mitochondria remain a mystery, but we do believe that the carrier may be specific
for oxaloacetate. The extraordinarily low half-saturation of oxaloacetate
transport (∼20 μM) would make it possible for a very active malate–
oxaloacetate shuttle to occur between the mitochondrial and the cytosolic
compartments of a plant cell under physiological conditions (Chen and
Heldt, 1983). Recent measurements in our laboratory using pea leaf
mitochondria suggest an uptake rate of close to 1000 nmoles/min/mg
protein. Finally, carbon input to the TCA cycle could occur in the form of
cytosolic oxaloacetate, thus disrupting conventional operation of the
TCA cycle.

7.   Glutamate and Aspartate Transport

Aspartate transport from mitochondrial matrices to the cell cytosol is a
necessary step in metabolic pathways such as gluconeogenesis and the
transport of reducing equivalents from the cell to the mitochondria in
animals. Aspartate is transported by a specific carrier that catalyzes an
exchange of glutamate for aspartate across the membrane (Azzi *et al.*,
1967). When a source of energy is available, in the form of ATP (ATP
hydrolysis) or electron transport, the exchange is virtually unidirectional, glutamate entering the mitochondrial matrix space and aspartate
leaving, because of the electrogenic nature of the exchange. The uncoupling of mitochondria or the presence of valinomycin and K^+ blocks
aspartate efflux and promotes aspartate entrance. An electrochemical
potential gradient of protons is responsible for the directional aspect of
the transport process. It has been demonstrated that a proton is cotrans-
ported on the carrier with glutamate, that the proton has a binding site
separate from either of the other substrates, and that release of the
proton from the binding site on the inner side of the membrane may be
rate-limiting for transport under certain conditions (Tischler *et al.*, 1976).
Since glutamate and aspartate are both monocarboxylic anions at neutral
pH, the process can be described as the exchange of neutral glutamic
acid for the aspartate anion and is, therefore, 100% electrogenic. The

driving force for the glutamate–aspartate exchange is provided by the electrochemical potential difference of the proton ($\Delta\bar{\mu}H$), and at equilibrium the following equation should hold:

$$RT\ln([Asp^-]_{in}\,[Glu^-]_{out}/[Asp^-]_{out}\,[Glu^-]_{in})$$
$$= F\Delta\Psi - 2.3RT\,\Delta pH = \Delta\bar{\mu}\,H \qquad (2.10)$$

Consequently, influx of glutamate into the mitochondrion in exchange with aspartate is energy consuming, because the net movement of negative charges from the matrix to the external side of the inner mitochondrial membrane dissipates a preexisting membrane potential (negative inside) and requires metabolic energy for its replenishment. The glutamate–aspartate carrier is intrinsic to the operation of the malate–aspartate cycle for transport of reducing equivalents from cytosolic NADH into the mitochondrion (animal mitochondria unlike plant mitochondria do not oxidize external NADH). In this process, the glutamate that enters the mitochondrion on the glutamate–aspartate carrier transaminates with oxaloacetate in the matrix space, and both aspartate and α-ketoglutarate exit from the mitochondrion. The malate–aspartate shuttle requires, therefore, the coordinated operation of several transport systems and intra- and extramitochondrial enzymes. This series of reactions effectively established a cyclic flow of oxaloacetate between the cytosol and the mitochondrion, and by acting in a concerted manner, ensures that the transport of reducing equivalents by oxaloacetate is rigorously controlled. It is important to think that this shuttle takes place against a concentration gradient of 1–2 orders of magnitude, since the cytosolic $NADH/NAD^+$ ratio is always much lower than the cytosolic mitochondrial matrix value.

Glutamate can in fact be transported across the mitochondrial membrane in mammalian cells by two separate systems. The first is by a glutamate–aspartate antiport. The second, is by a carrier-mediated glutamate–hydroxyl exchange diffusion (or a glutamate–hydrogen ion symport) and is electroneutral and freely reversible (Meyer and Vignais, 1973). This carrier is inhibited by lipid soluble thiol reagents. Based on this observation, a proteolipid that has a high affinity for glutamate and that exhibits the main properties expected from a glutamate carrier has been extracted from pig heart mitochondria (Gautheron and Julliard, 1979). The glutamate carrier has been incorporated in liposomes to reconstitute a glutamate–hydroxyl ion exchange (Gautheron and Julliard, 1979).

The penetration of glutamate into isolated plant mitochondria was investigated using the ammonium salt–swelling technique (Day and Wiskich, 1977b). Glutamate entry required the presence of both phosphate and a permeable dicarboxylic anion and was inhibited by 2-*n*-butylmalonate. Thus, glutamate transport in plant mitochondria seems to operate in a manner identical to the tricarboxylic anion transporter and is in marked contrast to results observed with mammalian mitochondria. In addition, Day and Wiskich (1977b) suggested that plant mitochondria may lack the specialized transport systems associated with the malate–aspartate cycle of animal mitochondria and that dicarboxylate–glutamate exchange may operate *in vivo* for the export of glutamate from the mitochondria. During the course of glutamate metabolism triggered by oxaloacetate in Percoll-purified potato tuber mitochondria, however, Journet *et al.* (1982) have suggested that the translocation of metabolites across the mitochondrial inner membrane plays an impor-tant role in the overall process and that glutamate enters the matrix in exchange for aspartate. Consequently, the results presented by Journet *et al.* (1982) strongly suggest that plant mitochondria are also able to import reducing equivalents in the form of glutamate and malate. In animal mitochondria, the entry of aspartate in exchange for glutamate is very slow and does not, in fact, occur in energized mitochondria. In plant mitochondria, however, data indicate that exchange in the opposite direction (entry of aspartate and efflux of glutamate) also occurs when aspartate is metabolized by the transamination pathway. For instance, during glycine oxidation by spinach leaf mitochondria, O_2 consumption shows a strong and transient inhibition upon addition of aspartate plus α-ketoglutarate (Fig. 2.26; Journet *et al.*, 1980). During the course of inhibition, aspartate and α-ketoglutarate are stoichiometrically transformed into malate and glutamate. Furthermore, added aspartate severely inhibits malate oxidation in the presence of glutamate (Wiskich and Day, 1979). Consequently, these results demonstrate that, in contrast with the situation observed in animal mitochondria, aspartate can enter a plant mitochondrion sufficiently rapidly to sustain a rapid transfer of reducing equivalents out of the mitochondrion. In other words, the glutamate aspartate exchange is freely reversible and appeared not to be electrogenic. In fact, Proudlove and Moore (1982) have recently demonstrated that aspartate, glutamate, serine, and glycine all permeate the inner membrane of the mitochondrion isolated from both etiolated and green plant tissues. These authors suggested that influx of each amino acid occurs by diffusion, because substrate saturation was not observed and there was no indication of specific inhibition or a requirement for a compensatory or counter-ion for uptake. Obviously, a more detailed knowledge of glutamate and aspartate transport in plant mitochondria is desirable.

It thus seems that plant cells, therefore, for an obscure reason, have several mechanisms for very rapidly transferring the reducing equiv-

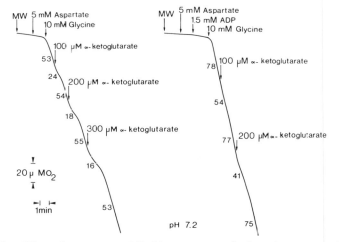

Fig. 2.26. Effect of aspartate and limiting amounts of α-ketoglutarate on glycine oxidation by mitochondria isolated from spinach leaves. Numbers on the traces refer to nmoles O_2 consumed/min/mg protein. Note that, after the addition of a small amount of α-ketoglutarate to mitochondria that are oxidizing glycine in the presence of 5 mM aspartate, a clear inhibition of the respiration rate occurs which is gradually reversed. Enzymatic analysis showed that, during the time of inhibition, α-ketoglutarate gradually disappears and is converted to malate and glutamate. Abbreviation: MW, washed mitochondria. [From Journet *et al.* (1981).]

alents produced during glycolysis into the mitochondrion, in order to circumvent the impermeable nature of the inner membrane to NADH. These are the malate–aspartate shuttle, the oxaloacetate–malate shuttle, and the Ca^{2+}-dependent NADH dehydrogenase located on the outer surface of the inner membrane. Obviously, more research is needed to understand the regulation of these mechanisms and to provide more accurate means of determining the intra- and extramitochondrial NAD/NADH levels, especially if one considers that mitochondria could generate NADH in the cytoplasm and, therefore, supply reductant for nitrate reduction (for a review see L. Beevers and Hageman, 1980). It is very likely that the presence of O_2 permits the mitochondrion to compete for NADH, thus inhibiting nitrate reduction. To advance the field, the most important piece of information needed concerns the *in vivo* concentrations of oxaloacetate and other components of the aspartate–aminotransferase system, not merely in the mitochondrion but also in the cytosol. Although confusing, the results presented here emphasize the great flexibility exhibited by all of the plant mitochondria isolated so far. Indeed, plant mitochondria are able to export or import reducing equivalents in the form of oxaloacetate, aspartate, α-ketoglutarate, malate, and glutamate. One example of this is the glycine metabolism in leaf mitochondria, in which the reoxidation of NADH produced in the matrix upon operation of the glycine decarboxylase complex can be reoxidized equally well by either the respiratory chain or the substrate shuttles (in which case, the NADH is transferred out of the mitochondrion and glycine decarboxylation will not be directly linked to the energy status of the mitochondria).

8. Miscellaneous Transport

a. NAD$^+$ Transport.   The mitochondrial inner membrane is generally considered to be impermeable to nicotinamide ("pyridine") nucleotides (VonJagow and Klingenberg, 1970). However, we have recently shown that isolated intact plant mitochondria actively accumulate NAD$^+$ from the external medium (Neuburger and Douce, 1978; Tobin *et al.*, 1980), leading to a substantial increase in the matrix concentration of the cofactor and stimulating dehydrogenases and electron-transport activity. Plant mitochondria apparently possess a specific NAD$^+$ carrier, since NAD$^+$ uptake is concentration dependent (Tobin *et al.*, 1980) and exhibits Michaelis–Menten kinetics (Neuburger and Douce, 1978; Tobin *et al.*, 1980). Furthermore, the rate of NAD$^+$ transport is strongly temperature-dependent (Neuburger and Douce, 1978), and the analogue N-4-azido-2-nitrophenyl-4-aminobutyryl-3' NAD$^+$ (NAP$_4$-NAD$^+$; Fig. 2.27) inhibits (almost completely) net NAD$^+$ import (Neuburger and Douce, 1983; Fig. 2.28). It should be emphasized that this transport does not reflect membrane damage during organelle isolation; the mitochondria in question have been rigorously tested and found to be greater than 90% intact.

Neuburger and Douce (1983) have also found that NAD$^+$ effluxes from intact isolated mitochondria maintained in a medium that avoids rupture of the outer membrane. The arguments in favor of this diffusion are: (a) when NAD$^+$-linked substrates are used, O$_2$ uptake becomes increasingly dependent on added NAD$^+$ (Fig. 2.29), (b) the concentration of NAD$^+$ in the mitochondria progressively decreases during aging of the mitochondria (Fig. 2.30), and (c) addition of NAD$^+$ to the suspending medium restores the NAD$^+$ content of the matrix space and prevents the decrease observed in the rate of substrate-dependent O$_2$ uptake over 48 hr (Fig. 2.29). The rate of NAD$^+$ efflux from the matrix space is strongly temperature-dependent and is inhibited by the analogue inhibitor of NAD$^+$ transport, indicating that a protein is required for net flux in either direction (Neuburger *et al.*, 1985). These data raise the question of the binding of NAD$^+$ to the NADH oxidases (complex I, rotenone-insensitive NADH oxidase) and the NAD$^+$-linked TCA cycle dehydrogenases (see Fig. 2.13). Although there is some uncertainty as to

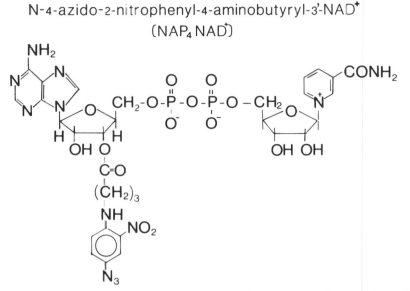

Fig. 2.27. Structure of the inhibitor N-4-azido-2-nitrophenyl-4-aminobutyryl-3'-NAD+ (NAP$_4$-NAD+) of the NAD carrier.

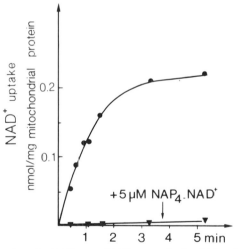

Fig. 2.28. Effect of NAP$_4$-NAD+ on NAD+ uptake by 24-hr-old mitochondria isolated from potato tubers. [^{14}C]NAD+ uptake was determined by silicone oil centrifugation–filtration. The pH of the assay medium was 7.2 and the temperature was 25°C. Labeled NAD+ was added at 30 μM and NAP$_4$-NAD+ was added at 5μM. Symbols: ●, control (without NAP$_4$-NAD+) and ▼, NAP$_4$-NAD+.

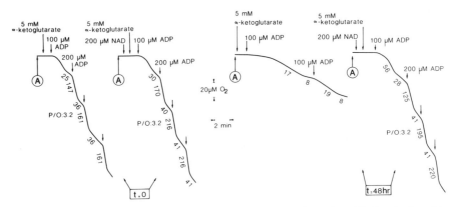

Fig. 2.29. Rates of α-ketoglurate oxidation during aging of purified mitochondria isolated from potato tubers. For each time period, the percentages of mitochondrial integrity were: $t = 0$ hr, 97.5%; $t = 48$ hr, 95%. In this particular experiment, the total amount of NAD^+ present in freshly prepared potato mitochondria and 48-hr-old mitochondria were, respectively, 4.4 and 0.3 nmoles/mg of mitochondrial protein. The values along the traces refer to nmoles of O_2 consumed/min/mg of protein. Note the marked decline of state-3 rates of α-ketoglutarate oxidation during mitochondrial aging and the recovery of the initial rate when the suspending medium contains 200 μM NAD^+. Since most of the plant mitochondria isolated so far are depleted of their endogenous thiamine pyrophosphate, it is necessary to add this cofactor to the medium in order to trigger α-ketoglutarate oxidation. Likewise, in order to be sure that the endogenous level of CoA is sufficient to saturate α-ketoglutarate dehydrogenase, the electrode medium also contains 0.5 mM CoA. Symbols: (A): 0.36 mg protein, 0.3 mM TPP, 3 mM malonate, and 0.5 mM CoA. [From Neuburger and Douce (1983) and reprinted by permission from *Biochem. J.* **216,** p. 445. Copyright © 1983. The Biochemical Society, London.]

the relative amounts of bound versus free NAD^+ and NADH, the fact that intact purified mitochondria progressively lose their NAD^+ content and that this leads to a dramatic decrease in the O_2 uptake rates strongly suggest that most of the NAD^+ is not firmly bound to the inner membrane or to the various dehydrogenases. It is highly plausible, therefore, that leakage of NAD^+ out of the mitochondrion occurred during their isolation (see, for example, Malhotra and Spencer, 1971; Hoekstra and Van Roekel, 1983).

The physiological role of this NAD^+ carrier still remains uncertain, but we can speculate that *in vivo* the matrix NAD pool is maintained against a concentration gradient by virtue of a one-way movement of nicotinamide nucleotides. This net uptake can also explain how the total nicotinamide-nucleotide content is established and maintained during mitochondrial proliferation. Consequently, the NAD^+ carrier could play its part in mitochondrial biogenesis. It is also possible that this carrier has an important regulatory function *in vivo*, by allowing manip-

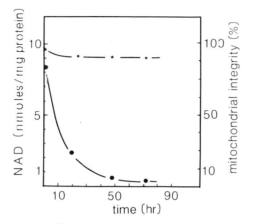

Fig. 2.30. Changes in NAD+ content of the mitochondria (●) and percentage integrity (★) during aging of purified mitochondria isolated from potato tubers. Note the absence of correlation between NAD+ concentration in the matrix and the percentage integrity. The integrity of the mitochondria was rapidly evaluated by cytochrome c oxidation using an oxygen electrode (Neuburger *et al.*, 1982). The concentration of NAD+ found in mitochondria isolated from potato tubers is variable and erratic. Assuming a volume of 2.5 μl/mg mitochondrial protein, values between 0.25 and 4 mM have been found. On the average, the concentration of NAD+ in mitochondria from sprouting tubers is higher than that of dormant tubers. [From Neuburger and Douce (1983) and reprinted by permission from *Biochem. J.* **216**, p. 447. Copyright © 1983. The Biochemical Society, London.]

ulation of the matrix NAD+ concentration and thus regulating the internal rotenone-insensitive pathway. In support of this suggestion, we have found that the NAD+ carrier also functions in the efflux of NAD+ from isolated mitochondria (Neuburger *et al.*, 1985). We believe that the NAD+ transporter may play a general role in the coarse regulation of plant respiration during transition from dormancy to a stage of active plant growth and may be particularly important during processes such as seed germination.

b. Coenzyme A Transport. Plant mitochondria isolated from a number of tissues also have a relatively low endogenous coenzyme A content, and these mitochondria are capable of actively accumulating coenzyme A (Fig. 2.31). This net uptake seems to be catalyzed by a specific transport system and can lead to impressive increases in the coenzyme A content of the matrix, consequently stimulating matrix enzymes such as NAD+-linked malic enzyme (Neuburger *et al.*, 1983; Fig. 2.32). The properties of coenzyme A uptake are very similar to those described previously for NAD+ transport. Since some similarity in the structure of the two molecules exists, we thought it possible that they

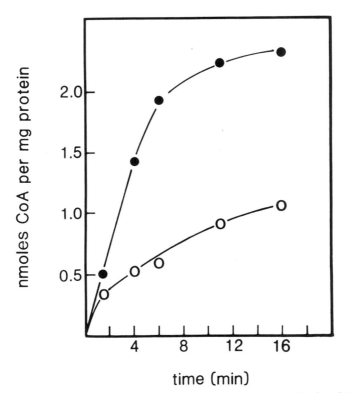

Fig. 2.31.   Time course of [³H]coenzyme A uptake by potato mitochondria. Freshly prepared mitochondria were incubated with 30 μM coenzyme A for the times shown, and the reactions terminated by centrifugation through silicon oil. Succinate (10 mM) and ATP (0.2 mM; required for the activation of succinate dehydrogenase) were included in the reaction medium that also contained 0.8 mg of mitochondrial protein. Symbols: ●, control; ○, 1.5 μM FCCP included in the reaction medium. [From Neuburger *et al.* (1984).]

might share a common carrier. This is clearly not so, however, since NAP₄-NAD has virtually no effect on coenzyme A transport whereas it strongly inhibits NAD⁺ influx. Interestingly, Day *et al.* (1984a) have shown that exogenous coenzyme A can be a potent regulator of malate oxidation by isolated plant mitochondria. Coenzyme A has a similar effect on the malic enzyme in intact mitochondria as it does with the solubilized enzyme, although more time is required to see the effect.

It is now apparent that plant mitochondria possess at least three distinct carriers for the net uptake of cofactors (Table 2.15). In addition to the previously known NAD⁺ and ADP carriers, we now have evidence for a third carrier that catalyzes the uptake of coenzyme A. It is also evident that thiamine pyrophosphate can enter isolated plant mitochon-

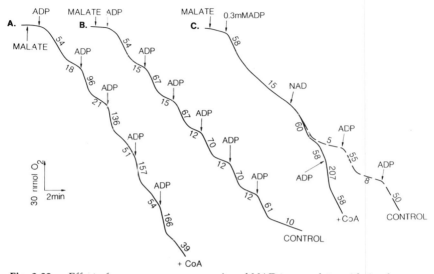

Fig. 2.32. Effect of exogenous coenzyme A and NAD⁺ on malate oxidation by potato mitochondria. The external pH was 7.3 in all experiments. The standard reaction medium included 0.2 mM NAD⁺ in (A) and (B); and in (C), the same quantity of NAD⁺ was added as indicated. Control [(B) and (C), dashed line] indicates that no coenzyme A was present; +CoA [(A) and (C), solid line] indicates that 0.5 mM coenzyme A was included in the reaction medium; in addition, other additions, as shown, were 10 mM malate and 0.1 mM ADP. Numbers on the traces refer to nmoles O_2 consumed/min/mg protein. [From Day *et al.* (1984).]

dria, since this cofactor is an essential addition for the oxidation of pyruvate and α-ketoglutarate (Lance *et al.*, 1965); apparently, the isolated organelles are depleted of endogenous thiamine pyrophosphate (presumably during their isolation, although this has yet to be demonstrated) but rapidly accumulate it when it is provided externally. Whether or not thiamine pyrophosphate can also use the coenzyme A carrier remains to be demonstrated. Jung and Laties (1979) have shown the need for a continued energy supply to maintain anions in the matrix (see also Romani *et al.*, 1974), and perhaps the same conditions are necessary to maintain NAD⁺, coenzyme A, and thiamine pyrophosphate at high levels.

c. Ca²⁺ Transport. For mammalian mitochondria there is a concensus that Ca^{2+} is transported electrophoretically down a gradient of electrical potential created by respiration or ATP hydrolysis (J. B. Chappell and Crofts, 1966). The transport of Ca^{2+} across the inner mitochondrial membrane is usually ascribed to a Ca^{2+}-binding carrier. In these mitochondria, Ca^{2+} uptake is linked to a stoichiometric increase in O_2

TABLE 2.15

Net Uptake of Cofactors by Plant Mitochondria

Species accumulated	Molecular weight	Parameters	Inhibitors
ADP (AMP, ATP)	427	K_m = 3.7 mM V_{max} = 37 nmoles/min/mg Requires P_i, Mg^+, and an energy source	Mersalyl
NAD^+	663	K_m = 0.3 mM V_{max} = 2 nmoles/min/mg Requires energy source; uptake depends on endogenous NAD content	NAP_4^+-NAD^+
Coenzyme A	768	K_m = 0.2 mM V_{max} = 6.5 nmoles/min/mg Requires energy; insensitive to NAP_4^+-NAD^+	None known
Thiamine pyrophosphate (TPP)	479	Not yet studied, but uptake shown by its effect on matrix enzymes	—

consumption, proton extrusion, and a depolarization of the membrane potential (for a review see Nicholls and Åkerman, 1982). However, there is scant evidence that plant mitochondria transport Ca^{2+} by a similar uniport. All studies agree that Ca^{2+} uptake (when it occurs) is linked to phosphate uptake. In contrast with animal mitochondria, however, Ca^{2+} does not act as a permeative cation in that it will not accumulate with acetate (Hanson and Hodges, 1967; C. H. Chen and Lehninger, 1973; Day et al., 1978b). There is also an opinion that Ca^{2+} uptake by plant mitochondria may be the artifactual result of preparative damage to the inner membrane (A. L. Moore and Bonner, 1977). Conflicting results such as these may have their origin in experimental procedures, but as yet one cannot distinguish between the loss of a calcium–phosphate carrier and the induction of Ca^{2+} permeability. In fact, the results of Åkerman and Moore (1983) do confirm the fact that fully intact plant mitochondria take up Ca^{2+} from the external medium in an energy-linked manner in the presence of P_i, although the uptake mechanism is slow (i.e., a rate which is two orders of magnitude lower than the mammalian counterpart). In addition, the absolute requirement for phosphate and insensitivity to ruthenium red, as well as the inability of other weak anions to promote uptake, strongly suggest that the mechanism of uptake is different from the uniport mechanism present in mammalian mitochondria. The low rates of transport reported cast some doubt on the postulated role of plant mitochondria in the regulation of cytosolic Ca^{2+} (Dieter and Marmé, 1980).

3

The Function of Plant Mitochondrial Matrix

The mitochondrial matrix is the site of TCA cycle enzymes that provide reducing equivalents to the electron transport chain for ATP synthesis and also, via ancillary reactions, provide numerous substrates for biosynthetic reactions in the cytoplasm.

Bonner and Voss (1961) and Wiskich and Bonner (1963) reported that in the presence of nonlimiting concentrations of substrates, such as succinate, malate or α-ketoglutarate, phosphate, and O_2, the respiratory activity of intact plant mitochondria was effectively controlled by the availability of ADP: this phenomenon is called respiratory control (Chance and Williams, 1956; Fig. 2.14). In the absence of ADP, respiratory activity is low (state 4) and reflects the slow rate of endogenous energy dissipation, i.e., the slow rate of proton diffusion. Following the addition of ADP, the rate of respiration increases considerably (state 3) and is limited either by the activity of the TCA cycle dehydrogenases or the activity of the electron chain. When the ADP is almost completely esterified, the rate of respiration declines to the state 4 rate. Two important parameters can be calculated from this type of experiment, namely the ADP/O ratio (the number of ATP molecules synthesized from ADP and phosphate per oxygen atom reduced) and the respiratory control ratio (i.e., the respiration rate that occurs at substrate saturation in the presence of ADP and the respiration rate under the same condition if ADP is lacking). The oxidation of α-ketoglutarate, other NAD^+-linked substrates, and succinate yields ADP/O ratios of approximately 3.2, 2.4, and 1.6, respectively. It should be noted that even at a low-levels of respiratory control a high degree of coupling can occur. This is particularly true with plant mitochondria. The majority of uncoupling agents, such as DNP and FCCP, induce an increase in the rate of respiration, and the final rate obtained is identical to the rate obtained in the presence of ADP. Likewise, intact purified plant mitochondria rapidly reduce ferricyanide in the presence of various substrates (succinate, α-ketoglutarate, and malate) and KCN (to inhibit cytochrome aa_3). The addition of a small quantity of ADP gives an increase in the rate of electron transport. This fast rate declines after all the ADP has been phosphorylated. The inner membrane represents an impenetrable barrier to the ferricyanide (Douce et al., 1972), and under these conditions it can react with only the cytochrome c that is localized on the external face of the inner membrane. Interestingly, when the mitochondria are damaged (osmotic shock, sonication), the rate of reduction of ferricyanide is no longer stimulated by ADP and shows little inhibition with antimycin A. In this case, the ferricyanide reacts directly with complex I or II and perhaps the various TCA cycle dehydrogenases. In fact, we have always observed that in preparations of intact purified mitochondria, the rate of reduction of ferricyanide in the presence of TCA cycle substrates (but not with NADH; see Coleman and Palmer, 1971) is strongly sensitive to antimycin A (Douce et al., 1972). All these phenomena can be fully explained by the chemiosmotic hypothesis formulated by Mitchell (1980), according to which oxidative phosphorylation is driven by a proton gradient across the membrane.

I. THE TRICARBOXYLIC ACID CYCLE AND ITS REGULATION

The remarkable work of Krebs and Johnson (1937) established that the TCA cycle plays a key role in aerobic respiration. In essence, the cycle is a conversion block coupling the catabolism of acetyl units to the generation of redox equivalents according to the following schema:

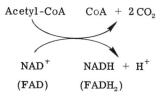

The TCA cycle also serves as an important source of carbon skeletons for synthetic processes, especially the formation of amino acids that takes place mainly in the plastidial compartment. During active cell division, the supply of carbon may be more important than the supply of ATP (Hunt and Fletcher, 1976). One cannot fail to be impressed and intrigued by the complexity of the TCA cycle, particularly when one realizes how small the volume is in which so much takes place. Instinctively, one feels that such a system is highly "organized," and perhaps the TCA cycle enzymes exist within the matrix of the mitochondrion as a multi-enzyme complex next to or on the inner surface of the inner membrane. Perhaps the regulatory potential of TCA cycle enzymes in plant mito-chondria may be realized by the apportionment of the enzymes between their soluble and bound forms (see, for example, Laties, 1983; Tezuka and Laties, 1983). The advantage of such an arrangement is obvious in terms of being able to maintain a high flux of substrates through the cycle with a relatively small number of intermediate molecules, the apparent concentrations of which in the cycle environment would be high. All the components of this cycle are kinetically very heterogeneous quantities, differing in turnover times, concentrations, and their connections with outside pools.

It was established very early that plant mitochondria contain the complete complement of enzymes for the oxidation of pyruvate via the TCA cycle (H. Beevers, 1961; Fig. 3.1). Since the early fifties, when the first successful isolation of mitochondria was achieved (Millerd *et al.*, 1951; Laties, 1953; Davies, 1953), mitochondrial preparations that show a high degree of both coupling and respiratory control have been isolated from a large variety of plant tissues, including underground storage organs, roots, various parts of seedlings, flowers, fleshy fruits, and leaves (for a review see Wiskich, 1980), and all of these have been thoroughly examined for their ability to carry out all of the reactions of the TCA cycle. For example, all of the plant mitochondria isolated so far are able to oxidize, more or less rapidly, pyruvate, citrate, α-ketoglutarate, succinate, and malate (see, for example, Wiskich and Bonner, 1963; Lance *et al.*, 1965; Ikuma and Bonner, 1967a,b; Ku *et al.*, 1968; Douce *et al.*, 1977). Tightly

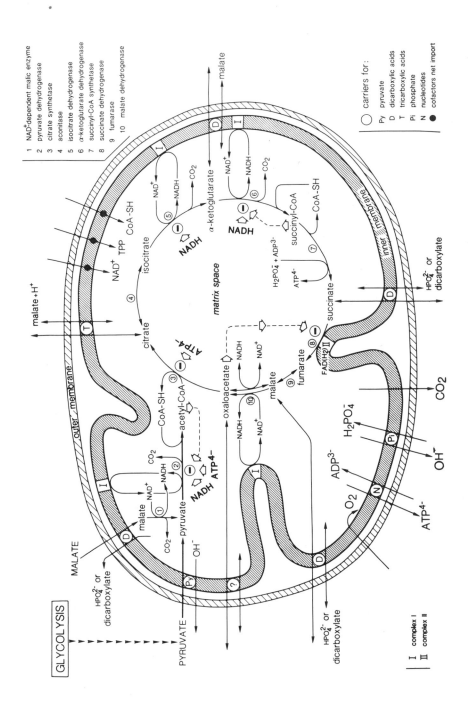

coupled mitochondria isolated from yeast and fungi (for an excellent review see Lloyd, 1974) reveal many characteristics similar to plant mitochondria. Both types of mitochondria oxidize, with good respiratory control and rates, representative acids of the TCA cycle as well as malate (Fig. 3.1) and externally added NADH (Fig. 2.14).

Three major mechanisms mediating control of the TCA cycle reactions may be considered as follows: (a) allosteric control of the initial reactions of the TCA cycle by ATP or ADP, (b) control by the oxidation–reduction state of pyridine nucleotides, and (c) control of the transport of intermediates across the mitochondrial membrane. Hence, there is a competition for intramitochondrial TCA cycle substrates between efflux from the mitochondrion and further metabolism in the TCA cycle. It is also possible that in plant mitochondria a membrane-bound protein could catalyze the transfer of electrons from the respiratory chain to a soluble protein modulating factor (reductively activated); which in turn controls the activity of mitochondrial enzymes such as aconitase (Mohamed and Anderson, 1983). In essence, the velocity with which this cycle turns is mainly determined by three quantities: (a) the availability of substrate, (b) the redox state ($NADH/NAD^+$), and (c) the energy state (ATP–ADP, energy charge, succinyl-CoA–CoA, acetyl-CoA–CoA; Fig. 3.1). These three controlling factors interact in a highly coordinated and cooperative manner (Williamson and LaNoue, 1975; Reich and Sel'kov, 1981), and these controls are complemented by input and output according to demand and offer of metabolites (Fig. 3.1).

Fig. 3.1. Metabolic regulation of the tricarboxylic acid (TCA) cycle in higher plant mitochondria. The heart of the cycle lies in the formation of citrate from oxaloacetate and acetyl-CoA and the regeneration of oxaloacetate from citrate. It is clear that flux through citrate synthase (3) in the intact mitochondria is controlled by the intramitochondrial concentration of either oxaloacetate or acetyl-CoA. This scheme indicates that there exists a concerted action of malate dehydrogenase (10), NAD^+-linked malic enzyme (1), and cytoplasmic glycolysis to provide pyruvate and oxaloacetate in the anaplerotic function of the TCA cycle. In addition, data obtained with intact plant mitochondria indicated the possibility of an important role for succinyl-CoA and NADH in coordinating flux through the separate spans from oxaloacetate plus acetyl-CoA to α-ketoglutarate and from α-ketoglutarate to malate in the TCA cycle. Furthermore, metabolic anions escape at different rates from the matrix during incubation of isolated mitochondrial suspensions until a steady state is reached, when the rate of efflux for each species becomes equal to the rate of influx. It is clear, therefore, that the TCA cycle could serve as an important source of carbon skeletons for synthetic processes and that input and output are very likely regulated according to supply and demand of metabolites. Finally, this scheme emphasizes the fact that isolated intact plant mitochondria actively accumulate NAD^+, thiamine pyrophosphate (TPP), and coenzyme A (CoA-SH) from the external medium, leading to a substantial increase in the matrix concentration of cofactors and stimulating TCA cycle dehydrogenases, NAD^+-linked malic enzyme, and electron transport activity.

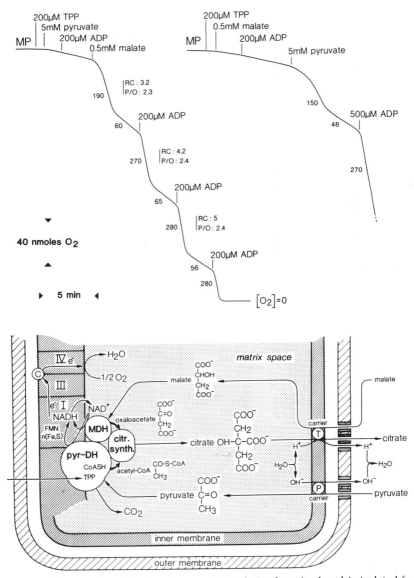

Fig. 3.2. Rates and mechanism of pyruvate oxidation by mitochondria isolated from higher plants. The reaction medium contained: 0.3 M mannitol; 5 mM $MgCl_2$; 10 mM KCl; 10 mM phosphate buffer, pH 7.2; and 0.1% (W/V) defatted bovine serum albumin. Since most of the plant mitochondria isolated so far are depleted of their endogenous thiamine pyrophosphate (TPP), it is necessary to add this cofactor in the medium in order to trigger pyruvate oxidation. Likewise, pyruvate oxidation requires the presence of small amounts of malate (or another TCA cycle intermediate) in the incubation medium. Under these circumstances, two dehydrogenases are operating, namely the pyruvate dehydrogenase

Therefore, the transport processes take control of events or contribute to them. The coordination is achieved by kinetic control of those enzymes that are associated with a large negative free energy change, namely citrate synthase, isocitrate dehydrogenase, and α-ketoglutarate dehydrogenase. Changes of the intramitochondrial NADH–NAD$^+$ ratio appear to provide the primary signal for coordination of flux through the electron transport chain with the irreversible steps of the TCA cycle. Cooperation manifests itself as a large acceleration factor although each of the controlling quantities changes only slightly. Unfortunately, a number of difficulties and discrepancies in the experimental data indicate that it is necessary to be cautious in the interpretation of results and particularly in the extrapolation of findings with isolated enzymes and mitochondria to the intact tissue. The greatest problem arises in regard to the knowledge of intramitochondrial concentrations of metabolites, both because of the uncertainty of the matrix activity coefficients and because binding to matrix enzymes of those metabolites present in low concentrations occurs.

A. Pyruvate Oxidation

Pyruvate oxidation in plants, for the most part, has been assumed to be via the pyruvate dehydrogenase complex and citrate synthase and has usually been measured using isolated mitochondria and small amounts of pyruvate and TCA cycle intermediates, such as malate (H. Beevers, 1961; Fig. 3.2) or oxaloacetate [Avron (Abramsky) and Biale, 1957; D. A. Walker and Beevers, 1956; Fig. 3.3]. This oxidation requires thiamine pyrophosphate and sometimes NAD$^+$ (Neal and Beevers, 1959; Lance et al., 1965; Bowman and Ikuma, 1976). It appears, therefore, that the binding and dissociation of thiamine pyrophosphate to the pyruvate dehydrogenase complex may have a marked effect on the respiration of plant mitochondria, although the exact mechanism of this control is not yet known.

The pyruvate dehydrogenase complex has been isolated from eukaryotic cells as functional units with molecular weights in the millions and is about 20–25 nm in diameter. This complex contains three functional enzymes required for the irreversible conversion of pyruvate to

(pyr-DH) and the malate dehydrogenase (MDH). Pyruvate oxidation requires a concerted interaction between MDH, pyr-DH, and citrate synthase (citr. synth.). It is very likely that all these enzymes exist within the matrix space as a multienzyme complex next to the inner surface of the inner membrane. Note that the pyruvate carrier (P) is responsible for the flux of pyruvate into the mitochondrial matrix and the tricarboxylate carrier (T) is responsible for the efflux of citrate.

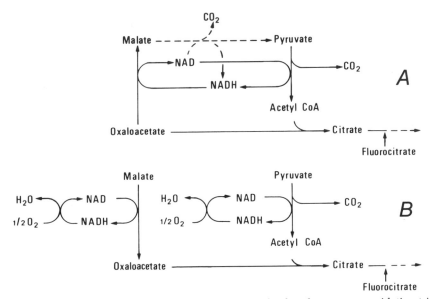

Fig. 3.3. Schemes of the metabolic processes involved in the pyruvate oxidation triggered by oxaloacetate in potato tuber mitochondria. (A) Added oxaloacetate readily penetrates the matrix space through the inner membrane. Part of the oxaloacetate, at least one half, reacts with acetyl-CoA under the control of citrate synthase. Simultaneously, the other part of the oxaloacetate is converted into malate under the control of malate dehydrogenase, a conversion that depends on the reduced pyridine nucleotides generated by the TPP-linked pyruvate dehydrogenase complex. In these conditions, O_2 consumption by the respiratory chain is inhibited. (B) When all the oxaloacetate is consumed, previously accumulated malate is rapidly oxidized into oxaloacetate. In these conditions NADH from malate dehydrogenase and pyruvate dehydrogenase is available for oxidation by the respiratory chain.

acetyl-CoA; pyruvate dehydrogenase itself, dihydrolipoyl trans-acetylase, and the flavoprotein dihydrolipoyl dehydrogenase (Reed *et al.*, 1978; Fig. 3.4). The pyruvate dehydrogenase component has a molecular weight of about 154,000 and possesses the subunit composition $\alpha_2\beta_2$. The molecular weights of the two subunits are about 41,000 and 36,000, respectively. The core enzyme, dihydrolipoyl transacetylase, consists of 60 apparently identical polypeptide chains of molecular weight about 52,000. The dihydrolipoyl dehydrogenase has a molecular weight of about 110,000 and contains two identical polypeptide chains and two molecules of FAD. The properties of these complexes in non-plant tissues have recently been reviewed by Reed *et al.* (1978). Because regulation of the flux of carbon through the pyruvate dehydrogenase system plays an essential role in the control of energy metabolism and

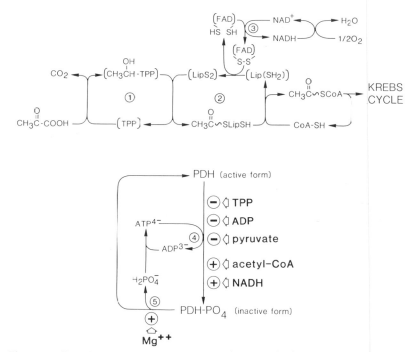

Fig. 3.4.   Reaction sequence in pyruvate oxidation and schematic representation of the covalent modification of pyruvate dehydrogenase and its control by various metabolites. Symbols and abbreviations: 1) pyruvate dehydrogenase; 2) dihydrolipoyl transacetylase; 3) dihydrolipoyl dehydrogenase; 4) PDH-kinase; 5) PDH-phosphatase; TPP, thiamine pyrophosphate; LipS2 and Lip(SH2), lipoyl moiety and its reduced form; CoA-SH, coenzyme A; and FAD, flavine adenine dinucleotide. [Adapted from Reed (1981).]

because of the other possible fates for pyruvate, it is obvious that regulatory control is exercised at this site. Two mechanisms for modulation of the activity of the mammalian pyruvate dehydrogenase complex have been well documented, (a) product inhibition and (b) a phosphorylation–dephosphorylation cycle (Fig. 3.4). The activity of the complex is inhibited by NADH and acetyl-CoA, and these inhibitions are competitive with respect to CoA and NAD^+, respectively (Reed *et al.*, 1978). These observations have led to suggestions that the activity of the pyruvate dehydrogenase complex may be regulated *in vivo*, at least in part, by the intramitochondrial acetyl-CoA–CoA and $NADH–NAD^+$ molar ratios. Phosphorylation and concomitant inactivation of the complex are catalyzed by an MgATP-dependent kinase, and dephosphorylation and concomitant reactivation are catalyzed by an Mg^{2+}-dependent phosphatase (Fig. 3.4). The site of this covalent regulation is the pyruvate dehydrogenase component of the complex. Interconversion of the active and inactive phosphorylated forms of pyruvate dehydrogenase is a dynamic process.

The pyruvate dehydrogenase complex has been purified and characterized from higher plants and appears to be similar in its reaction mechanism to the complexes from mammalian tissues, microbial systems, and fungi (Rubin et al., 1978). In addition, castor bean mitochondrial pyruvate dehydrogenase has been shown to be tightly bound to the inner membrane (Reid et al., 1975). The phosphorylation–dephosphorylation reactions of the pyruvate dehydrogenase complex have also been demonstrated in numerous plant tissues (Rubin and Randall, 1977; Thompson et al., 1977; Ralph and Wojcik, 1978; Randall et al., 1981). For example, the pyruvate dehydrogenase complex from pea leaf mitochondria was rapidly deactivated in the presence of ATP, whereas activation was inhibited by the phosphatase inhibitor F^- (Randall et al., 1981).

Citrate synthase catalyzes the irreversible condensation of acetyl-CoA and oxaloacetate and is regulated directly by the relative availability of either of its two substrates, acetyl-CoA or oxaloacetate (Fig. 3.2).

$$\text{Oxaloacetate} + \text{acetyl-CoA} + H_2O \rightarrow \text{citrate} + \text{CoA} \tag{3.1}$$

Like its mammalian counterpart, the plant mitochondrial enzyme is inhibited by ATP^{4-} (not MgATP) (Axelrod and Beevers, 1972; Greenblatt and Sarkissian, 1973). This inhibition is competitive with respect to acetyl-CoA (ATP and acetyl-CoA compete for an adenine binding site on the enzyme) and noncompetitive with respect to oxaloacetate. The kinetic characteristics of plant citrate synthase indicate a strong affinity of the enzyme for both of its substrates, oxaloacetate and acetyl-CoA. Thus, the apparent K_m of plant citrate synthase for oxaloacetate at saturating acetyl-CoA concentrations is in the range of 1–5 μM. Givan and Hodgson (1983) argued that in pea leaf mitochondria the great majority of the acetyl-CoA produced by pyruvate oxidation will likely enter into citrate formation and that the acetyl units will thereby pass into the TCA cycle. In addition, their data preclude any substantial hydrolytic release of CoA from acetyl-CoA, as suggested before by Liedvogel and Stumpf (1982). The intramitochondrial oxaloacetate concentration measured in plant mitochondria is extremely low and is expected to be determined primarily by the malate concentration and the $NADH–NAD^+$ ratio via equilibrium with malate dehydrogenase. Consequently, oxaloacetate availability in state 3 will increase primarily as a result of the lowered $NADH–NAD^+$ ratio. Since the intramitochondrial oxaloacetate concentration measured in plant mitochondria under various metabolic concentrations is extremely low (below 1 μM, R. Douce, unpublished data), it is probable that the effective oxaloacetate concentration available to citrate synthase is in the regulatory region of its K_m, unless oxaloacetate is compartmentalized due to protein binding or possibly to a concerted interaction between malate dehydrogenase, pyruvate dehydrogenase, and citrate synthase (Srere, 1980; Beeckmans and Kanarek, 1981). Interactions between malate dehydrogenase and citrate synthase have been proposed as a means of efficiently utilizing or drawing off the oxaloacetate produced by malate dehydrogenase. It is clear, therefore, that ligand interactions with citrate synthase and separate factors controlling the intramitochondrial concentrations of oxaloacetate and acetyl-CoA make the overall regulation of citrate synthase rather complex.

B.   Citrate Oxidation

It is more relevant, as suggested by Duggleby and Dennis (1970a,b) to consider citrate and isocitrate *in vivo* as a "single citrate pool" in which the citrate–isocitrate ratio remains fairly constant because of the aconitase reaction (between 10 : 1 and 70 : 1 depending on the Mg concentration). The results presented by Journet and Douce (1983) demonstrated that plant mitochondria retain a good rate of citrate-dependent O_2 consumption insofar as the medium contains thiamine pyrophosphate and NAD^+ (Fig. 3.5). In addition, ATP has a strong action upon citrate oxidation. Thus, by preincubating the mitochondria with ATP, the maximal rate of citrate-dependent O_2 uptake is attained more rapidly. In contrast to mammalian mitochondria, freshly isolated plant mitochondria contain very low levels of endogenous exchangeable metabolites. Consequently, citrate must be taken up in the mitochondria by energy-dependent net anion accumulation. It is clear, therefore, that uptake and retention of citrate are probably dependent upon the proton-motive force arising from ATP hydrolysis (Birnberg *et al.*, 1982; Jung and Laties, 1979). During the course of citrate oxidation in the absence of ADP (state 4), mitochondria almost exclusively produce α-ketoglutarate (Journet and Douce, 1983). On the contrary, under state 3 conditions plant mitochondria produce α-ketoglutarate, succinate, and malate. Consequently, *in vitro* the reactions of the citric acid cycle do not operate essentially as a cycle because of carbon loss as α-ketoglutarate and succinate. Because the rate of O_2 uptake is approximately equal to the mean rate of α-ketoglutarate accumulation plus twice the rate of succinate formation and three times the rate of malate formation, it is clear that three dehydrogenases are operating during the course of citrate oxidation in state 3, namely the isocitrate dehydrogenase, the α-ketoglutarate dehydrogenase, and the succinate dehydrogenase.

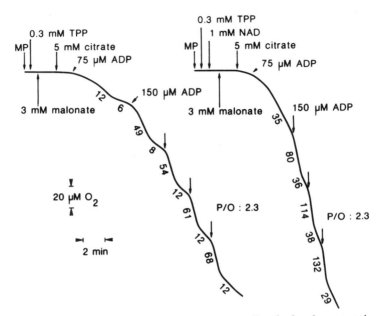

Fig. 3.5.   Effect of NAD$^+$ on citrate oxidation by NAD$^+$-depleted potato tuber mitochondria. The concentrations given are the final concentrations in the reaction medium.
The arrows on the traces indicate successive additions of 150 μ*M* ADP. The numbers on
the traces refer to nmoles O$_2$ consumed/min/mg protein. Note the strong increase of state
3 O$_2$ uptake rates and the decrease of the lag period when NAD$^+$ is added. Inasmuch as
the medium contains 3 m*M* malonate, the succinate dehydrogenase is fully inhibited.
[From Journet and Douce (1983).]

As noted by Plaut (1970), mitochondrial NAD-linked isocitrate dehydrogenase is inhibited by NADH. Further studies with purified enzymes
from mammalian tissues showed that inhibition by NADH was competitive with respect to NAD$^+$ ($K_i \simeq 20$ μ*M*) and was, therefore, dependent on the NADH–NAD$^+$ ratio. The NAD$^+$-linked isocitrate dehydrogenase of higher plants was first isolated and partially purified from pea
mitochondria by Davies (1953). It is also inhibited by the increasing mole
fraction of NADH (Cox, 1969; Duggleby and Dennis, 1970a,b) and is
activated by isocitrate and citrate (Cox and Davies, 1969, 1970; Duggleby
and Dennis, 1970a,b). Consequently, isocitrate dehydrogenase activity
in intact mitochondria is mainly regulated by intramitochondrial
NADH–NAD$^+$ ratios and citrate concentrations. Since the transition
from resting (state 4) to active (state 3) respiration is associated with an
oxidation of the mitochondrial pyridine nucleotides (Douce and Bonner,
1972; Fig. 3.6), it is concluded that feedback inhibition to the NAD$^+$-
linked isocitrate dehydrogenase is secondary to phosphate acceptor con-

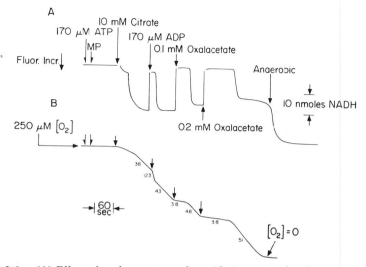

Fig. 3.6. (A) Effect of oxaloacetate on the oxidation state of endogenous NAD^+ in purified mung bean hypocotyl mitochondria oxidizing citrate. (B) Effect of oxaloacetate on citrate oxidation. The arrows correspond to the addition in (A), above. The numbers on the trace refer to nmoles O_2 consumed/min/mg protein. [From Douce and Bonner (1972).]

trol of electron transport. In support of this suggestion, Journet and Douce (1983) have demonstrated that oxaloacetate increased the rate of citrate oxidation in state 4. In this process oxaloacetate, which rapidly penetrates the inner membrane, is immediately converted into malate by malate dehydrogenase at the expense of intramitochondrial NADH.

According to Birnberg *et al.* (1982), potential utilization of citrate as a mitochondrial substrate from the organic pool is most unlikely, since the majority (or all) of the extramitochondrial citrate is synthesized in the mitochondria originally. Oxidation of this citrate would constitute an apparently futile cycle. They speculate, therefore, that there might be endogenous controls polarizing the malate–citrate exchange for citrate efflux (i.e., in exchanging mitochondrial citrate for cytoplasmic malate *in vivo*). In support of this suggestion, it appears that malate is usually present at higher concentrations than citrate in the cytoplasm (Beevers *et al.*, 1966) and is much more rapidly oxidized than citrate (Lance, 1972).

C. α-Ketoglutarate Oxidation

Plant mitochondria oxidize α-ketoglutarate insofar as the medium contains thiamine pyrophosphate and ADP (Fig. 3.7). In the absence of ADP and thiamine pyrophosphate, α-ketoglutarate is very poorly oxidized (Bowman *et al.*, 1976; Journet *et al.*, 1982). In fact, it is well known

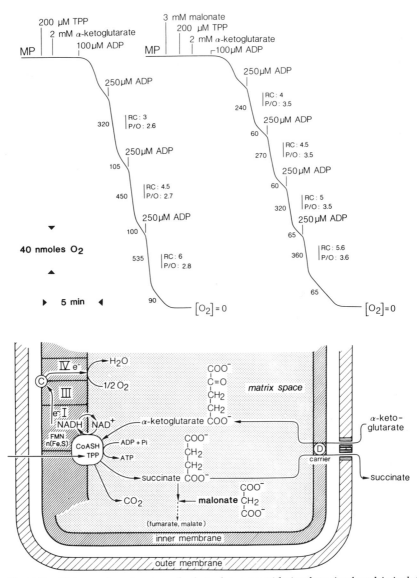

Fig. 3.7. Rates and mechanism of α-ketoglutarate oxidation by mitochondria isolated from higher plants. The reaction medium contained: 0.3 M mannitol; 5 mM MgCl$_2$; 10 mM KCl; 10 mM phosphate buffer, pH 7.2, and 0.1% (W/V) defatted bovine serum albumin. Since most of the plant mitochondria isolated so far are depleted of their endogenous thiamine pyrophosphate (TPP), it is necessary to add this cofactor in the medium in order to trigger α-ketoglutarate oxidation. Likewise in the absence of ADP, α-ketoglutarate is very poorly oxidized (the succinyl-CoA synthetase catalyzing a substrate level phosphorylation requires ADP). Addition of malonate (a potent inhibitor of succinate oxidation) to the incubation medium increases the ADP/O ratio during the course of α-ket-

that the oxidation of α-ketoglutarate requires thiamine pyrophosphate (oxidative decarboxylation) and ADP (phosphorylation at the substrate level). Analysis of the reaction products shows that during the course of α-ketoglutarate oxidation, plant mitochondria excrete malate; in addition, a transient accumulation of succinate occurs (Fig. 3.8). Under these conditions, the rate of $O(n$ atom) uptake is equal to the rate of succinate accumulation plus twice the rate of malate formation, and two dehydrogenases are operating, namely, the α-ketoglutarate dehydrogenase and the succinate dehydrogenase (Journet *et al.*, 1982). When the same experiment is carried out in the presence of malonate, the rate of α-ketoglutarate disappearance remains unchanged and the mitochondria almost exclusively excrete succinate (Fig. 3.8). One dehydrogenase is, therefore, operating, namely the α-ketoglutarate dehydrogenase.

α-Ketoglutarate dehydrogenase complexes have been isolated from animal mitochondria as functional units with molecular weights of about 2×10^6 (Reed and Cox, 1966). Like the pyruvate dehydrogenase complex, they contain three functional enzymes required for the irreversible conversion of α-ketoglutarate to succinyl-CoA; α-ketoglutarate dehydrogenase itself, dihydrolipoyl transsuccinylase, and the flavoprotein dihydrolipoyl dehydrogenase. This complex, therefore, is considered to be very similar in some respects to the pyruvate dehydrogenase complex.

$$\text{α-Ketoglutarate} + NAD^+ + CoA \rightarrow \text{succinyl-CoA} + NADH + H^+ + CO_2 \quad (3.2)$$

All of the α-ketoglutarate dehydrogenase complexes isolated from mammalian mitochondria contain thiamine pyrophosphate, flavin adenine dinucleotide, lipoic acid, and a divalent cation, all of which are required for the oxidation of α-ketoglutarate by NAD^+ (Massey, 1960). The metal ion is presumed necessary for the binding of thiamine pyrophosphate to the α-ketoglutarate dehydrogenase, and the lipoyl moiety is covalently bound to the lipoyl transsuccinylase portion of the complex, where it is linked to the ε-amino group of a lysine residue by an amide bond. In the mitochondria, the turnover of α-ketoglutarate dehydrogenase is strongly dependent on product removal that is achieved by succinyl-CoA synthetase catalyzing a substrate-level phosphorylation requiring ADP (J. M. Palmer and Wedding, 1966) and, therefore, related to energy demand.

The α-ketoglutarate dehydrogenase–lipoyl transsuccinylase complex

---

oglutarate oxidation. Note that the α-ketoglutarate carrier (D) is responsible for the high flux of α-ketoglutarate into the mitochondrial matrix. During the course of α-ketoglutarate oxidation, net exchange between added α-ketoglutarate and intramitochondrial succinate occurs. Isolated plant mitochondria very often display depressed initial state 3 rates of α-ketoglutarate oxidation which rise to a maximum with several consecutive state 3–state 4 cycles. The depressed state 3 rates result largely from inadequate substrate uptake (the entry of α-ketoglutarate is activated by small amounts of malate).

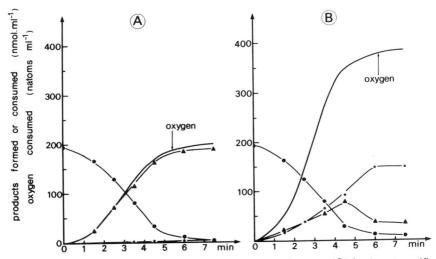

Fig. 3.8. Oxidation of a limiting amount of α-ketoglutarate (●) by intact purified potato tuber mitochondria, appearance of succinate (▲) and malate (★), and O₂ consumption. The standard assay solution was used with 0.85 mg mitochondrial protein/ml, 0.3 m *M* TPP, and 1.5 mM ADP. Final pH was 7.2. The reaction was initiated by addition of 195 μ *M* α-ketoglutarate. (A) Assay in presence of 2 mM malonate; (B) assay without malonate. Note that the rapid disappearance of α-ketoglutarate is not affected by the presence of malonate. Symbols: ● α-ketoglutarate, ▲ succinate, and ★ malate. [From Journet *et al.* (1982).]

of higher plants, first isolated and partially purified from cauliflower mitochondria by Poulsen and Wedding (1970), appears to catalyze a reaction like that shown for mammalian enzymes and to utilize the same array of substrates and cofactors (Craig and Wedding, 1980a,b). Highly purified preparations of α-ketoglutarate dehydrogenase complex from plant mitochondria are, however, completely dependent on added thiamine pyrophosphate, but similar preparations of the mammalian enzyme show little or no response to exogenous thiamine pyrophosphate. According to Poulsen and Wedding (1970), the absence of bound thiamine pyrophosphate in the plant complex has made it possible to show that the cofactor for this enzyme appears to be Mg–thiamine pyrophosphate (this complex dissociates readily) rather than Mg and thiamine pyrophosphate separately. Again, this dissociation of a required cofactor, which is also observed at the level of isolated mitochondria, may represent one method of controlling the activity of α-ketoglutarate dehydrogenase in plants. The same thing holds true with the pyruvate dehydrogenase complex. α-Ketoglutarate dehydrogenase is inhibited by both of its products, NADH and succinyl-CoA (Williamson and LaNoue, 1975). Consequently, *in vivo*, flux through α-ketoglutarate dehydrogenase is regulated by the succinyl-CoA–CoA and NADH–NAD$^+$ ratios. Finally, AMP acts as a positive effector on maximum velocity at all pH values (Craig and Wedding, 1980a,b). This observation places the plant enzyme in a different category from the mammalian α-ketoglutarate dehydrogenase (Hansford, 1972). It is unlikely, however, that this would be an effective control mechanism as far as the TCA cycle is concerned, because the rate of α-ketoglutarate oxidation is strongly dependent on the cycling of CoA.

D. Succinate Oxidation

All of the plant mitochondria isolated so far oxidize succinate very rapidly. Succinate dehydrogenase in mitochondria isolated from plant tissues is usually found to be in a deactivated state, and preincubation of the mitochondria with ATP overcomes this problem (Oestreicher *et al.*, 1973; Fig. 3.9). Succinate dehydrogenase, as pointed out before, is inhibited by oxaloacetate. However, we have found that the degree of inhibition of succinate oxidation by oxaloacetate in intact potato tuber mitochondria depends upon the respiratory state of the mitochondria, with a greater inhibition occurring during state 3 respiration than during state 4. The difference in the inhibition is related to conditions whereby added oxaloacetate can be metabolized. A pool of NADH is maintained during state 4 (Douce and Bonner, 1972; Fig. 3.10) which acts as a cofactor in the reduction of oxaloacetate by malate dehydrogenase. This pool is depleted during state 3 respiration. Furthermore, endogenously produced oxaloacetate during rapid state 3 oxidation of succinate inhibited succinate-supported respiration at low-succinate concentrations (0.5 mM) but not at low-succinate plus glutamate nor at high-succinate concentrations (10 mM). Consequently, we believe that oxaloacetate could play an important role in the regulation of plant succinate dehydrogenase if one considers that a molar equivalent of oxaloacetate suffices for extensive loss of succinate dehydrogenase activity (Singer *et al.*, 1972).

E. Miscellaneous Oxidations

1. Glutamate Oxidation

The NAD-linked glutamate dehydrogenase activity from both root and shoot tissues of higher plants is associated with the mitochondria and is predominantly concerned with deamination of glutamate to α-ketoglutarate.

Isolated plant mitochondria are capable of synthesizing glutamate under certain defined but nonphysiological conditions (Davies and Teix-

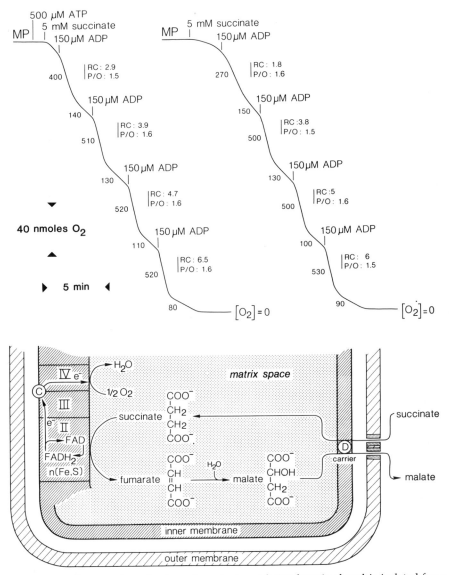

Fig. 3.9. Rates and mechanism of succinate oxidation by mitochondria isolated from higher plants. The reaction medium contained: 0.3 M mannitol; 5 mM MgCl$_2$; 10 mM KCl; 10 mM phosphate buffer, pH 7.2, and 0.1% (W/V) defatted bovine serum albumin. In the presence of 150 μM ATP the maximal rate of succinate oxidation is attained more rapidly than in its absence. (Succinate dehydrogenase in mitochondria isolated from plant tissues is usually found to be in a deactivated state and preincubation of the mitochondria with ATP overcomes this problem.) Note that the dicarboxylate carrier (D) is responsible for the high flux of succinate into the mitochondrial matrix. During the course of succinate oxidation, a net exchange between added succinate and intramitochondrial malate occurs.

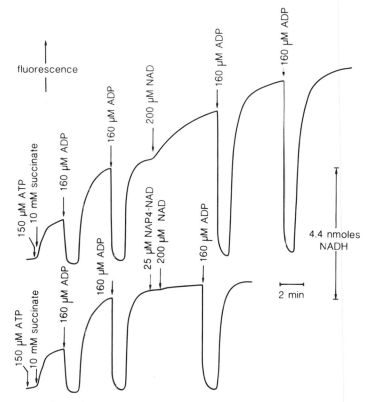

Fig. 3.10. Effect of exogenous NAD$^+$ and NAP$_4$-NAD$^+$ on the oxidation state of endogenous NAD in purified potato tuber mitochondria oxidizing succinate and aged for 24 hr. The standard assay medium contained 150 μM ATP, 10 mM succinate, and 0.22 mg of mitochondrial protein; the pH was 7.2, and the temperature 25°C. The final volume of the reaction mixture was 2 ml. To prevent the possible reduction of exogenous NAD$^+$, the medium contained 1 mmole pyruvate and 10 μg of lactate dehydrogenase. Note that the addition of succinate to potato tuber mitochondria aged for 24 hr resulted in a prompt reduction of endogenous nicotinamide nucleotides, provided that the medium contained ATP. It is well-established that net electron transport from succinate to NAD$^+$ (reversed electron transport) is driven by ATP hydrolysis (Ernster and Lee, 1967). The subsequent addition of ADP caused a transient oxidation of nicotinamide nucleotides (state 3–state 4 transition). With succinate as substrate, however, the state-4 level of nicotinamide nucleotide reduction was consistently greater after a state-3–state-4 cycle than before such a respiratory cycle. Addition of NAD$^+$ caused an increase in the extent of nicotinamide nucleotide fluorescence up to a new steady-state level. Further addition of ADP produced a rapid reoxidation of the total pool of mitochondrial NADH. Addition of NAP$_4$-NAD$^+$ prevented exogenous NAD$^+$ from increasing internal nicotinamide nucleotide fluorescence. [From Neuburger and Douce (1983) and reprinted by permission from *Biochem. J.* **216**, 449. Copyright © 1983. The Biochemical Society, London.]

eira, 1975). Since in the presence of citrate (as both the carbon source and the hydrogen donor) washed mitochondria from *Pisum* produce considerable amounts of glutamate upon addition of NH_4^+ (one half saturation concentrations of NH_4^+ of 1.6–3.6 mM were determined), Nauen and Hartmann (1980) considered the possibility that there might exist a special link between NAD-specific isocitrate dehydrogenase and glutamate dehydrogenase in plant mitochondria. In addition, they believe that plant mitochondria are adapted to assimilate NH_4^+, at least at elevated levels of intracellular NH_4^+. A major objection to a role for mitochondrial glutamate dehydrogenase in ammonia assimilation, however, is the enzyme's high K_m for ammonia (10–80 mM; Stewart *et al.*, 1980).

All of the plant mitochondria isolated so far oxidize glutamate poorly unless the medium contains thiamine pyrophosphate, in order to trigger α-ketoglutarate oxidation. Furthermore, mitochondria isolated from a large variety of plant tissues oxidize glutamate in the presence of malate (for a review see Hanson and Day, 1980). This oxidation is probably due to a combination of glutamate dehydrogenase, malate dehydrogenase, and glutamate-oxaloacetate transaminase localized in the matrix space (Journet *et al.*, 1982; Fig. 3.11). With the aim of clarifying the mechanisms of glutamate metabolism in intact plant mitochondria, Journet *et al.* (1982) detailed the effect of oxaloacetate on glutamate metabolism in potato tuber mitochondria. They demonstrated that, during the course of glutamate oxidation triggered by oxaloacetate, part of the oxaloacetate is metabolized by transamination whereas the other part is rapidly converted to malate, and this conversion is dependent on the reduced pyridine nucleotide generated by the thiamine pyrophosphate-linked α-ketoglutarate dehydrogenase localized in the matrix space. During the course of oxaloacetate disappearance, part of the α-ketoglutarate formed may derive from the action of glutamate dehydrogenase. This pathway is not important under these conditions, however, since the sum of the rates of α-ketoglutarate and succinate accumulation (most of the succinate was formed directly from α-ketoglutarate) is approximately equal to the rate of aspartate production (Fig. 3.12). When all the oxaloacetate is consumed, a rapid O_2 uptake is initiated and α-ketoglutarate and succinate are rapidly reabsorbed and oxidized by the mitochondria (Fig. 3.12). All of these results strongly emphasize the importance of α-ketoglutarate oxidation during the course of glutamate metabolism in intact plant mitochondria. The organization and integrated function of plant mitochondrial glutamate dehydrogenase need further clarification. The physiological significance of glutamate dehydrogenase is difficult to assess, however, without knowing the substrate concentrations and the pH at the site of the enzyme within the mitochondria of individual

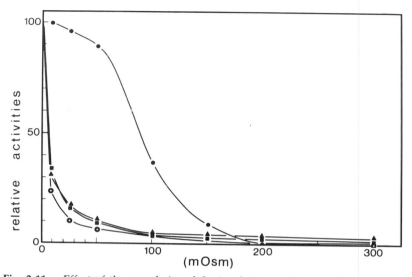

Fig. 3.11. Effect of the osmolarity of the incubation medium on the appearance of succinate–cytochrome *c* oxidoreductase, glutamate–oxaloacetate transaminase, malate dehydrogenase, and fumarase activities in intact purified potato tuber mitochondria. Aliquots of the mitochondrial suspension (1.2 mg mitochondrial proteins) were added to mannitol solutions of various osmolarities, obtained by mixing in definite proportions of medium I (0.3 *M* mannitol, 0.1% (W/V) BSA, 2 m*M* MgCl$_2$, and 10 m*M* phosphate buffer, pH 7.2) and medium II (0.1% (W/V) BSA, 2 m*M* MgCl$_2$, and 2 m*M* Mops buffer, pH 7.2). After 1-min incubation at 4°C, the osmolarity of the medium was adjusted to 0.3 Osm; the final volume was 3 ml. The succinate–cytochrome *c* oxidoreductase activity was determined. Simultaneously, part of the mitochondria were centrifuged at 10,000 *g* for 10 min, and the glutamate–oxaloacetate transaminase, the malate dehydrogenase, and the fumarase activities were determined in the supernatant. The maximal values of glutamate–oxaloacetate transaminase (580 nmoles oxaloacetate formed/min/mg mitochondrial protein), malate dehydrogenase (55 μmoles NADH consumed/min/mg mitochondrial protein), and fumarase (540 nmoles fumarate formed/min/mg mitochondrial protein) activities were obtained after addition of Triton X-100 (0.02%; W/V) to the incubation medium. The total succinate–cytochrome *c* oxidoreductase activity was 880 nmoles cytochrome *c* reduced/min/mg mitochondrial protein. The enzymatic activities are expressed as percentages of their maximal values. Symbols: ● succinate–cytochrome *c* oxidoreductase; □ malate dehydrogenase, ▲ glutamate–oxaloacetate transaminase; and ○ fumarase. [From Journet *et al.* (1982).]

organisms. Further studies should include the possibility of control by the NADH–NAD$^+$ ratio and the differential control of the reversible enzyme reactions by divalent cations (Garland and Dennis, 1977).

2. Proline Oxidation

The mitochondrial oxidation of proline has been observed in isolated organelles from a variety of higher plant mitochondria (Boggess *et al.*, 1978; Huang and Cavalieri, 1979). Recently, the enzymes involved with

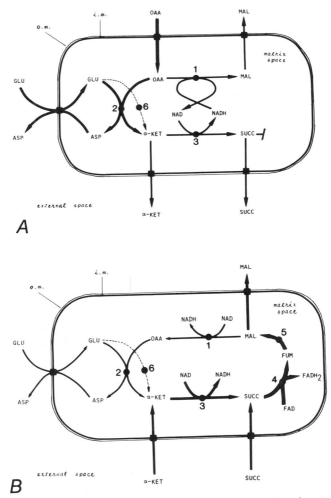

Fig. 3.12. Schemes of the metabolic processes involved in the glutamate oxidation triggered by oxaloacetate in potato tuber mitochondria. (A) Added oxaloacetate readily penetrates the matrix space through the inner membrane. Part of the oxaloacetate (OAA), at least one half, reacts with glutamate (GLU) under the control of the glutamate–oxaloacetate transaminase (2); α-ketoglutarate thus produced is either oxidized by the α-ketoglutarate dehydrogenase (3) or excreted into the external space; since the succinate dehydrogenase (4) is strongly inhibited by oxaloacetate, succinate (SUCC) accumulates in the external space. Simultaneously, the other part of the oxaloacetate is converted into malate (MAL) under the control of malate dehydrogenase (1), a conversion that depends on the reduced pyridine nucleotides generated by the TPP-linked α-ketoglutarate dehydrogenase. In these conditions, O_2 consumption by the respiratory chain is inhibited. The activity of the glutamate dehydrogenase (6) is relatively negligible. (B) When all the oxaloacetate is consumed, the inhibition of succinate dehydrogenase by oxaloacetate is released, and the

proline oxidation have been partially characterized as to their sub-mitochondrial location and electron transport characteristics (Boggess *et al.*, 1978; Elthon and Stewart, 1982; Huang and Cavalieri, 1979; Stewart and Lai, 1974). The first enzyme in this process is proline dehydrogen-ase, which catalyzes the conversion of proline to Δ'-pyroline-5-carbox-ylic acid (P5C). The second enzyme is P5C dehydrogenase, which cata-lyzes the conversion of P5C to glutamate. Proline dehydrogenase activity (pH optimum around 7.2) does not involve NAD^+ reduction, and thus electrons from proline enter the respiratory chain directly. On the other hand, P5C dehydrogenase (pH optimum around 6.4) requires NAD^+ reduction. These two enzymes are bound to the matrix side of the inner membrane (Elthon and Stewart, 1982).

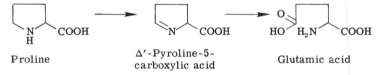

Proline Δ'-Pyroline-5-carboxylic acid   Glutamic acid

The accumulation of proline in several species of plants is a well established response to salt- or water-stress. Proline accumulation re-sults primarily from stimulated synthesis and the concomitant inhibition of proline oxidation (Stewart *et al.*, 1977). In support of this interesting suggestion, the state 3 rate of proline oxidation by mitochondria from water-stressed maize shoots decreased as a logarithmic function of the increased seedling water-stress of between 5 and 10 bars (Sells and Koeppe, 1981). Accumulated proline may serve as a neutral osmoticum and as a reserve of nitrogen; upon relief of stress, accumulated proline is rapidly oxidized, suggesting that it serves a role as an energy reserve (Stewart *et al.*, 1977).

II. THE MECHANISM OF MALATE OXIDATION

A. Mechanism Involved

The unfavorable equilibrium of the reaction catalyzed by malate dehy-drogenase (L-malate/oxaloacetate $\simeq 10^4$) necessitates the rapid removal

---

previously accumulated succinate and α-ketoglutarate can be rapidly oxidized. Malate thus produced is excreted from the matrix space. With time, malate is slowly oxidized into oxaloacetate, which reacts immediately with glutamate. Three dehydrogenases are in-volved in that slow process. Abbreviations and symbols: 5, fumarase; ■, carriers; o.m., outer membrane; i.m., inner membrane; and ASP, aspartate. [From Journet *et al.* (1982).]

of the product, oxaloacetate, in order for malate oxidation to proceed. For this reason, interactions between mitochondrial malate dehydrogenase and citrate synthase or aspartate transaminase have been proposed as a means of efficiently utilizing or drawing off the oxaloacetate produced by mitochondrial malate dehydrogenase during operation of the TCA cycle.

Nonetheless, plant mitochondria, in contrast to mammalian mitochondria, readily oxidize malate without the necessity of removing oxaloacetate, because they possess a specific NAD^+-linked malic enzyme and a specific oxaloacetate carrier. Pyruvate and/or oxaloacetate are the major products formed during the course of malate oxidation with lesser amounts of citrate produced, insofar as the medium contained thiamine pyrophosphate (Lance et al., 1965; Coleman and Palmer, 1972; Brunton and Palmer, 1973; Macrae and Moorhouse, 1970; Hulme and Rhodes, 1968). Citrate, of course, is derived from oxaloacetate by condensation with acetyl-CoA in the presence of thiamine pyrophosphate (Lance et al., 1967). The acetyl-CoA derives from the pyruvate, which is formed during malate oxidation and accumulates when pyruvate dehydrogenase activity is limited by the reduced availability of thiamine pyrophosphate. Since crude mitochondrial extracts exhibit no appreciable oxaloacetate decarboxylase activity, even at low pH values, it is clear that during the course of malate oxidation pyruvate is formed directly from malate using a mitochondrial NAD^+-requiring malic enzyme, rather than by a route involving the malate dehydrogenase reaction followed by decarboxylation of oxaloacetate as suggested before by Lance et al. (1967). In the absence of thiamine pyrophosphate, O_2 uptake with malate as the substrate is attributed solely to malate dehydrogenase and/or NAD^+-linked malic enzyme [L-malate–NAD^+ oxidoreductase (decarboxylating) EC1.1.1.39]. The latter enzyme was discovered in plant mitochondria by Macrae and Moorhouse (1970) and is not found in most animal mitochondria; the NAD^+-malic enzyme of higher plants is always located in the mitochondria (Hirai, 1978; Hatch and Kagawa, 1974; Dittrich, 1976). Macrae (1971a,b), in a classical experiment, first isolated the malic enzyme from plant mitochondria and demonstrated that it is specific for L-malate, has an absolute requirement for Mn^{2+} or Mg^{2+}, and is characterized by a low substrate affinity. The enzyme is inhibited by NADH and stimulated by low concentrations of CoA (Macrae, 1971a; Valenti and Pupillo, 1981; Neuburger and Douce, 1980; Day et al., 1984a). In addition, bicarbonate inhibits malate oxidation by malic enzyme (Hatch et al., 1974; K. S. R. Chapman and Hatch, 1977; Neuburger and Douce, 1980), and this inhibition is relieved by NAD^+ and CoASH (Neuburger and Douce, 1980; Hatch et al., 1974). It is clear, therefore, that NAD^+-linked malic enzyme is subject to controls of various types, particularly allosteric signals from reactions that precede or follow the reaction it catalyzes. Grover and Wedding (1982) have shown that NAD^+-malic enzyme is capable of existing in three different molecular weight forms that, on the basis of a subunit molecular weight of 60,000, appear to be a dimer, a tetramer, and an octamer. The 264,000-form of the enzyme, which they take to be a tetramer composed of four identical subunits, is the most active form of the enzyme, having a low K_m for substrate and the highest V_{max}. Such a conclusion is in agreement with the properties of the mitochondrial NADP-dependent malic enzymes isolated from animal mitochondria (see, for example, Nagel and Sauer, 1982). The dimer with its high K_m and low V_{max} is the least active form. High concentrations of malate or citrate favor tetramer formation. Given the fact that plant cells often accumulate large quantities of malate (Ranson, 1965), it seems probable that this effect of malate on the aggregation state of malic enzyme has regulatory significance (Grover and Wedding, 1982). It is to be expected that malic enzyme should be tightly controlled unless high levels of organic acids are present in the cytosol, as when stored acids are mobilized. The relative effects of malate, CoA, enzyme concentration, and ionic strength on the enzyme aggregation equilibrium are only partially understood, however, and remain to be examined in greater detail.

Macrae (1971b) also demonstrated that the rates and products of malate oxidation for cauliflower bud mitochondria will be altered in response to relatively small changes in the pH of the incubation medium. Similar changes do not occur with α-ketoglutarate and succinate oxidation. The effect of pH on malate oxidation is interpreted in terms of the pH–activity profiles of the NAD^+-requiring malic enzyme and malate dehydrogenase. Thus, during the course of malate oxidation, addition of a known amount of NaOH, which causes a rapid increase of the pH from 6.5–7.5, immediately triggers oxaloacetate formation and stops pyruvate production. Conversely, addition of a known amount of HCl, which causes a rapid fall of the pH from 7.5–6.5, immediately triggers pyruvate formation (Neuburger and Douce, 1980; Tobin *et al.*, 1980) and stops oxaloacetate production (Fig. 3.13). The ratio of the products, oxaloacetate and pyruvate, during the course of malate oxidation reflects the balance of the two malate-oxidizing enzymes (Neuburger and Douce, 1980; J. M. Palmer *et al.*, 1982). When the activity of the NAD^+-linked malic enzyme is weakened (high bicarbonate concentration, low CoA concentration, alkaline pH; Fig. 3.14), oxaloacetate is preferentially excreted and there is a decrease in the rate of malate oxidation as the reaction proceeds, owing to the accumulation of oxaloacetate and the

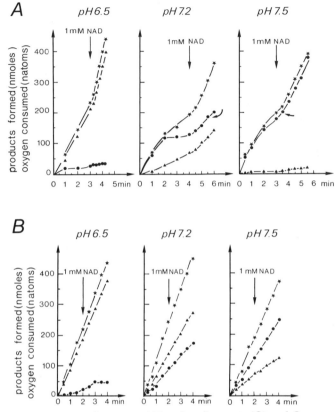

Fig. 3.13.   Production of pyruvate (▲) and oxaloacetate (●) and O_2 consumption (★) during malate oxidation by mitochondria from potato tubers (A) and mung bean hypocotyl mitochondria (B) as a function of pH. The standard assay solution was used with 0.4 mg mitochondrial protein/ml, 15 mM malate, 1 mM ADP, and 3 mM sodium arsenite. As indicated, 1 mM NAD+ was added. In this particular experiment the total amount of NAD+ present in freshly prepared potato tuber mitochondria was at least five times lower than that of isolated mung bean hypocotyl mitochondria (1.1 versus 5.7 nmoles/mg mitochondrial protein). This experiment clearly indicates that as the pH increases, the fraction of O_2 uptake appearing as oxaloacetate increases very rapidly. In addition, potato tuber mitochondria oxidizing malate respond to NAD+ addition with increased oxidation rates, whereas mung bean hypocotyl mitochondria do not. This is traced to a low-endogenous content of NAD+ in potato mitochondria which promotes the take up of added NAD+. [From Tobin et al. (1980).]

unfavorable equilibrium of the reaction catalyzed by malate dehydrogenase (Fig. 3.15). Under state 4 conditions, matrix NADH levels also increase (Tobin et al., 1980), malate dehydrogenase and, therefore, O_2 uptake are more severely restricted, and malate dehydrogenase may be reversed (Fig. 3.15). One of the reasons potato mitochondria are able to

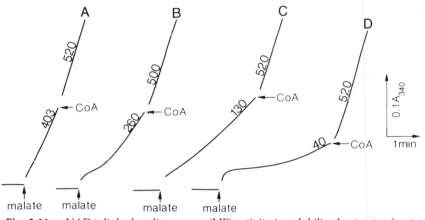

Fig. 3.14. NAD$^+$-linked malic enzyme (ME) activity in solubilized extracts of potato tuber mitochondria. (A) pH 7.2; (B) pH 7.2 plus 5 mM HCO$_3^-$; (C) pH 7.5; and (D) pH 7.5 plus 5 mM HCO$_3^-$. Where indicated, 10 mM malate and 25 μM coenzyme-A were added. [From Day *et al.* (1984).]

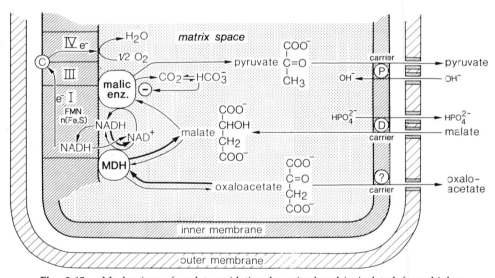

Fig. 3.15. Mechanism of malate oxidation by mitochondria isolated from higher plants. This scheme indicates that NAD$^+$-linked malic enzyme (malic enz) and malate dehydrogenase (MDH), located in the matrix space, compete at the level of the pyridine nucleotide pool and that the NAD$^+$-linked malic enzyme provides NADH for the reversal of the reaction catalyzed by the malate dehydrogenase. Note that the dicarboxylate carrier (D) is responsible for the high flux of malate into the mitochondrial matrix. Oxaloacetate can rapidly traverse the inner membrane via a specific carrier (malate efflux or influx and oxaloacetate influx or efflux occur on separate carriers). Pyruvate is probably excreted via the pyruvate carrier (P). Finally, bicarbonate directly produced in the matrix space is a potent inhibitor of the NAD$^+$-linked malic enzyme.

consume O_2 at significant rates at high pH conditions with malate as a substrate (in the absence of a system to remove oxaloacetate) is that they excrete oxaloacetate in the external medium. Phthalonate is a potent inhibitor of oxaloacetate uptake and efflux in plant mitochondria, consequently, adding phthalonate to plant mitochondria respiring malate at alkaline pH conditions induces a marked inhibition of O_2 uptake. On the other hand, when the activity of the NAD^+-linked malic enzyme is powerful (low bicarbonate concentration, high CoA concentration, acidic pH; Fig. 3.14), oxaloacetate concentration is maintained at a low level and pyruvate is rapidly excreted. In other words, with all the plant mitochondria isolated so far, whenever the NAD^+-linked malic enzyme activity is weakened, the rate of oxaloacetate production is higher than that of pyruvate. At pH values intermediate between 6.6 and 7.5, malate oxidation rates were initially slow but gradually increased with time. At these pH values, state 4 rates are biphasic with a slow initial phase followed by a faster second phase (Fig. 3.16). This biphasic pattern of state 4 malate oxidation has been investigated by J. M. Palmer *et al.* (1982) and reflects the concerted operation of malate dehydrogenase and malic enzyme. When both enzymes are active and some oxaloacetate accumulates, the respiratory chain and malate dehydrogenase compete for the NADH generated during malate oxidation by malic enzyme.

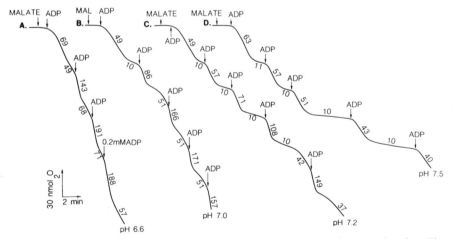

Fig. 3.16.   Effect of external pH on malate oxidation by potato tuber mitochondria. The external pH was adjusted by adding small aliquots of HCl or KOH to the standard reaction medium. Other additions, as shown, were 10 mM malate and 0.1 mM ADP (except where indicated otherwise). NAD^+ (0.2 mM) was included in the reaction medium in all experiments. Mitochondria integrity was 98% and 0.4 mg mitochondrial protein was used in each experiment. Numbers on traces refer to nmoles O_2/min/mg protein.

During state 4, NADH oxidation via the respiratory chain is restricted (by lack of ADP) and oxaloacetate previously accumulated during state 3 reenters the mitochondria and is reduced via malate dehydrogenase at the expense of NADH produced by the operation of malic enzyme; consequently, O_2 uptake is initially slow. When all the oxaloacetate has been removed, more NADH becomes accessible to the respiratory chain and O_2 uptake rates increase. Under these conditions the rotenone-insensitive internal NADH dehydrogenase may be engaged (J. M. Palmer *et al.*, 1982). This interpretation of the biphasic state 4 rates is supported by analyses of the products of malate oxidation which have shown that pyruvate accumulates while oxaloacetate levels decrease during the initial phase (J. M. Palmer *et al.*, 1982; Day *et al.*, 1984a). Consequently, it is clear that both enzymes are competing at the level of the pyridine nucleotide pool and that the regulation *in vivo* of malate dehydrogenase can be readily accounted for by equilibrium effects alone (Neuburger and Douce, 1980; Fig. 3.15). In fact, in plant mitochondria malate dehydrogenase is present in a great excess capable of oxidizing NADH at the rate of 50,000 nmoles/min/mg protein compared with 300 nmoles/min/mg protein for NAD^+-linked malic enzyme (Neuburger, 1980), and there is, therefore, good reason to believe that the equilibrium can respond rapidly to changes in the concentration of any one of the substrates.

The rate of malate oxidation by intact plant mitochondria can frequently be stimulated by the addition of NAD^+, a coenzyme which has not been thought to be able to traverse the inner membrane of the mitochondria (Coleman and Palmer, 1972; Day and Wiskich, 1974a,b; Marx and Brinkmann, 1979; Fig. 3.13). A great deal of speculation into the mechanism of malate oxidation has arisen as a result of this NAD^+-stimulation of malate oxidation. Palmer and co-workers have suggested that two pathways of malate oxidation exist in plant mitochondria. One pathway is controlled by malate dehydrogenase in the matrix space and linked to an NADH dehydrogenase on the inner face of the inner membrane. This NADH dehydrogenase (complex I) is thought to be rotenone sensitive and coupled to three sites of phosphorylation. The second pathway is thought to be controlled by malic enzyme present in the intermembrane space and linked to the external NADH dehydrogenase on the outer surface of the inner membrane, in the presence of added NAD^+. This NADH dehydrogenase is rotenone-resistant and coupled to only two sites of phosphorylation. This hypothesis is no longer valid, however, for the following reasons: (a) the stimulation of malate oxidation by NAD^+ could be due to the stimulation of malic enzyme and malate dehydrogenase from a small percentage of "broken" mitochondria (Mannella, 1974; J. M. Palmer, 1980; Day et al., 1983); (b) a stimulation of rotenone-resistant and antimycin-sensitive electron transport from malate to ferricyanide by NAD^+ is also observed (Brunton and Palmer, 1973; J. M. Palmer and Arron, 1976); (c) Day et al. (1979) and ap Rees et al. (1983) sequentially disrupted the membranes of isolated potato tuber mitochondria using digitonin and concluded that malate dehydrogenase and malic enzyme were both matrix enzymes; and (d) if the various NAD^+-linked dehydrogenases are in different compartments, then one might expect to see differential inhibition of oxaloacetate. Douce and Bonner (1972) and Wiskich and Day (1982), however, have indicated that O_2 uptake declines equally with all substrates when oxaloacetate is added; that is the NADH generated in the matrix upon oxidation of different NAD^+-linked substrates is equally accessible to malate dehydrogenase and oxaloacetate. In addition, Day and Wiskich (1974a,b) indicated that malate had to penetrate the matrix in order to be oxidized even in the presence of rotenone. They speculated, therefore, that NAD^+ was acting as an external substrate for a transmembrane transhydrogenase that received its reducing equivalents from internal NADH (Day and Wiskich, 1978). Again several results are not in agreement with the involvement of a transmembrane transhydrogenase. This claim was investigated by J. M. Palmer (1980) who concluded that malate-dependent reduction of external NAD^+ was due to contamination by malate dehydrogenase from broken mitochondria. In addition, NAD^+ stimulates the rotenone-resistant oxidation of NAD^+-linked substrates by Jerusalem artichoke mitochondria in the presence of EGTA, which severely inhibits the activity of the external NADH dehydrogenase. Furthermore, we have observed that whenever the total amount of mitochondrial NAD^+ was low (i.e., below 2 mM), the rate of malate oxidation by plant mitochondria was considerably stimulated by exogenous NAD^+ (Tobin et al., 1980). In addition, if the NADH was reacting directly with the flavoprotein on the external face of the inner membrane of intact mitochondria, one would expect a decrease in the ADP/O ratio during the course of malate oxidation in the presence of NAD^+. This is not seen unless the mitochondria are damaged (Neuburger and Douce, 1980) or NAD^+ is contaminated with AMP. In fact, as pointed out by J. M. Palmer et al. (1982), all these results are most easily explained by assuming that the added NAD^+ stimulates malate oxidation by entering the matrix space (Neuburger and Douce, 1978, 1983; Tobin et al., 1980; Day et al., 1983) where it can be reduced by the NAD^+-linked dehydrogenases and reoxidized by the rotenone-sensitive NADH dehydrogenase (complex I) or the rotenone-insensitive bypass located on the inner face of the inner membrane. In a recent reinvestigation of NAD^+

effects, Day *et al.* (1983) have demonstrated conclusively that the stimulation of respiration in intact plant mitochondria by added NAD^+ is due to the uptake of NAD^+ into the matrix space. The consequential increase in the concentration of endogenous NAD^+ stimulates the activity of all NAD-linked enzymes (Figs. 2.29 and 3.1), thus providing more reducing power to the respiratory chain and stimulating the rotenone-insensitive pathway. The stimulation by NAD^+ of NAD^+-linked malic enzyme is stronger than that of the NAD-linked TCA cycle dehydrogenases because the affinity of the latter enzymes for NAD^+ is higher than that of the former (Cox and Davies, 1967; Massey, 1960; Rubin and Randall, 1977; Macrae, 1971a,b).

B. Physiological Significance

The physiological significance of mitochondrial NAD^+-linked malic enzyme is unclear. This enzyme is present in all of the plant mitochondria isolated so far and may play a key role in the organic acid metabolism. When stored reserves of malate within the vacuole [the available evidence suggests that malate accumulated in vacuoles is synthesized by a sequence involving phosphoenolpyruvate carboxylase and malate dehydrogenase (Osmond, 1976) and rapid transfer to the vacuole after synthesis] are mobilized, NAD^+-linked malic enzyme allows their complete oxidation via conversion of the malate to pyruvate. Pyruvate is then converted to acetyl-CoA, which in turn can be completely oxidized in the normal reactions of the TCA cycle. In other words, malic enzyme allows the conversion of C_4 acids into acetyl-CoA, the normal respiratory substrate (Beevers, 1961) without the necessity of supplying pyruvate from glycolysis (J. M. Palmer, 1976). In support of this suggestion, Grover and Wedding (1982) have shown that high levels of malate and citrate act to maintain malic enzyme in its active tetrameric form. It is also possible that the NAD^+-linked malic enzyme could play a significant role in the metabolism of fast growing cells by producing carbon skeletons for biosynthetic purposes (e.g., amino acid synthesis). Under these circumstances there may be an advantage in having a nonphosphorylating electron transport chain to enable this role to be fulfilled in the presence of a high level of ATP (J. M. Palmer, 1976). Recently, Rustin *et al.* (1980b) have suggested that malic enzyme is functionally linked not only to the rotenone-insensitive NADH dehydrogenase but also to the alternative path in etiolated tissues. Unfortunately, under no conditions tested could J. M. Palmer *et al.* (1982), Wiskich and Day (1982), and Gardeström and Edwards (1983a) detect the operation of malate oxidation that was specifically linked to nonphosphorylating electron transport pathways in mitochondria from various tissues, including etiolated tissues and the mesophyll cells of *Panicum miliaceum*, an NAD^+-malic enzyme type C_4 plant. Furthermore, Neuburger (1980) and Wiskich and Day (1982) have presented evidence that the NAD^+-linked malic enzyme and TCA cycle dehydrogenases of plant mitochondria have equal access to internal respiratory-linked NADH dehydrogenases and to the alternative pathway (see also Moreau and Romani, 1982). This fits well with the generally held concept of a concerted action of malate dehydrogenase and malic enzyme to provide both pyruvate and oxaloacetate in the anaplerotic function of the TCA cycle. Under these conditions, the action of the two enzymes must be coordinated to balance these metabolites. Such coordination would be difficult if the enzymes were associated with different respiratory chain components oxidizing NADH at different rates (Wiskich and Day, 1982).

Mitochondrial malic enzyme is also important to the metabolism of plants having the C_4 dicarboxylic acid pathway of photosynthesis (C_4 plants) in which CO_2 is initially fixed in mesophyll cells by phosphoenolpyruvate carboxylase into four-carbon dicarboxylic acids. These C_4 acids are then transferred into the bundle sheath cells where they are decarboxylated. In one group of C_4 pathway species (*Atriplex spongiosa, Amaranthus edulis, Amaranthus retroflexus, Portulaca oleracea, Panicum miliaceum*) designated as NAD^+-malic enzyme type (Hatch *et al.*, 1975), C_4 acid decarboxylation in bundle sheath cells is catalyzed via a very active mitochondrially located NAD^+-malic enzyme (Hatch and Kagawa, 1974; Gutierrez *et al.*, 1974; Rathnam, 1978; Hatch *et al.*, 1982) (Fig. 3.17; Table 3.1). In fact, C_4 species are divisible into three clear groups based on the predominance of one of three different decarboxylating enzymes. For those in which NAD^+-malic enzyme predominate (*Atriplex spongiosa, Amaranthus edulis*), Hatch *et al.* (1982) recorded maximum activities in the range of 7–22 μmoles/min/mg chlorophyll for whole leaf extracts. The activities in C_3 plants are low (in the range of 0.3–0.7 μmoles/min/mg chlorophyll). The C_4 acid decarboxylation in the mitochondria of these species is believed to proceed by the sequence of reactions catalyzed by aspartate aminotransferase, NAD^+-malate dehydrogenase, and NAD^+-malic enzyme, respectively (Kagawa and Hatch, 1975; Fig. 3.17). The CO_2 produced is fixed in the Benson–Calvin cycle to make starch and sucrose. Thus in this C_4-type, the mitochondrial NAD^+-linked malic enzyme is directly involved in the carbon fixation cycle. A particular difficulty in the study of C_4 plant mitochondria is that two distinctly different cell types, mesophyll and bundle sheath cells, exist within the leaves. Therefore, mitochondria must be isolated from

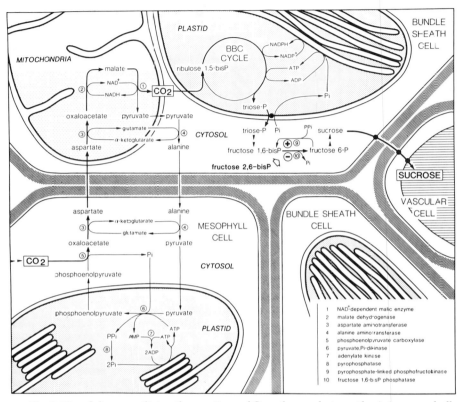

Fig. 3.17. Scheme outlining the reactions of C_4 pathway photosynthesis in mesophyll and bundle sheath cells and the intracellular location of these reactions in NAD-malic enzyme-type C_4 plants. In these plants aspartate is decarboxylated via a series of reactions (aminotransferase-mediated interconversions coupled by a glutamate–α-ketoglutarate cycle; conversion of oxaloacetate to malate and malate decarboxylation) located in bundle sheath cell mitochondria (Hatch and Osmond, 1976). Significantly, this group has a notably high frequency of mitochondria in bundle sheath cells (Hatch *et al.*, 1975). Sucrose is synthesized either in the bundle sheath cell (this figure) or in the mesophyl cell.

both of these cell types with sufficiently low cross contamination. Mitochondria from leaf tissue (mesophyll and bundle sheath fractions) of *Panicum miliaceum* have been successfully purified using Percoll density gradient centrifugation (Gardeström and Edwards, 1983b). Malate oxidation in mesophyll mitochondria is sensitive to cyanide and shows good respiratory control, as does mitochondria from the leaves of C_3 species and etiolated tissues. In bundle sheath mitochondria, however, malate oxidation is largely insensitive to cyanide and shows no respiratory control. On the other hand, this oxidation is strongly inhibited by

TABLE 3.1

A Survey of NAD^+-Linked Malic Enzyme Activity of Mitochondria from Leaves

Groups and species	Rate (μmoles pyruvate formed/ min/mg chlorophyll)
C_4–NAD-malic enzyme type species	
Atriplex spongiosa (BS)[a]	5–8
Panicum milaceum (BS)	4
Atriplex spongiosa (MC)[a]	0.7
Panicum miliaceum (MC)	0.7
C_4–NADP-malic enzyme type species	
Zea mays (BS)	0.03
Sorghum sudanense (BS)	0.01
C_4–PEP-carboxykinase type species	
Chloris gayana (BS)	0.2
Panicum maximum (BS)	0.2
C_3–Species	
Spinacia oleracea	0.03
Triticum aestivum	0.02

[a] Source of mitochondria: BS, bundle sheath strands; MC, mesophyll cells. [From Kagawa and Hatch (1975).]

salicyl hydroxamic acid, showing that the alternative oxidase is involved. Consequently, in bundle sheath mitochondria, malate oxidation, linked to an uncoupled alternate pathway, may allow decarboxylation to proceed without the restraints that might occur via coupled electron flow through the cytochrome chain. However, decarboxylation with aspartate as donor of carbon to the C_3 pathway via decarboxylation through NAD^+-malic enzyme need not be linked to the electron transport chain, since the pyridine nucleotide would be recycled through malate dehydrogenase and NAD^+-malic enzyme (Kagawa and Hatch, 1975; Fig. 3.17).

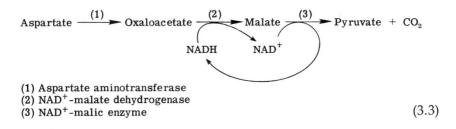

(1) Aspartate aminotransferase
(2) NAD^+-malate dehydrogenase
(3) NAD^+-malic enzyme

(3.3)

CAM plants (crassulacean acid metabolism) fix CO_2 at night into malic acid, which is stored in the vacuole. During the subsequent light period, malate is decarboxylated in the cytoplasm, and the CO_2 that is released is refixed by photosynthesis. Therefore, the biochemistry of CAM photosynthesis has many similarities with C_4 photosynthesis, the main difference being that, in CAM, the separation of the synthesis of C_4 acids from their decarboxylation is achieved by temporal rather than spatial separation. Decarboxylation is achieved either via phosphoenolpyruvate carboxykinase or malic enzyme depending on the particular plant (Osmond, 1976; Ting and Gibbs, 1982). A number of CAM plants (for example, members of the crassulaceae) have been shown to possess two malic enzymes; one NADP-linked and the other NAD^+-linked (Dittrich, 1976). The NADP-malic enzyme is localized in the cytoplasm (Spalding *et al.*, 1979). The NAD^+-linked malic enzyme, on the other hand, is a mitochondrial enzyme (Day, 1980; Arron *et al.*, 1979a,b). For instance, mitochondria isolated from *Kalanchoë daigremontiana* (Day, 1980) and *Saedum praealtum* (Arron *et al.*, 1979a,b) decarboxylate malate at significant rates compared to *in vivo* deacidification rates in the light (100 μmoles/hr/mg original chlorophyll). According to Arron *et al.* (1979a,b), pyruvate so formed is transported out of the organelles and used during a reversal of glycolysis in the cytoplasm and chloroplast rather than being metabolized to CO_2 and ATP within the mitochondria. It is clear, therefore, that mitochondria with a higher capacity to oxidize malate through malic enzyme need to have a similar capacity to transport malate and pyruvate, although the characteristics of the transport have not been studied. As pointed out by Day (1980) and Edwards *et al.* (1982), the extent to which pyruvate is exported from the mitochondria versus being metabolized through the TCA cycle *in vivo*, is unknown. The operation of the TCA cycle in light may be controlled by the demand for ATP in the cytoplasm during deacidification. In addition, Day (1980) believes that cytoplasmic pH may regulate malate decarboxylation by mitochondria. This may be important *in vivo* since the cytoplasmic malate concentration probably increases early in the light period and, consequently, cytoplasmic pH may fall. Both of these conditions, as shown earlier, favor mitochondrial malate decarboxylation. Under these conditions it may be advantageous to have malic enzyme specifically linked to nonphosphorylating respiratory pathways, as suggested by Rustin *et al.* (1980b). It should be noted, however, that in CAM leaf mitochondria the capacity of the alternative pathway is probably not great enough to support *in vivo* rates of malate decarboxylation, being no more than 20% of the capacity of the cytochrome oxidase pathway (Arron *et al.*, 1979a,b;

Day, 1980). It is also possible that the cytoplasmic NADP-linked malic enzyme may operate in conjunction with the NAD^+-linked malic enzyme in these tissues.

III. THE MECHANISM OF GLYCINE OXIDATION

A. Characterization

Benson and Calvin (1950) first reported that glycolate, glycine, and serine were formed in sequence during photosynthesis by higher plants. Experiments in which [^{14}C]glycolate was supplied to green leaves in the light demonstrated its conversion to glycine and serine (Tolbert and Cohan, 1953). It is now well established that glycolate is formed inside the chloroplasts via phosphoglycolate from ribulose-1,5-bisphosphate. Both CO_2 and O_2, the substrates for the ribulose-1,5-bisphosphate carboxylase–oxygenase behave as competitive inhibitors of the oxygenase and carboxylase reactions, respectively (for an excellent review see Lorimer and Andrews, 1981). Fixation of O_2 by ribulose-1,5-bisphosphate carboxylase is the primary event in the metabolic sequence, leading to uptake of O_2 and, eventually, the evolution of CO_2 (photorespiration). The glycolate so formed is then thought to leave the chloroplast and enter the peroxysome where, by the action of glycolate oxidase and a transaminase, glycine is formed (Kisaki and Tolbert, 1969). The glycine in turn migrates to the mitochondria where it is converted to serine and CO_2. Loss of the carboxyl group of glycine takes place in the synthesis of one molecule of serine from two molecules of glycine. Serine then returns to the peroxisomes, where via the action of a transaminase and hydroxypyruvate reductase, glycerate is formed. Finally, glycerate reenters the chloroplast where it is phosphorylated by glycerate kinase to give glycerate 3-P (C_2 cycle; Fig. 3.18). The conversion of glycine to serine in green leaf mitochondria is currently considered to be the major source of CO_2 released during photorespiration (Kisaki *et al.*, 1971). Enzymes capable of catalyzing each of these steps have been described and are present in sufficient activity to handle the measured flux of carbon through the cycle (Tolbert, 1980; Keys, 1980; Lorimer and Andrews, 1981).

The ability to oxidize glycine via the respiratory chain is specific for mitochondria from leaf tissues having the C_3 pathway of photosynthesis (C_3 plants; Fig. 3.19). Woo and Osmond (1977) reported the isolation of bundle sheath mitochondria from NAD^+-malic enzyme type C_4 plants which slowly decarboxylated glycine, although no oxidative data were given. On the other hand, Rathnam (1978) and Ohnishi and Kanai (1983) reported that mitochondria isolated from mesophyll protoplasts were incapable of oxidizing glycine while bundle sheath mitochondria slowly oxidized glycine. The capacity of isolated mitochondria from *Sedum praealtum* (Arron *et al.*, 1979a,b) and *Kalanchoë daigremontiana* (Day, 1980), two crassulacean metabolism plants, to respire glycine in the presence of ADP is low (much slower than that of other substrates), again in agreement with a lowered *in vivo* capacity of the glycolate pathway in these species. In fact, plants with C_4 and CAM photosynthesis are potentially able to suppress ribulose-1,5-bisphosphate oxygenation, and, therefore, photorespiration, by concentrating CO_2 at the site of ribulose-1,5-bisphosphate carboxylase–oxygenase (Chollet, 1974; Spalding *et al.*, 1979; Canvin, 1979).

No glycine oxidation activity is present in mitochondria from non-green tissues. Thus, in the case of spinach the activity is present in mitochondria from the green parts of the leaves but not in mitochondria from roots, stalks, or leaf veins (Gardeström *et al.*, 1980). Arron and Edwards (1980a,b) showed that in etiolated sunflower cotyledons, the development of mitochondrial glycine oxidation is light-induced in the same manner as the development of glycolate oxidase but is independent of chlorophyll synthesis. One can thus conclude that the capacity to very rapidly oxidize glycine seems to be specific to mitochondria from photosynthetically active tissues of C_3 plants.

B. Mechanism Involved

Earlier studies with mitochondrial fractions have shown that 2 moles of glycine are converted to 1 mole each of serine, NH_3, and CO_2 in a complex reaction catalyzed by glycine decarboxylase (also named glycine synthase, EC 2.1.2.10) and serine hydroxymethyltransferase (EC 2.1.2.1.) (Tolbert, 1971; Neuburger and Douce, 1977; Rathnam, 1978; Arron *et al.*, 1979b; see also the excellent and penetrating review of Keys, 1980). Kisaki *et al.* (1971) using spinach leaf mitochondria found that NAD^+ was necessary as an electron acceptor. Bird *et al.* (1972a) using *Nicotiana tabacum* leaf mitochondria reported that the reaction required aerobic conditions, was linked to the electron transport chain, and was coupled to the synthesis of two ATP molecules. Interestingly, mitochondria isolated from leaves oxidize glycine with a stoichiometry of CO_2 evolution to O_2 uptake of 2 : 1 (Bird *et al.*, 1972b; Arron *et al.*, 1979a,b). In addition, Bird *et al.* (1972a) demonstrated that ferricyanide could act as the final electron acceptor instead of O_2 and that serine synthesis was strongly inhibited by antimycin A whether O_2 or ferricyanide was the

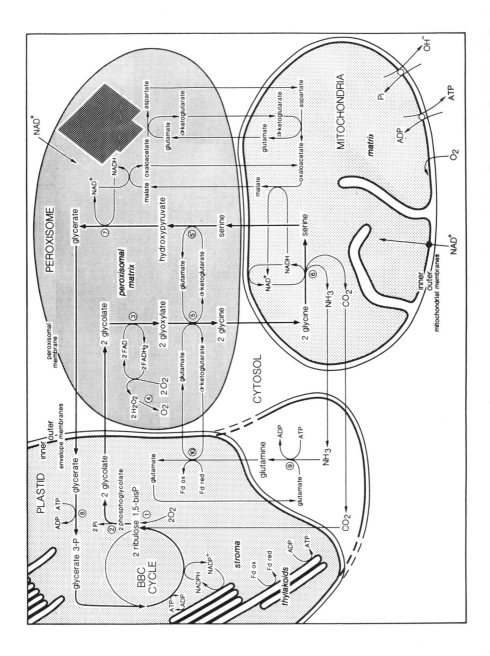

final electron acceptor. However, this coupling (i.e., ATP synthesis) was not observed in isolated pea leaf mitochondria (Clandinin and Cossins, 1972), probably because they were badly damaged. Douce *et al.* (1977) first reported a method to isolate functionally intact mitochondria from the C_3 plant spinach and demonstrated that glycine oxidation is in fact coupled to the synthesis of three ATP molecules (Fig. 3.19). Since then, several methods have been described to purify spinach leaf mitochondria, including phase partition (Bergman *et al.*, 1980), Percoll density gradient centrifugation (Jackson *et al.*, 1979a,b), and use of protoplasts (Nishimura *et al.*, 1982). All these authors confirmed that the glycine oxidation is coupled to three phosphorylation sites, indicating that NADH produced during the course of glycine oxidation is reoxidized by complex I (i.e., the rotenone-sensitive NADH dehydrogenase; Fig. 2.13). In fact, it can be calculated that the following reaction:

$$1/2 \; O_2 \; (g) + 2 \; glycine \rightarrow serine + CO_2 \; (g) + NH_3 \; (g) \qquad (3.4)$$

is accompanied by a free energy change of -42.2 kcal (Bird *et al.*, 1972b). This would be adequate for synthesis of three molecules of ATP for each molecule of serine formed. In this system, the role of the respiratory chain appears to be the regeneration of NAD^+ (in addition to synthesis of ATP). In support of this suggestion, Woo and Osmond (1976) and Journet *et al.* (1980), following a previous observation (Douce and Bonner, 1972), have shown that the glycine decarboxylating activity was observed in spinach leaf mitochondria supplemented with oxaloacetate. The addition of oxaloacetate in this system appears to ensure the oxidation of endogenous NADH, via mitochondrial malate dehydrogenase

---

Fig. 3.18.   The glycolate and glycerate pathways of photorespiration. In this scheme, reducing equivalents formed inside the mitochondria during the course of glycine oxidation are transported to the peroxisome for the reduction of hydroxypyruvate. This could be achieved by metabolic shuttles such as malate–oxaloacetate or malate–aspartate shuttles (Journet *et al.*, 1981). The reoxidation of NADH produced in the matrix upon operation of the glycine decarboxylase complex can be reoxidized equally well by the respiratory chain. The reassimilation of CO_2 and NH_3 formed during the course of glycine oxidation also is indicated (Keys *et al.*, 1978). The close association observed *in vivo* between mitochondria, chloroplasts, and peroxisomes could facilitate the direct transport of metabolites between cell organelles. *In vitro*, plant mitochondria prove to be biochemically very flexible organelles. Symbols and abbreviations: 1) ribulose-1,5-bisphosphate carboxylase–oxygenase; 2) P-glycolate phosphatase; 3) glycolate oxidase; 4) catalase; 5) 5'-aminotransferase; 6) glycine decarboxylase and serine hydroxymethyl transferase; 7) NADH-hydroxy pyruvate reductase; 8) glycerate kinase; 9) glutamine synthetase; 10) glutamate synthase; Fd ox, oxidized ferredoxin; and Fd red, reduced ferredoxin.

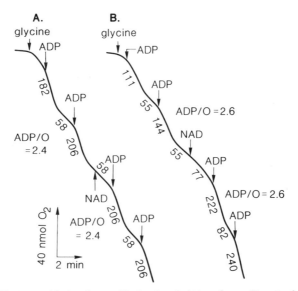

Fig. 3.19.   Glycine oxidation by purified spinach (A) and pea (B) mitochondria. Oxygen consumption was measured with 0.16 (A) and 0.18 (B) mg mitochondrial protein. Additions as indicated were 10 mM glycine, 140 μM ADP, and 200 μM NAD$^+$. Numbers on traces refer to nmoles O_2/min/mg protein. Note that exogenous NAD$^+$ stimulates the rate of glycine oxidation in pea leaf mitochondria but is without effect on those from spinach, because pea and spinach leaf mitochondria have different endogenous NAD$^+$ contents.

(see also A. L. Moore *et al.*, 1977; Day and Wiskich, 1981a; Fig. 3.20). The use of two separate carriers during the operation of the malate–oxaloacetate shuttle, one with a high affinity and specificity for oxaloacetate (Day and Wiskich, 1981b), would enable it to operate in the desired direction even when cytosolic malate concentrations are substantially higher than those of oxaloacetate (a situation likely to occur given the high activity of cytosolic malate dehydrogenase). Likewise, Journet *et al.* (1980) have demonstrated that during glycine oxidation by spinach leaf mitochondria, O_2 consumption showed a strong and transient inhibition upon addition of aspartate plus α-ketoglutarate (Fig. 2.26). During the course of the inhibition, aspartate and α-ketoglutarate were stoichiometrically transformed into malate and glutamate. They concluded, therefore, that oxaloacetate formed by transamination is reduced by the malate dehydrogenase which allows the regeneration of NAD$^+$ for glycine oxidation and, thus, bypasses the respiratory chain. The latter systems may involve either an oxaloacetate–malate shuttle or a malate–α-ket-

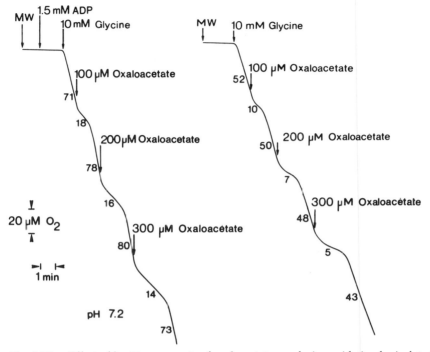

Fig. 3.20. Effect of limiting amounts of oxaloacetate on glycine oxidation by isolated mitochondria from spinach leaves. Numbers on traces refer to nmoles O_2 consumed/min/mg protein. [From Journet *et al.* (1980).]

oglutarate–aspartate–glutamate shuttle between mitochondria and peroxisomes *in vivo* (Woo and Osmond, 1976; Journet *et al.*, 1980; Fig. 3.18) and have the advantage of being unaffected by phosphate potential. As a matter of fact, a consideration to be taken into account regarding the possibility of *in vivo* glycine oxidation via the respiratory chain, is that this process necessarily leads to the production of ATP, which must be recycled back as ADP at a rate sufficient to account for the potential rate of glycine-dependent oxidative phosphorylation. This problem could, of course, be negated if glycine oxidation was linked to the non-phosphorylating alternate pathway. The results of Douce *et al.* (1977) and Dry (1984), however, indicate that for pea and spinach there would be insufficient alternate pathway activity to support *in vivo* rates of photorespiratory glycine oxidation.

Since turnover of organic acids via the TCA cycle continues in the light (Graham, 1980), it is important to determine to what extent glycine oxidation and the TCA cycle interact or interfere with each other. Day

TABLE 3.2

NH$_3$ Release during Glycine Oxidation by Pea Leaf
Mitochondria[a]

Additions	Oxygen consumption[b]	NH$_3$ release[b]
Experiment 1		
Glycine	40	78
Plus ADP, state 3	76	157
State 4	32	67
Experiment 2		
Glycine	39.0	65.5
Plus OAA	9.5	165.5
Experiment 3		
Glycine plus ADP	69	139
Plus malate	133	141
Plus succinate	144	139
Experiment 4		
Glycine	96	198
Plus α-ketoglutarate	105	224

[a] Oxygen consumption and NH$_3$ release were measured simultaneously. Concentrations of reagents were
10 mM glycine, 1 mM oxaloacetate (OAA), 0.3 mM ADP,
10 mM malate, and 10 mM succinate. [From Day and
Wiskich, (1981b).]
[b] Units: nmoles/min/mg protein.

and Wiskich (1981b), Bergman (1983), and Dry *et al.* (1983a) have shown
that pea and spinach leaf mitochondria preferentially oxidize glycine
when confronted with a mixture of glycine and TCA cycle substrates
(Table 3.2). Even under state 4 (ADP-limited) conditions, when competition among substrates for electron transport could be expected to be
most severe, glycine oxidation was not diminished. It therefore appears
that the electron transport chain of pea or spinach leaf mitochondria has
an absolute preference for the NADH generated from glycine oxidation.
The mechanism by which glycine is preferentially oxidized via the respiratory chain is yet to be elucidated but relates to a stronger interaction
between complex I and complex III than that between complex II (and
the external NADH dehydrogenase) and complex III and to the ability of
glycine decarboxylase to compete effectively at the level of matrix
NAD$^+$. This preferred oxidation of glycine may represent an important
mechanism for the regulation of mitochondrial functions in the light.

Glycine decarboxylase represents a large proportion of the matrix protein in green leaf mitochondria (Fig. 3.21), however, and this might easily explain the fact that glycine oxidation has preferred access to the mitochondrial electron transport chain. In support of this suggestion, no similar system of preferential glycine oxidation is present in animal mitochondria. In this case, the glycine decarboxylase complex is present in a small amount (Hampson *et al.*, 1983). In addition, we have found that glycine decarboxylase, in contrast with a previous suggestion (Sarojini and Oliver, 1983), is not membrane bound since it is easily removed after gentle sonication of the mitochondria (Fig. 3.21).

The NH_3 (g) and CO_2 (g) produced in the mitochondria during the course of glycine oxidation readily diffuse out of the mitochondria. The CO_2 released is refixed by ribulose-1,5-bisphosphate carboxylase in the chloroplast (Schaefer *et al.*, 1980). Indeed, the C_2 carbon oxidation cycle is absolutely dependent on the C_3 carbon reduction cycle, because the substrate for the C_2 cycle, ribulose-1,5-bisphosphate, is supplied directly from the C_3 carbon reduction cycle (Fig. 3.18). On the other hand, the rate of photorespiratory NH_3 released is an order of magnitude larger than the net rate of inorganic nitrogen assimilation by leaves (1–5 μmoles/hr/mg chlorophyll). Consequently, in steady-state operation, this NH_3 must be refixed or else the plant would not survive the rapid deficiency of organic nitrogen during photorespiration (Keys *et al.*, 1978). The ammonia produced by the action of glycine decarboxylase was assumed to be assimilated into glutamate by mitochondrial glutamate dehydrogenase. As shown by Keys *et al.* (1978), however, this scheme provides an unsatisfactory explanation, because addition of α-ketoglutarate to spinach leaf mitochondria oxidizing glycine has no effect on the rate of NH_3 release. In fact, it is well known that glutamate dehydrogenase exhibits a low affinity for NH_3. The presence of glutamine synthetase and glutamate synthase which catalyze equations (3.5) and (3.6), respectively,

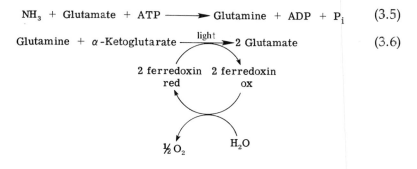

$$NH_3 + Glutamate + ATP \longrightarrow Glutamine + ADP + P_i \qquad (3.5)$$

$$Glutamine + \alpha\text{-Ketoglutarate} \xrightarrow{\text{light}} 2\ Glutamate \qquad (3.6)$$

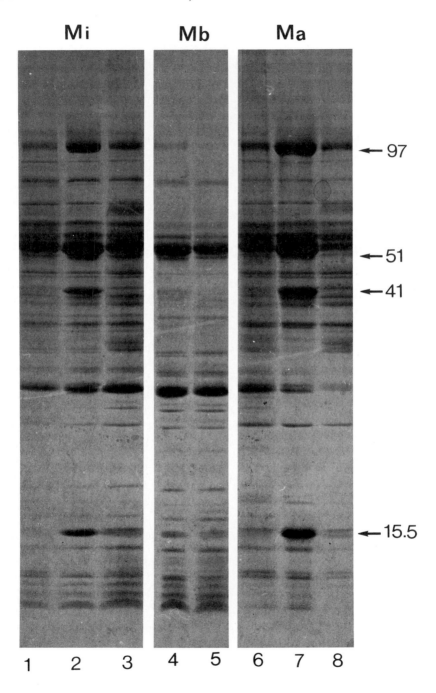

affords a better explanation for ammonia assimilation, because the glutamine synthetase has a high affinity for ammonia with a K_m of 19 μM (O'Neal and Joy, 1974; Fig. 3.18). In addition, using [^{15}N]glycine Keys *et al.* (1978) demonstrated that the rate of $^{15}NH_3$ release to the atmosphere by wheat leaves could be enhanced fivefold by application of methionine sulfoximine, a specific inhibitor of glutamine synthetase. Keys *et al.* (1978) have, therefore, proposed a photorespiratory nitrogen cycle involving glutamine synthetase and glutamate synthase (probably localized in the chloroplasts). In support of this suggestion, J. W. Anderson and Done (1977) have demonstrated that isolated intact pea chloroplasts catalyzed a very rapid (ammonia plus α-ketoglutarate)-dependent O_2 evolution. The reaction had a high affinity for ammonia (a characteristic of glutamine synthetase) and was sensitive to azaserine and methionine sulfoximine (inhibitors of glutamate synthase and glutamine synthetase, respectively). The ATP required for equation (3.5) was assumed to be supplied by photophosphorylation (see also Dry and Wiskich, 1983; J. W. Anderson and Walker, 1983). Jackson *et al.* (1979b) have suggested that ammonia can also be refixed in the mitochondria by a hypothetical mitochondrial glutamine synthetase. Several recent reports, however, have clearly demonstrated that glutamine synthetase is not present in the mitochondria and is, therefore, an extramitochondrial enzyme (Wallsgrove *et al.*, 1979, 1980; Nishimura *et al.*, 1982). In spinach, the cytosolic glutamine synthetase seems to be completely missing, the chloroplastic enzyme being the single detectable isoform (Hirel *et al.*, 1982). Glutamine, on the other hand, is only very slowly taken up by chloroplasts (Barber and Thurman, 1978). On the basis of this and the cellular distribution of glutamine synthetase, it appears most likely that the NH_3 released in the mitochondria during photorespiration is reassimilated in the chloroplasts by the glutamine synthetase–glutamate

---

Fig. 3.21. Electrophoretic analysis of mitochondria (Mi), membrane (Mb), and matrix (Ma) polypeptides from purified pea mitochondria by LDS–PAGE at 4°C with a 9–15% acrylamide gradient. Protein load in each slot was equivalent to 100 μg of protein (slots 1–8). Lanes 1, 2, and 3, total polypeptides from etiolated leaf, green leaf, and hypocotyl mitochondria, respectively; lanes 4 and 5, total polypeptides from the membranes of green leaf and etiolated leaf mitochondria, respectively; and lanes 6, 7, and 8, total polypeptides from the matrix of etiolated leaf, green leaf, and hypocotyl mitochondria, respectively. The positions of marked molecular weight polypeptides specific to green leaf mitochondria are indicated on the right. Note that the membrane polypeptide patterns are remarkably similar (slots 4, 5) but substantial differences are observed in the various matrix fractions. The matrix of green leaf mitochondria is distinguished by the presence of four pronounced bands at 97,000, 51,000, 41,000, and 15,500 which are absent in the other tissues (slots 6, 7, 8).

synthase pathway (Bergman et al., 1981). Close interaction between chloroplasts and mitochondria must, therefore, also be postulated to avoid loss of NH_3 (Fig. 1.5).

Mitochondria isolated from C_3 plant leaves should possess a glycine transporter that is capable of moving glycine from the cytosol into the matrix, the site of glycine decarboxylase. According to Cavalieri and Huang (1980), Day and Wiskich (1980), and Proudlove and Moore (1982), influx of glycine is not related to the energy status of the mitochondria and appears diffusional. Day and Wiskich (1980) have suggested that glycine permeates isolated pea leaf mitochondria as the neutral amino acid. The ability of amino acids to form rings with intramolecular hydrogen bonds may permit them to rapidly penetrate the inner membrane. On the other hand, G. H. Walker et al. (1982) gave some evidence to suggest that glycine entry into spinach leaf mitochondria is mediated by a specific carrier. Using spinach leaf mitochondria isolated in a Percoll density gradient and taking into account the water space and sucrose space of the organelles, Yu et al. (1983) demonstrated that the transport of glycine, serine, and perhaps proline, is predominantly by diffusion at amino acid concentrations above 0.5 mM and is carrier mediated at concentrations of amino acids below 0.5 mM. The carrier-mediated system appears to be an active uptake system, since accumulation of amino acids in the matrix against a concentration gradient occurs, since the uptake is partly dependent on the supply of external energy, and since the uptake is inhibited by electron transport inhibitors and protonophores. Since glycine and serine are all substrates of the photorespiratory pathway, an investigation of any interrelationship between their mechanisms of transport across the mitochondrial inner membrane might yield insight into the regulation of photorespiration.

Glycine decarboxylase catalyzes the oxidative decarboxylation of glycine, producing CO_2, NH_3, methylene tetrahydrofolate, and the reduction of NAD^+ (Fig. 3.22). The enzyme also catalyzes an exchange between the carboxyl carbon of glycine and bicarbonate (Clandinin and Cossins, 1975) and is able to reduce (possibly indirectly) dichlorophenol indophenol (A. L. Moore et al., 1980). Glycine hydroxamate (Lawyer and Zelitch, 1979) and aminoacetonitrile (Usuda et al., 1980; Oliver, 1981) competitively inhibit the partial reaction of glycine–bicarbonate exchange (Gardeström, 1981; Gardeström et al., 1981). The plant glycine decarboxylase is located behind the mitochondrial membrane (Woo and Osmond, 1977; A. L. Moore et al., 1977). Both the decarboxylase and bicarbonate exchange activities are lost when the inner membrane is ruptured (Woo and Osmond, 1976). The loss of activity can depend on several factors. One or several components, like proteins or cofactors, may leak out from the mitochondria (Gardeström, 1981). It is also possible that the very fragile complex involved in glycine cleavage is partially disrupted during the course of inner mitochondrial membrane rupture. This has hampered the study of the plant enzyme and very little is known about the composition, precise intramitochondrial location, and properties of the enzyme. However, glycine decarboxylase has been successfully solubilized from pea leaf mitochondria as an acetone powder. The presence of an essential lipoamide in the plant complex can be inferred by its sensitivity to arsenite (Sarojini and Oliver, 1983). In contrast, glycine synthase (glycine decarboxylase) from animals and bacteria has been studied in detail (Fig. 3.22). It consists of four components that have tentatively been named P-protein (a pyridoxal phosphate containing protein), H-protein (a lipoic acid-containing protein), T-protein (a protein that catalyzes the tetrahydrofolate-dependent step of the reaction), and L-protein (a lipoamide dehydrogenase). The P-protein shows an absorption spectrum characteristic of the pyridoxal phosphate enzyme. It gives two distinct absorption maxima at 335 and 428 nm (Hiraga and Kikuchi, 1980a,b). The glycine cleavage system in animals is also

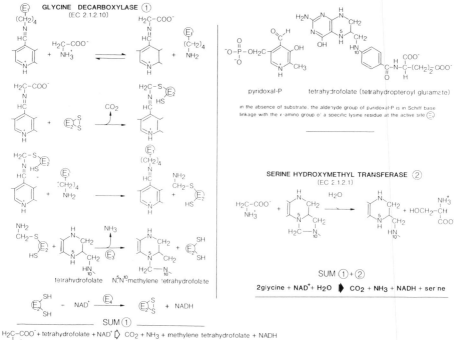

Fig. 3.22. Reaction sequence in glycine oxidation.

confined to mitochondria, possibly as an enzyme complex that is loosely bound to the mitochondrial inner membrane (Hiraga *et al.*, 1972). Consequently, the general features of the glycine cleavage reaction have a considerable similarity to those of the pyruvate and α-ketoglutarate dehydrogenase complexes. In these reactions, the disulfide group of the protein-bound lipoic acid plays a key role as both electron acceptor and acyl carrier (Figs. 3.4, 3.22). However, the keto acid dehydrogenases (decarboxylases) actively catalyze decarboxylation of respective keto acids without the participation of the lipoic acid containing acyltransferases. In contrast, glycine decarboxylase hardly exhibits significant enzymatic activity unless it is coupled with H-protein (Hiraga and Kikuchi, 1980a). Interestingly, leaf mitochondria differ significantly in polypeptide composition compared to mitochondria from nonphotosynthetic tissues like petiole and root. These differences are more pronounced in the matrix fractions. The matrix of leaf mitochondria contains at least four polypeptides (15,900; 41,700; and 101,000) which are absent, or present in only small amounts, in the matrix of petiole or root mitochondria (Ericson *et al.*, 1983; D. Day, R. Douce, and M. Neuburger, unpublished data; Fig. 3.21). One of the polypeptides (15,900 protein) which binds [^{14}C]glycine could be involved in the exchange of $^{14}CO_2$ into the carboxylic moiety of glycine (Ericson *et al.*, 1983).

Serine hydroxymethyltransferase catalyzes the reversible interconversion of serine and glycine as follows (Fig. 3.22):

$$\text{glycine} + \text{methylene tetrahydrofolate} \rightleftarrows \text{serine} + \text{tetrahydrofolate} \qquad (3.7)$$

The mitochondrial serine hydroxymethyltransferase is yellow, exhibiting an absorption maximum at 428 nm due to the pyridoxal P bound as a Schiff base. However, the spectral properties of the enzyme–glycine complex show absorption maxima at 343, 425, and 495 nm (Schirch and Peterson, 1980) and increasing the pH causes a rather large increase in the amount of complex absorbing at 498 nm. Recently, Gardeström (1981), utilizing mitochondria from spinach leaves, and Rustin and Alin (1983), with mitochondria from *Kalanchöe blossfeldiana* (CAM-type plant) leaves, have clearly shown that the glycine spectrum (glycine anaerobic minus oxidized difference spectra) compared to the spectra of the other substrates (NADH, succinate, and malate) is characterized by two additional peaks at 493 and 503 nm. According to Gardeström (1981), the peaks at 493 and 503 nm could represent different enzyme-pyridoxal P–glycine complexes in the glycine–serine interconversion. Potassium cyanide competitively inhibited serine hydroxymethyltransferase at a low concentration (Gardeström *et al.*, 1981) and completely reduced the absorption peak at 493 nm. Consequently, according to Gardeström

(1981), the peak at 493 nm is probably associated with serine hydroxymethyltransferase. In contrast to glycine decarboxylase, serine hydroxymethyltransferase is present in both photosynthetic and non-photosynthetic cells (Cossins, 1980). According to Woo (1979), however, this enzyme is almost exclusively a mitochondrial enzyme in leaves of C_3 plants. The activity of serine hydroxymethyltransferase was observed to be cocompartmented with glycine decarboxylation and malate dehydrogenase behind the mitochondrial inner membrane. The integration of this enzyme system *in vivo* would presumably favor the transfer of carbon (C_1 unit) from glycine to serine during photorespiration. In addition, in green leaves the rapid reoxidation of NADH and the immediate utilization of NH_3 and CO_2 during the course of glycine oxidation continuously shifts the equilibrium toward serine formation, even though the reactions are readily reversible *in vitro*.

Finally, if the glycolate pathway is strictly cyclic (C_2 cycle) as suggested by Lorimer and Andrews (1981), then all methylene tetrahydrofolate formed upon glycine cleavage must be used for serine synthesis. In other words, methylene tetrahydrofolate arising from glycine during operation of the glycolate pathway does not gain access to the general C_1 pool. It has been clearly shown, however, that the methylene tetrahydrofolate formed could be important in the C_1 metabolism of photosynthesizing and etiolated cells (Cossins, 1980), depending on the cell's requirement for C_1 units. In growing plant cells for instance, glycine cleavage in the mitochondria could serve as an important source of methylene tetrahydrofolate for biological synthesis leading to purines, thymidilate, methionine, and perhaps the formylation of methionyl tRNA (Cossins, 1980). In support of this suggestion, Clandinin and Cossins (1972) found that [2-^{14}C]glycine contributed carbon to the pools of formyl and methylene tetrahydrofolate in mitochondria of actively germinating seeds. In this case, however, the rate of glycine decarboxylation is much slower than that observed in mitochondria from C_3 plant leaves. Photorespiratory serine has been reported to be an intermediate in sucrose synthesis in leaves (Waidyanatha *et al.*, 1975). It has also been suggested that serine and glycine could participate in protein synthesis (Tolbert, 1980). There is also evidence for the conversion of photorespiratory serine to malate via phosphoenolpyruvate in *Vicia faba* leaves (Kent, 1979). Much additional work is needed to establish whether photorespiratory glycine, serine, and methylene tetrahydrofolate participate in other metabolic reactions to a significant degree.

4

Mitochondrial Oxidative Activities in Relation to in Vivo Metabolism

The mechanism of regulation of cellular respiration with its accompanying phosphorylation reactions is of particular physiological importance. Respiration is the most important process for the production of cytoplasmic ATP and at the same time provides the major sink for phosphate, ADP, and NADH, the metabolites that are also reactants in most of the other metabolic pathways of the cell. The rates of mitochondrial oxidative phosphorylation must be rigorously coordinated to meet the ATP demands of the cytoplasm, and how this is achieved *in vivo* is a crucial issue in the field of cellular bioenergetics. Understanding the interrelationships of mitochondrial and cytosolic metabolism has proven extraordinarily controversial. This is partially attributable to the spatial separation of the two compartments (mitochondrial matrix space and cytoplasm) which necessitates the operation of many vectorial translocation systems across the inner mitochondrial membrane. The variety of possible mechanisms, coupled with an inability to account quantitatively for all of the species (especially in plant cytoplasm) crossing the membrane in a given experiment, has led to many different views of the intra- and extramitochondrial metabolic relationships. This is particularly true concerning the role of the adenine nucleotide translocase in cellular energy metabolism.

The rate at which a cell respires can theoretically be limited by any of a number of factors, depending on the condition assigned: (a) availability of substrate, (b) availability of inorganic phosphate or of ADP for oxidative phosphorylation, (c) supply of O_2 to the respiratory chain, and (d) the overall capacity of the respiratory chains in a cell for the sequence of redox reactions leading to the reduction of O_2 to water. It is also possible that the rate of respiration may be controlled by more than one reaction (see Fig. 2.16).

The flow of substrates across the mitochondrial membrane during respiration is highly controlled, and probably none of the metabolically important substrates except CO_2 and O_2 are able to escape the stringent control exerted by the various specific carriers. One must distinguish, however, when considering rate control by carriers, between rate limitation by the capacity of the carrier and by the availability of the substrate, such as phosphate, nucleotides, and organic acids (Stubbs *et al.*, 1978). Thus, the capacity of the carrier or the rate of supply of the substrate can limit the rate of translocation.

It is now widely accepted that O_2 can move through the gas spaces of plant tissues (for review see Vartapetian *et al.*, 1978) and that the rate of O_2 uptake by a tissue is determined by diffusion and utilization. In cases in which diffusion can be eliminated, the rate of respiration is independent of the O_2 concentration, except at extremely low concentrations (Davies, 1980). Thus at low O_2 concentrations, it has been shown that the value of the ATP/ADP ratio or of the energy charge in lettuce seeds was controlled by the O_2 partial pressure (Raymond and Pradet, 1980; Saglio *et al.*, 1983). There are no data to indicate that molecular oxygen is a positive modulator of the enzymes involved in the early stages of glucose oxidation. Thus, oxygen must limit the terminal respiratory electron acceptors, namely cytochrome oxidase or the alternate oxidase in tissue where this enzyme is present and functioning. It is also possible that O_2 may limit "oxidases" with relatively low affinities for oxygen. [Plant tissues can survive beautifully for a long period of time at very low O_2 concentrations (for review see Morisset *et al.*, 1982), indicating that plant tissues may possess an as yet unknown mechanism to compensate for a partial restriction of the cytochrome oxidase.]

I.   MITOCHONDRIAL RESPIRATORY CONTROL

It is generally accepted that *in vivo* the cell is usually not respiring at maximum capacity. Thus, addition of an uncoupler to the medium which induces a passive permeability in the mitochondrial membrane to protons, increases O_2 uptake rates by cells or tissues (Plantefol, 1922, 1932; H. Beevers, 1961; Fig. 4.1). However, this stimulation by un-couplers is not always evident in plant tissues that are limited by sugars (Saglio and Pradet, 1980; Azcón-Bieto *et al.*, 1983a,b). Chance and Williams (1955), in their classical studies, paid particular attention to the "resting state" (state 4), in which lack of ADP limits respiration, and to the "active state" (state 3), in which ADP is present in excess. It is likely that cells *in vivo* will generally be respiring in states between 4 and 3, as defined by Chance and Williams (1955) (Azcón-Bieto *et al.*, 1983b).

Although most investigators now agree that there is thermodynamic control of mitochondrial respiration in the resting state and that an obvious candidate for control of respiration in state 4 is the passive permeability of the mitochondrial membrane to protons, opinions differ with regard to the control of respiration in the intermediate states and in the active state. Chance and Williams (1956) initially proposed that rates

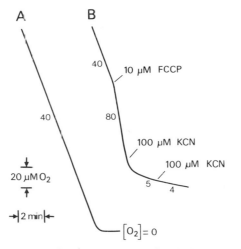

Fig. 4.1.   Oxygen consumption by sycamore cells. Cells harvested from the culture medium (23 mg wet weight/ml) were directly introduced into the 1-ml electrode chamber. Numbers along the traces represent the O_2 uptake rates (nmoles O_2/hr/mg wet weight). Measurement temperature: 25°C. The concentrations given are the final concentrations in the reaction medium. Abbreviations: FCCP, carbonyl cyanide *p*-trifluoromethoxyphe-nylhydrazone.

of respiration graded between state 3 and state 4 were a function of ADP availability. Klingenberg (1961) subsequently varied not only the ADP concentration but also that of ATP and phosphate. He concluded that the extramitochondrial phosphorylation potential ($[ATP][ADP])/[P_i]$ (i.e., that the mass action ratio of the reaction $ADP + P_i \rightleftarrows ATP$), was the parameter that determined the immediate rates of O_2 consumption. This hypothesis has received much experimental and theoretical support, mainly based on thermodynamic considerations of ATP synthesis (for a review see Tager *et al.*, 1983). Thus, Krebs and co-workers (Stubbs *et al.*, 1972) and D. F. Wilson *et al.* (1983) have postulated that, during phosphorylation-limited respiration, a near equilibrium relationship is maintained between the steady states of reduction of the redox components of the phosphorylation reaction, as specified in the following equation:

$$NADH + 2 \text{ cytochrome } c^{3+} + 2 ADP + 2P_i \rightleftarrows NAD^+ + 2 \text{ cytochrome } c^{2+} + 2 ATP$$
$$(4.1)$$

In other words, under physiological conditions the coupled reactions of ATP synthesis and electron transport are at near equilibrium for the first two "sites of energy transduction." The equilibrium constant K_{eq} for this reaction is defined as:

$$K_{eq} = [NAD^+]/[NADH] \times [c^{2+}]^2/[c^{3+}]^2 \times [ATP]^2/[ADP]^2[P_i]^2 \qquad (4.2)$$

for a given pH, in which the first two terms express the oxidation–reduction state of the mitochondrial pyridine nucleotide and cytochrome c couples, respectively, and the third term represents the extramitochondrial phosphorylation–state ratio. [Forman and Wilson (1982) showed that in rat liver mitochondria reduction of molecular oxygen by cytochrome c oxidase is the only irreversible step in oxidative phosphorylation and, hence, may be the primary site of respiratory control.] In fact, reversibility of the mitochondrial respiratory chain from cytochrome c to NAD^+ and *vice versa* was first observed in the early 1960s (Klingenberg and Shollmeyer, 1960), and it was later demonstrated that the net electron transfer from cytochrome c to NAD^+ could be driven by ATP hydrolysis. Consequently, as pointed out by Forman and Wilson (1982), if a reaction sequence is near equilibrium, all intermediate steps must similarly approach equilibrium. This suggests that none of the partial reactions of oxidative phosphorylation, such as adenine nucleotide translocation, are markedly displaced from equilibrium. Stubbs *et al.* (1978) have shown on theoretical grounds that because near-equilibrium exists between the electron transport chain and the cytosolic phosphorylation potential, ADP translocation, which is an intermediate step, cannot be rate limiting. Stubbs *et al.* (1978) also noted that since the cytosolic ADP content of rat liver is 300–600 μM (115–250 μM in intact liver cells; Akerboom *et al.*, 1978) and the K_m of the translocator for ADP is 1–10 μM (Vignais, 1976), it seems likely, therefore, that the adenine nucleotide translocator has the capacity to account for known rates of oxidative phosphorylation at the low levels of cytosolic free-ADP postulated here.

On the other hand, Slater *et al.* (1973) postulated that respiratory control was simply a function of the ATP/ADP ratio and somewhat independent of P_i. This theory has likewise achieved considerable attention (for a review see Tager *et al.*, 1983) and is founded on kinetic considerations of the adenine nucleotide translocase. These authors explained the effect of the external ATP/ADP ratio as a kinetic limitation of the overall process of oxidative phosphorylation by adenine nucleotide translocation, suggesting that the translocator is out of equilibrium. Indeed, by prefering uptake of ADP and release of ATP, the energization of the membrane modulates the carrier so that the ADP/ATP ratio was found to be considerably higher in the cytoplasmic than in the mitochondrial compartment. Consequently, at a high extramitochondrial ATP/ADP ratio, ADP translocation by isolated mitochondria is probably inhibited by ATP. Vignais and Lauquin (1979), however, have pointed out that despite these findings a role for the nucleotide translocator in *in vivo* regulation of oxidative phosphorylation is unlikely, since the ATP/ADP ratios measured in animal tissues are well below those needed to cause significant inhibitions of mitochondrial respiration *in vitro*.

Erecinska *et al.* (1977) experimentally perturbed the phosphate concentration of intact liver and yeast cells and from this study they speculated that the rate of respiration changes in parallel with changes in [ATP]/([ADP][Pi]) but not in parallel with changes in the ATP/ADP ratio. These considerations strongly suggest that the nucleotide translocator does not limit the rates of oxidative phosphorylation under physiological conditions. However, this work has been criticized because the phosphate concentration was varied by incubation with fructose. This treatment resulted in a four- to sixfold decrease in phosphate concentration, down to less than 1 mM. Davis and Davis-Van Thienen (1978) suggested that this concentration of phosphate is the same as or lower than K_m for phosphate translocation. Therefore, limitation of respiration by nonphysiologically low concentrations of phosphate may obscure the regulatory role of the adenine nucleotide translocator.

Recently, Jacobus *et al.* (1982) have confirmed the hypothesis first proposed by Chance and Williams more than 25 years ago and cast considerable doubt on both of the hypotheses discussed earlier for the mechanism of respiratory control i.e., either thermodynamic control by the external phosphorylation potential or kinetic control by the extra-mitochondrial ATP/ADP ratio. Instead, they strongly suggested that the most plausible explanation of respiratory control *in vivo* is actually the availability of ADP and the kinetics of its transport by the adenine nucleotide translocase. Their results convincingly showed that, under steady state conditions with isolated mitochondria, respiration is directly controlled by the concentrations of extramitochondrial ADP, with little or no positive correlation with either the extramitochondrial phosphorylation potential or the extramitochondrial ATP/ADP ratio. In addition, their data showed that the $K_{i_{ATP}}$ for respiration is not in the 100–200 μM range, as suggested before, but is actually closer to 30 mM. This very low ability of ATP to inhibit state 3 respiration is additional evidence why theories of respiratory control formulated on a simple ATP/ADP ratio are invalid. Jacobus *et al* (1982) also pointed out that recent nuclear magnetic resonance (NMR) data with ^{31}P from the *in vivo* rat brain suggest that the concentration of free ADP in tissues is much lower than that estimated from the values of ADP measured previously in cellular acid extracts (Ackerman *et al.*, 1980). These new values fall in control ranges for which only very low rates of mitochondrial respiration are measured.

Finally, Tager and co-workers (Tager *et al.*, 1983; Groen *et al.*, 1982) have recently applied the control theory of Heinrich and Rapoport (1974) to the inhibition of mitochondrial respiration by various inhibitors of electron and energy transfer. Based on this analysis, they have suggested that control of respiration is distributed among several of the partial reactions of oxidative phosphorylation. In the case of adenine nucleotide translocation, the glucose–hexokinase system was used to vary the respiratory rate of suspensions of liver mitochondria and carboxyatractyloside was used as an inhibitor of the translocase. Analysis of their data indicated that, at low-levels of hexokinase, the translocase did not contribute to control of the respiratory rate, while at higher levels of hexokinase (approaching maximal respiratory rates), it contributed up to 30% of the total control, with remaining control distributed among other partial reactions. In other words, in any sequence of reactions (in which every step depends on the substrate supplied by the preceding step), inhibition of any single step is liable to cause an inhibition of the overall process. Indeed, the extramitochondrial phosphate concentration and the supply of hydrogen, which provide the other substrates for oxidative phosphorylation, could also play significant roles in controlling respiration and must be considered in addition to the ADP concentration.

Very few studies have been made of the mechanism of respiratory control in plant mitochondria. Stitt *et al.* (1982) demonstrated that when wheat leaf protoplasts were kept in the dark, a high cytosolic ATP/ADP quotient was found, while the mitochondrial ATP/ADP quotient was about 20 times lower. Inhibition of electron transport, or uncoupling, apparently lowered the cytosolic ATP/ADP quotient. This shows that *in situ* plant mitochondria resemble animal mitochondria in that the membrane potential generated by electron transport not only drives ATP synthesis within the mitochondrial matrix but also its active transport into the extramitochondrial space i.e., the cytoplasm. A. L. Moore (1978) examined the effects of phosphorylation potential on succinate respiration in mung bean hypocotyl mitochondria by recording the respiratory rates before and after the addition of ADP at varying ATP concentrations and at constant phosphate concentrations. A high phosphorylation potential was found to be related to low rates of respiration. A. L. Moore (1978) concluded, therefore, that mitochondrial respiration is under thermodynamic control by the phosphorylation potential.*

Recently, Rebeillé *et al.* (1983; 1984), by using ^{31}P-NMR to discriminate the cytoplasmic from the vacuolar phosphate pool, have determined the phosphate concentration in the cytoplasmic compartment of *Acer pseudoplatanus* cells grown as cell suspensions. The value found was within the range 6–12 mM. Consequently, the cytosolic concentration of phosphate is higher than the K_m value of the mitochondrial phosphate translocator, indicating that *in vivo* this carrier is amply supplied. With the aim of determining the effect of endogenous phosphate concentration on the rate of O_2 consumption, experiments were done in order to decrease the endogenous phosphate pool. When phosphate was omitted from the nutrient solution, the endogenous phosphate pool decreased to 8% of the control value within only 72 hr. During the same time, the normal and uncoupled rates of O_2 consumption remained practically constant throughout the experiment. These results demonstrate that, at least in the case of isolated sycamore cells, the endogenous phosphate concentration does not limit the rate of respiration.

Dry and Wiskich (1982) studied the effect of changes in the external ATP/ADP ratio on the rates of respiration of malate plus glutamate in isolated pea leaf and cauliflower mitochondria. Various respiratory rates

---

*In fact, as pointed out judiciously by Pradet and Raymond (1983), plant physiologists will be concerned with energy charge (Bomsel and Pradet, 1968) and phosphorylation potential not because of their regulatory properties but rather because they can be used as indicators of cellular metabolic status. For example, in tissues having a high fermentative capacity, such as maize root tips, the energy charge–value remained high under anoxia (Saglio *et al.*, 1980). In the presence of 2 mM NaF (a potent inhibitor of glycolysis), on the contrary, the energy charge–value decreased sharply (Saglio *et al.*, 1983).

were generated using a creatine–creatine kinase system in the presence of varying levels of ATP. In this way, in contrast to the experiments carried out by A. L. Moore (1978), mitochondrial metabolism could be examined in the presence of different steady-state levels of ADP and ATP. ATP was found to have an inhibitory effect on the rate of O_2 uptake, but inhibition was both small (only noticeable at ATP/ADP ratios greater than 20) and influenced by the ADP concentration. Thus, the inhibition of O_2 uptake by ATP diminishes as the level of ADP increases. In other words, control of respiration in isolated plant mito- chondria occurred only with a very high ratio of ATP/ADP, exceeding about 20. In addition, the sensitivity of the control was found to be markedly influenced by the total concentration of ADP present, indicat- ing that respiration is regulated not so much by the external ATP/ADP ratio but more by the absolute concentration of ADP available for uptake into the mitochondria, in agreement with the work of Jacobus *et al.* (1982) in animals. They also demonstrated that decreasing the con- centrations of Mg^{2+} and phosphate by 10-fold had little effect on the respiratory response of the mitochondrion to changes in the ATP/ADP ratio. The observation concerning the absence of an Mg^{2+} effect is sur- prising, however, because only free-ATP or -ADP are translocated, the Mg^{2+} complexes being inactive. In addition, most of the ATP present in the cytoplasm is probably bound to Mg^{2+} (Siess *et al.*, 1982; Rebeillé *et al.*, 1985) and, therefore, must not affect the functioning of the ADP carrier.

The determination of free and complexed forms of adenine nu- cleotides in the cytosol of plant cells is a prerequisite to understanding the control *in vivo* of the rate of respiration. Unfortunately, particular care must be taken in adapting techniques shown to be successful with animal and microbial cells to plants. In comparison with animal and microbial cells, plant cells contain 2–10% of the amounts of nucleotides involved in general metabolism (Bieleski, 1964). Furthermore, the tech- niques used for killing and extracting plant tissues may cause changes in the amounts of nucleotides present. The use of hot acids or alkalis will hydrolyze some phosphate esters. Probably the most dangerous source of error is the ability of plant phosphatases to resist many of the com- monly used killing agents for long enough to cause significant alteration in the amounts of substrates recovered in the extracts. For example, adenylate energy charge–values of about 0.6 are obtained when maize roots are killed with boiling buffers (Lin and Hanson, 1974) but values close to 0.9 are obtained when the roots are frozen with liquid nitrogen in the presence of ether and then extracted in trichloroacetic acid (Saglio and Pradet, 1980). Likewise, the concentration of bound- and free-Mg^{2+}

in the cytoplasm of the plant cells is entirely unknown, making the calculation of cytosolic complexed forms of adenine nucleotides impossible.

II. CONTROL OF CARBOHYDRATE OXIDATION IN PLANT CELLS

Despite much recent progress in elucidating the chemical pathways involved in the respiration of higher plants, relatively little information is available concerning the control of carbohydrate oxidation in plant cells. A plant grows under rapidly changing environmental conditions and the rate of carbohydrate oxidation varies correspondingly. It is probable that the cells of most higher plants receive, via the phloem, the bulk of their organic carbon as sucrose synthesized in the cytoplasm of green cells. Indeed, photosynthesis in C_3 and CAM plant chloroplasts and in bundle sheath cell chloroplasts of C_4 plants involves conversion of CO_2, H_2O, and phosphate into triose phosphate (Edwards and Walker, 1983). Most of the triose phosphate molecules thus formed are exported from the chloroplasts via the phosphate translocator in exchange for phosphate and are used to synthesize sucrose in the cytosol (Bird *et al.*, 1974). During sucrose synthesis, the phosphate is released and can be cycled back into the chloroplasts. Some fixed carbon can also be temporarily retained in the chloroplasts as starch (Edwards and Walker, 1983; Preiss, 1984). Central to an understanding of plant respiration is an understanding of the mechanism by which sucrose is transferred from the mesophyll cells to the companion cells of the phloem. Two distinct pathways have been proposed for this transfer, symplastic and apoplastic transport. In symplastic transport, photosynthetate remains in the symplasm and passes through the plasmodesmata into the phloem (Cataldo, 1979). Although this type of transport is important in several tissues, it is not the commonly accepted route for photosynthetate transfer between mesophyll and phloem (Giaquinta, 1976). Instead, transfer of photosynthetate occurs across the mesophyll plasma membrane (J. M. Anderson, 1983), with subsequent uptake by the phloem companion cells (Giaquinta, 1976). Phloem loading of sucrose has been shown to be an energy-dependent process that probably involves the co-transport of protons and sucrose into the phloem (Delrot and Bonnemain, 1981). Hence, the driving force for sucrose uptake was proposed to be a proton gradient established across the plasmalemma by a vectorial ATPase (Giaquinta, 1979). Consequently, during the day if nothing inhibits or slows down the rate of photosynthesis, all of the plant cells are flooded with sucrose. Storage products in most of the plant cells are sucrose and perhaps malate in the vacuole and starch in the plastids, although sugar alcohols, fructans, and lipids are also used for respiratory purposes (ap Rees, 1977). We do not know how the nonphotosynthetic cells of higher plants convert translocated sucrose into starch (Preiss, 1982). The process is distinguished from starch synthesis in photosynthesizing cells by its dependence upon the transport of organic carbon into the amyloplasts (Fig. 4.2). It is possible that sucrose is converted to triose phosphate for entry into the amyloplast. Since amyloplasts lack any enzyme known to metabolize sucrose (sucrose synthetase, UDP-glucose pyrophosphorylase, or invertase), sucrose is not likely to be the compound that crosses the amyloplastic inner membrane (McDonald, and ap Rees, 1983). Conversion of sucrose to starch may involve hydrolysis of sucrose by the alkaline invertase, followed by glycolysis to triose phosphate in the cytosol, entry into the amyloplast, and synthesis therein of starch from triose phosphate (Fig. 4.3).

The respiration rate of many plant tissues falls off when they are deprived of a carbon source for a long period of time, and the reduced respiration rates resulting from such starvation can often be elevated by supplying the appropriate sugars (W. O. James, 1953). Recently Saglio and Pradet (1980) have clearly shown that O_2 uptake declines immediately after excision of maize root tips and that the addition of exogenous sugars induces a rapid rise in the respiratory rate while the energy charge remains constant. These results indicate that metabolic activity of root tips is highly reliant on sugar import and that carbohydrate reserves at the time of excision cannot compensate for the cessation of import. Likewise, exogenous sugars stimulate CO_2 efflux and/or O_2 consumption in the dark in the leaves of *Rumex acetosa* (Goldthwaite, 1974), *Avena sativa* (Tetley and Thimann, 1974), and wheat (Azcón-Bieto *et al.*, 1983a,b). Furthermore, Rebeillé *et al.* (1985) have demonstratd that the rate of respiration falls off rapidly in sycamore cells when they are deprived of a carbon source. Several lines of evidence suggest that there is a direct relationship between the rate of respiration and the carbohydrate status of plant tissues. For example, dark CO_2 efflux of mature wheat leaves (Azcón-Bieto and Osmond, 1983) and tomato leaves (Ludwig *et al.*, 1975) increases considerably after a long period of photosynthesis. At least in these cases, two groups of substrates contribute to the CO_2 efflux. Because 15–20% of the CO_2 evolved in the first 30 min of darkness is abolished if leaves are kept in low O_2 during the latter part of

Fig. 4.2. Electron micrograph of Percoll-purified cauliflower bud amyloplasts (×64,000). Note that big starch grains of variable sizes occur in the dense stroma. (Courtesy of Michèle Montlahuc.)

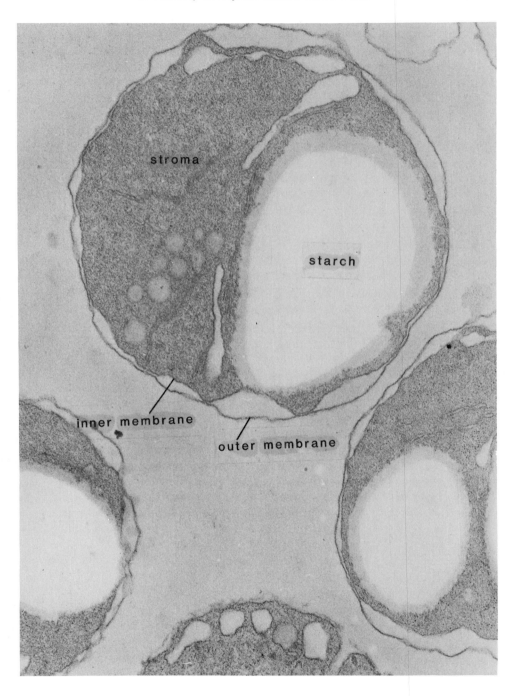

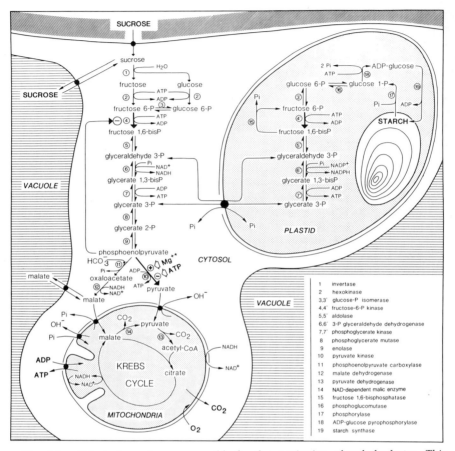

Fig. 4.3. Principal reactions responsible for the respiration of carbohydrates. This scheme indicates that the major storage products in most of the plant cells are sucrose and perhaps malate in the vacuole and starch in the plastids. Conversion of starch to pyruvate may involve phosphorolysis of starch by the phosphorylase (17), followed by glycolysis to triose phosphate in the plastids, entry into the cytoplasm via the phosphate translocator, and synthesis therein of pyruvate from triose phosphate. Starch synthesis and breakdown are probably under the control of phosphate and triose phosphate concentrations in the cytoplasm and plastid stroma (Preiss, 1984). The estimates of the maximum catalytic activities and of the free energy changes pinpoint the steps catalyzed by pyruvate kinase (10) and fructose-6-P kinase (4,4′) as nonequilibrium reactions and thus potential points of regulation. Finally, this scheme indicates that phosphoenolpyruvate could be converted to oxaloacetate and then to malate, because both phosphoenolpyruvate carboxylase (11) and malate dehydrogenase (12) activities are high in the cytoplasm. Malate thus formed is either stored in the vacuole or directly utilized by the mitochondrion for respiratory purposes.

the photosynthetic period, Azcón-Bieto and Osmond (1983) concluded that this CO_2 arose from accumulated photorespiratory substrates such as glycine. The remaining CO_2 efflux (i.e., after the first 30 min in the dark) is closely correlated with several carbohydrate fractions. This CO_2 presumably arose from TCA cycle and pentose phosphate pathway oxidation of carbohydrate-derived substrates. According to Azcón-Bieto and Osmond (1983), the linear relationship between the prior rate of photosynthesis and the subsequent rate of dark CO_2 efflux can be explained in terms of quantitative changes in cellular carbohydrates, common metabolites to both processes. Interestingly, the dark rate of O_2 uptake in bean leaves and leaf slices decreases during development. The rate of overall respiration, the activities of the cytochrome and alternative pathways, and the extent to which uncouplers stimulate respiration in bean leaf slices are positively correlated with endogenous free sugar levels during aging (Azcón-Bieto *et al.*, 1983a). The enhancement of leaf respiration by carbohydrates could not be related primarily to growth requirements, since mature leaves were used; this is in contrast with fast growing cells, such as sycamore cells. Alternatively, excess respiration may be used for synthesis of compounds (e.g., amino acids) in the leaf which can be utilized for growth in other parts of the plant and for providing energy for transport of assimilates (Ho and Thornley, 1978).

Carbohydrate oxidation usually occurs in the cytoplasm but plastids also contain starch, one of the respiratory substrates of plant tissues (Fig. 4.2). In fact, plastids make a major contribution to the total carbohydrate oxidation occurring in plant cells deprived of exogenous carbohydrate (Rebeillé *et al.*, 1985). Exogenous carbohydrate respiration begins with their conversion to glucose-6-P (the hexokinase step is a nonequilibrium reaction).* Glucose-6-P is then oxidized via glycolysis (Fig. 4.3) and the oxidative pathway, and the products are then metabolized via the TCA cycle (Fig. 4.3). The surplus of carbohydrates and part of the TCA cycle intermediates are utilized for biosynthetic purposes.

A. Respiration and Glycolysis

1. Glycolysis in the Cytosol

There is a good evidence to support the view that glycolysis is universal in higher plants (H. Beevers, 1961; Turner and Turner, 1980; ap Rees,

---

*Sucrose can be metabolized to hexose phosphates via either invertase and hexokinase or sucrose synthetase and UDP-glucose pyrophosphorylase since both these enzymes are readily reversible and since the cytosolic pyrophosphate content is high enough to allow UDP-glucose pyrophosporylase to convert UDP-glucose to glucose-phosphate *in vivo* (Edwards *et al.*, 1984).

1980). The basic pathway is similar to that described for animals, yeast, and bacteria. Investigations into the distribution of hexokinase activity in plant cells have concentrated on nonphotosynthetic tissues and show the activity to be distributed between both the soluble and particulate fractions (Marré et al., 1968). Some evidence indicates that hexokinase may have a similar distribution in phothosynthetic tissues (Baldus et al., 1981) and that the particulate activity may be associated with the outer mitochondrial membrane (Dry et al., 1983b; Tanner et al., 1983). One possible explanation for the mitochondrial location of this enzyme may be that it serves to maintain a close relationship with the ATP-generating system of oxidative phosphorylation. This would ensure that the hexokinase enzyme could respond rapidly to changes in the cellular demand for glucose-6-P, which is known to be a key intermediate in a number of different metabolic pathways including glycolysis, sucrose synthesis, and the pentosephosphate pathway. The reactions catalyzed by phosphoglucomutase, glucose phosphate isomerase, triose phosphate isomerase, NAD^+-specific glyceraldehyde 3-P dehydrogenase phosphoglyceromutase, and enolase are close to equilibrium. On the other hand, pyruvate kinase and phosphofructokinase reactions are far from thermodynamic equilibrium, as determined by estimations of substrates and products, providing valuable evidence for regulatory reactions (Fig. 4.3). Theoretically, the rate of glycolysis should be regulated to strictly match the cell's energy demand. In fact, there is strong evidence that the activities of phosphofructokinase and pyruvate kinase in vivo regulate carbon flow through glycolysis. Pyruvate kinase is at a major metabolic intersection and the reaction it catalyzes is far from equilibrium. The K_m values for phosphoenolpyruvate have been estimated at 20 μM and 17–55 μM for ADP (Turner and Turner, 1980). The enzyme that is Mg^{2+}-dependent is inhibited by ATP. Consequently, the overall concentration of cytoplasmic ATP, when very high, should inhibit pyruvate kinase activity. Variations in the levels of ATP will be reinforced in vivo by changes in the opposite direction of the substrate ADP. Phosphofructokinase, which utilizes MgATP as a substrate (Dennis and Coultade, 1967) also catalyzes a nonequilibrium reaction. The enzyme is strongly inhibited by phosphoenolpyruvate. For example, with the pea seed enzyme, total inhibition may be obtained with 4 μM phosphoenolpyruvate (Turner and Turner, 1980). In addition, low concentrations of free-ATP considerably enhance the inhibition given by phosphoenolpyruvate (Kelly and Turner, 1969). On the other hand, inorganic phosphate (P_i; 5mM) completely reversed the inhibition of phosphofructokinase by phosphoenolpryrunate (Dennis and Miernyk, 1982). Consequently, the overall concentration of cytoplasmic free-ATP and phosphoenolpyruvate will influence phosphofructokinase activity. It is clear, therefore, that glycolysis is directly or indirectly under the control of cytoplasmic ADP and ATP concentrations. Since both enzymes are stimulated by Mg^{2+}, it is possible that changes in the concentrations of cytoplasmic metallic ions, such as Mg^{2+}, may also play a part in the regulation of glycolysis. The same thing probably holds true for the cytoplasmic level of K^+ (Turner and Turner, 1980). Finally, a competition between oxidative phosphorylation and glycolysis for ADP and phosphate probably occurs *in vivo* (Racker, 1974).

Sites responsible for glycolytic control *in vivo* can be pinpointed by comparing intracellular concentrations of glycolytic intermediates under circumstances in which alterations in adenine nucleotide concentrations are brought about owing to anoxia or uncoupling. Facilitation of substrate flux, through a rate-limiting step in response to altered adenine nucleotide concentrations, will presumably result in a decrease in the intracellular concentration of the substrate of this reaction, accompanied by a relative increase in its product. Such a reaction is said to constitute a "crossover point" (Williamson, 1966). Barker *et al.* (1967) followed the changes in ADP, ATP, and a series of glycolytic intermediates in pea seeds transferred from air to nitrogen and *vice versa.* The rate of utilization of carbohydrates by peas in nitrogen was higher than that observed in air. When the seeds were transferred from air to nitrogen there was a rapid increase in ADP and a fall in ATP which was accompanied by a large decrease in phosphoenolpyruvate and by an increase in fructose-1,6-P_2. Barker *et al.* (1967) concluded that pyruvate kinase, through the cytoplasmic adenine nucleotide concentration, was of primary importance in controlling the rate of glycolysis. In addition, this work emphasized the probable significance of the change in the level of phosphoenolpyruvate on the fructose-6-P kinase reaction, because the change in phosophoenolpyruvate tended to precede the change in fructose-1,6-P_2. The data of Givan (1968) on the response of intact cells of *Acer pseudoplatanus* to anoxia also indicate that phosphofructokinase of higher plants is subject to metabolic control *in vivo*. The short term changes in hexose monophosphate and diphosphate levels occurring in response to anoxia are readily explained on the basis of a relief of ATP-inhibition by phosphofructokinase. The rise in fructose-1,6-P_2 observed by Barker *et al.* (1967) and Givan (1968) could be attributable to inhibition of the triose phosphate dehydrogenase, owing to an accumulation of NADH under anaerobic conditions. This interpretation would not be compatible with the data of Lynen (1963), however, who found that in yeast the diphosphate esters accumulated when aerobic glycolysis was accelerated in the presence of dinitrophenol, which did not interfere with NADH oxidation via the respiratory chain. Kobr and Beevers (1971) and Faiz-ur-Rahman (1974) have also confirmed that the fructose-6-P–fructose-1,6-P_2 and phosphoenolpyruvate–pyruvate reactions were the regulatory steps in castor bean endosperm and carrot disks. In view of the above findings, it seemed likely that the rate of glycolysis might be an important factor controlling the rate of O_2 uptake by plant cells. Likewise, it seemed probable that any respiratory response to uncouplers might be due in part to the acceleration of glycolytic flux that would supply the mitochondria with additional substrate (Givan and Torrey, 1968).

The role of ADP in controlling glycolysis has also been examined in a soluble extract of germinating pea seeds (Givan, 1974). A shortage of ADP appears to retard glycolysis, principally by restricting the conversion of phosphoenolpyruvate to pyruvate. Accumulation of phosphoenolpyruvate during a period of ADP shortage could produce a feed back inhibition of glycolysis at the phosphofructose kinase step. Upon addition of ADP to the extract, there is an immediate decrease in the concentration of phosphoenolpyruvate accompanied by an increase in pyruvate. Apparently, the pyruvate kinase step shows the most marked response to fluctuations in ADP availabiltiy. Furthermore, the postulated role of phosphate in regulating glycolysis should not be neglected in attempting to explain the overall factors responsible for *in vivo* glycolytic regulation. For example, phosphate stimulated phosphofructokinase in *Pisum sativum* seed extracts but had no significant effect on pyruvate kinase (Radzali and Givan, 1981; see also Lagunas and Gancedo, 1983). In fact, only phosphate changed in a consistent way, increasing in anaerobiosis when the Pasteur effect occurred (Martin *et al.*, 1982; Fig. 4.4). Finally, ap Rees *et al.* (1976) have shown that during the development of the spadix of *Arum maculatum* there is a large rise in the phosphofructokinase activity. This rise corresponds to a coarse control of the glycolytic sequence and favors the flooding of mitochondria with respiratory substrate during the respiratory crisis.

One characteristic of glycolysis in higher plants is the great flexibility of its control. Day and Lambers (1983) working with different root species having different alternative pathway activities concluded that the phosphofructokinase and pyruvate kinase reactions were displaced far from equilibrium, the degree of displacement being approximately equal in roots with little and roots with substantial alternative pathway engagement. In other words, glycolysis is controlled via the same key enzymes regardless of whether the alternative pathway is engaged or not. The data of Lambers *et al.* (1983) and Day and Lambers (1983) are an example of the flexibility of the regulation of respiration in plant root

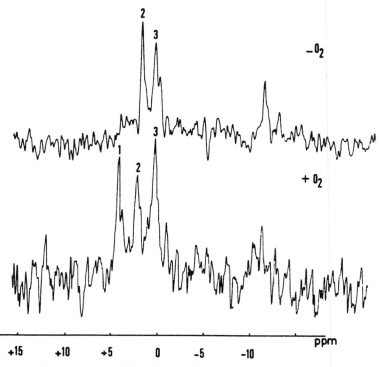

Fig. 4.4. Influence of air bubbling through the cell suspension on the relative amounts of glucose-6-P and P_i of *Acer pseudoplatanus* cells. Cells were placed in the nuclear magnetic resonance (NMR) tube and different ^{31}P NMR spectra recorded. Spectra were obtained with 127 scans ($-O_2$) or 396 scans ($+O_2$) at 1.36 sec/scan. Oxygen was injected ($+O_2$) by bubbling air through the cell suspensions with a peristaltic pump (20 ml/min). Peaks 1, 2, and 3 were assigned to glucose-6-P and cytoplasmic and vacuolar P_i, respectively. The main differences observed between the aerobic and anaerobic cells were the strong increase of peaks in the sugar-phosphate region (peak 1) and the decrease of the cytoplasmic P_i peak (Pasteur effect). [From Martin *et al.* (1982).]

systems. For example, in bean roots, the cytochrome pathway appeared to be unrestricted and glycolysis was regulated so that substrate supply to the mitochondria matched the capacity of the cytochrome chain; hence, the alternative pathway was not expressed. In wheat roots, glycolytic flux was sufficient to match the capacity of both the alternative and cytochrome pathways. In spinach and maize, however, the cytochrome pathway was controlled so that flux through it was less than that through glycolysis, with reducing equivalents therefore "spilling over" into the alternative pathway. This may reflect the involvement of the alternative pathway in an "energy overflow mechanism" (in this case the TCA cycle will function anaplerotically). Consequently, according to Day and Lambers (1983), the rate of glycolysis in plants is not always matched to the energy demands of the cell, and adenylate control of the cytochrome pathway can be tighter than that of glycolysis, allowing the alternative pathway to be expressed. Similar conclusions have been drawn concerning the regulation of leaf respiration (Azcón-Bieto et al., 1983b). Such flexibility might be achieved in parts, bypassing pyruvate kinase that is inhibited, as shown earlier, by ATP; phosphoenolpyruvate could be converted to oxaloacetate and then to malate, because both phosphoenolpyruvate carboxylase and malate dehydrogenase activities are high in the cytoplasms of all plant cells studied so far (ap Rees, 1980). In support of this suggestion, Popp et al. (1982) using wheat and lupin roots have shown that in short term labeling experiments with $H^{14}CO_3^-$, malate is predominantly labeled in the (C-4) carboxyl group, as would be expected from the operation of phosphoenolpyruvate carboxylase and/or phosphoenolpyruvate carboxykinase. In fact, there is appreciable evidence for the conversion of glycolytic phosphoenolpyruvate to C_4 acids in a range of tissues (ap Rees, 1980). The malate could then enter the mitochondrion via the dicarboxylate translocator, to be decarboxylated by NAD-linked malic enzyme. In this way, pyruvate would be generated within the mitochondrion to supplement that imported via the pyruvate carrier (Day and Hanson, 1977b). Such a situation could also prevent depletion of carbon in the mitochondrion so that the TCA cycle will function anaplerotically i.e., for synthetic purposes. Quite rapid release of ^{14}C, fixed from $^{14}CO_2$ in the dark, has been shown in the roots of Opuntia (Ting and Dugger, 1966), maize (Ting and Dugger, 1967), Plantago major, and Plantago lanceolata (Bryce, 1983). Furthermore, treatment of excised clubs of the spadix of Arum maculatum with 2-N-butylmalonate (a specific inhibitor of the dicarboxylate carrier) largely prevented the development of the rapid respiration responsible for thermogenesis and severely inhibited the dark fixation of $^{14}CO_2$ (ap Rees et al., 1983).

Recently, a pyrophosphate-linked phosphofructokinase (PP_i-PFK) has been shown in leaf cells (Carnal and Black, 1979, 1983; Cséke et al., 1982; Stitt et al., 1982b). This enzyme, which is activated by fructose-2,6-P_2 (Sabularse and Anderson, 1981), catalyzes the following reversible reaction: fructose-6-P + pyrophosphate $\leftrightarrows$ fructose-1,6-P_2 + phosphate (ΔG ° = −2.93 kJ/mole). The mechanism of activation of PP_i-PFK by fructose-2,6-P_2 generally involves both an increase in the affinity of the enzyme for fructose-6-P and an increase in the maximal velocity of the reaction (Sabularse and Anderson, 1981; Cséke et al., 1982; Van Schaftingen et al., 1982). Since PP_i-PFK is reversible, it can theoretically catalyze both glycolytic and gluconeogenic carbon flow. In fact, PP_i-PFK is present in the cytosol at all stages, in all tissues that have been investigated and in each the activity was 2 to 10 times greater than that of phosphofructokinase. Accordingly, it would seem that pyrophosphate-linked phosphofructokinase plays at least as important a role in cytoplasmic glycolysis of the leaf and perhaps of the nongreen cells (Van Schaftingen *et al.*, 1982; Kruger *et al.*, 1983) as does phosphofructokinase, provided that sufficient concentrations of pyrophosphate are present. Reports by Smyth and Black (1984) and Edwards *et al.* (1984) suggest that in certain nonphotosynthetic tissues this may be the case. The participation of PP_i-PFK in plant glycolysis has been demonstrated using extracts from pea seeds (Smyth *et al.*, 1984a). Under these conditions, since PP_i-PFK is freely reversible, the formation of fructose-1,6-P_2 from fructose-6-P, which is a control point in the regulation of sugar breakdown, is bypassed. Future experiments are necessary to determine if this is indeed the case, and if so, the source of the pyrophosphate needed by the enzyme must be identified (Czéke *et al.*, 1982). Furthermore the exact quantity of carbon flowing through either pathway remains to be quantitatively established in plants. Wu *et al.* (1983) have partially purified PP_i-PFK from germinating pea seeds. The enzyme exists in two forms (small and large), with apparently different molecular weights. In addition, these authors suggest that the rate of glycolytic flux in plants can be controlled by the concentration of fructose-2,6-P_2, through its regulating role on the dissociation–association of PP_i-PFK (fructose-2,6-P_2 is involved in converting a small form of PP_i-PFK into a large form). Finally, the experimental observations described by Wu *et al.* (1983) also suggest that the small form of pea cotyledon PP_i-PFK is probably the inactive species for the forward reaction.* These findings have demonstrated that fructose-2,6-P_2 functions in the plant cell cytosol to regulate carbon flow through activation of a glycolytic enzyme (PP_i-PFK) and inhibition of a glucogenic enzyme (fructose 1,6-bisphosphatase) (Czéke *et al.*, 1982; Stitt *et al.*, 1982b). It also appears possible that alterations in fructose-2,6-P_2 play a role in the partitioning

*Balogh *et al.* (1984) discovered the occurrence of a metabolite-mediated interconversion between PP_i-PFK and phosphofructokinase (PFK) activities of spinach leaf cytosol. Pyrophosphate and UDP-glucose promoted the conversion of cytosolic PFK activity to PP_i-PFK. Conversely, fructose-2,6-P_2 plus fructose-6-P promoted the conversion to PP_i-PFK activity to cystosolic PFK. The observed shifts in substrate specificity of the preparation were accompanied by changes in charge and in molecular weight. It is possible, therefore, that these enzymes may be visualized as regulatory antennae that serve to detect metabolite shifts and adjust carbon flow (gluconeogenic or glycolytic flux) accordingly (Cséke *et al.*, 1985). It remains to be seen how this PP_i-PFK/PFK interconversion is related to the interconversion of two different forms of PP_i-PFK described by Wu *et al.* (1983).

of carbon between sucrose and starch in green leaves (Stitt et al., 1983; Preiss, 1984; Stitt et al., 1984a,b) and in the synthesis of sucrose in castor bean endosperm (Kruger et al., 1983). Furthermore, Smyth et al. (1984b) anticipated that the amount of fructose-2,6-P_2 phosphatase (Cséke et al., 1983) or kinase (Czéke and Buchanan, 1983) activity may prove to be a key factor in hexose phosphate interconvertions inasmuch as PP_i-PFK can be activated 5- to 15-fold by micromolar fructose-2,6-P_2.

2. Glycolysis in Plastids

A significant proportion of the photosynthetate can be temporarily retained in the plastids as starch (Fig. 4.2). This compound might be seen as a buffer to sucrose metabolism. Starch accumulates in the light and is broken down in the subsequent dark period; in fact, net starch accumulation in the light represents a balance between synthesis and breakdown (Stitt and Heldt, 1981a; Kruger et al., 1983). In addition, Preiss (1982) considered the possibility that starch degradation is not regulated to any extent in leaves and that the starch content is determined solely by the regulation of synthesis, with the rate of degradation not varying much between night and day. The questions arise, therefore, as to how the starch is degraded [there is still no information on the relative contributions of amylases (α-amylase, β-amylase) and α-glucan phosphorylase to starch breakdown], in what form the products (triose phosphate, glucose) are exported from the plastids, and to what extent the starch is used for respiratory purposes. Since α-amylase is thought to be the initial enzyme which is active on starch granule (Preiss, 1982), it is very likely that amylolysis and phosphorolysis act in cooperation and not so much in competition.

The key enzymes of glycolysis, pyruvate kinase (Bird et al., 1973) and phosphofructokinase (Kelly and Latzko, 1975), have been detected in chloroplasts. Ammonium sulfate fractionation of an extract from the leaves of spinach, wheat, pea, and maize produced two isoenzymes of phosphofructokinase activity. Only the second isoenzyme was obtained from similar treatment of an extract of isolated chloroplasts (Kelly and Latzko, 1977). Plastidial and cytosolic isoenzymes of pyruvate kinase and phosphofructokinase have also been shown to be present in etiolated pea plants, castor bean endosperm (Dennis and Miernyk, 1982), and cauliflower florets (Journet and Douce, 1984). It seems likely, therefore, that the key enzymes of glycolysis are present in all of the plastids (Fig. 4.3; Table 4.1). The kinetic and regulatory properties of many of the isoenzymes have been characterized (Dennis and Miernyk, 1982), but few of them have been purified to homogeneity or examined structurally (see, however, Pichersky and Gottlieb, 1984; Miernyk and Dennis, 1984;

TABLE 4.1

Distribution of Enzymes of Carbohydrate Metabolism in Plastids and the Soluble Phase
of Developing Castor Oil Seeds[a]

Enzyme	Activity (μmoles/min/g fresh wt.)	Distribution (percentage of total activity)	
		Cytosol	Plastids
Acetyl-CoA carboxylase	0.78	12	88
Ribulose-1,5-P_2 carboxylase	3.70	52	48
Invertase	0.10	99	1
Sucrose synthase	0.01	79	21
UDPG-synthase	0.18	88	12
Hexokinase	0.34	69	21
Hexose-P isomerase	3.10	77	23
Phosphofructokinase	1.80	60	40
Aldolase	1.10	82	18
Glyceraldehyde-3-P dehydrogenase	7.31	86	14
Triose-P isomerase	23.35	91	9
3-P-Glycerate kinase	13.43	92	8
Phosphoglycerate mutase	4.75	80	20
Enolase	8.37	67	33
Pyruvate kinase	4.10	61	39
Glucose-6-P dehydrogenase	0.58	98	2
6-P-Gluconate dehydrogenase	1.80	31	69
Transketolase	0.38	44	56
Transaldolase	0.11	62	38

[a] [From Dennis and Miernyk (1982) and reproduced, with permission, from the *Annu. Rev. Plant Physiol.* **33,** 27–50. Copyright © 1982 by Annual Reviews Inc.]

Kurzok and Feierabend, 1984). Insofar as their genetics have been studied, each isoenzyme is coded by an independent nuclear gene (Gottlieb, 1982). However, Stitt and ap Rees (1979) have demonstrated that pea shoot chloroplasts can catalyze glycolysis only as far as glycerate-3-P. Further metabolism of the later via glycolysis would have to involve export (via the phosphate translocator) to the cytoplasm, because all of the data they obtained for phosphoglyceromutase indicate that this enzyme is absent from pea shoot chloroplasts. The observation that the phosphate translocator is also present in chromoplasts of the daffodil (Liedvogel and Kleinig, 1980) and amyloplasts of the cauliflower (Journet and Douce, 1984) is of central significance to the regulation of carbohydrate metabolism in nonphotosynthetic tissues. The results of Stitt and ap Rees (1979) also suggest that at least some of the glycerate-2-P formed from glycerate-3-P could be returned to the chloroplast for metabolism to pyruvate and fatty acids. Although the phosphate translocator transports it much less effectively than it does glycerate-3-P, the rate may still be enough to meet the demands *in vivo*. In support of this suggestion, Levi and Gibbs (1976) and Heldt *et al.* (1977) observed formation of glycerate-3-P from starch in intact chloroplasts held in darkness in the presence of P_i. Starch breakdown with rates above 10 µatoms carbon/mg chlorophyll/hr has been monitored in starch-loaded spinach chloroplasts and compares favorably with the rates in whole leaves (Stitt and Heldt, 1981b). Phosphate has a large influence on the starch metabolism of chloroplasts in the dark. When phosphate is omitted, accumulation of phosphorylated anions is prevented and 80% of the starch degradation leads to free sugar. More of the starch will now be degraded hydrolytically and exported as free sugars to the external medium. The response of chloroplastic starch mobilization to changing P_i concentrations in the medium occurs gradually over a range varying from 50 µM to 1 mM (Stitt and Heldt, 1981b). Since starch phosphorolysis in the dark is promoted by P_i and is inhibited by glycerate-3-P, it is clear that starch mobilization will be modified by the presence of glycerate-3-P and triose-P as well as P_i in the cytosol (Stitt and Heldt, 1981b). It is reasonable to suggest therefore, that as extrachloroplastic demand for respiratory substrates increased (during the night, for example) the triose-P and glycerate-3-P levels in the cytoplasm will fall and the P_i level will rise. This should increase the net flux of P_i into the chloroplasts and that of triose-P and glycerate-3-P out of the chloroplast and, hence, increase the net flux starch via glycolysis through to exported phosphorylated products. The same thing probably holds true in the case of nongreen cells, because Rebeillé *et al.* (1984b) working with sycamore cells, have shown that when cells are transferred to a saccharose-deficient medium, the cytoplasmic phosphorylated compounds decrease whereas cytoplasmic P_i increases symmetrically. Such a situation triggers starch breakdown.* Apparently, in these cells during saccharose deficiency, there is a switch from cytoplasmic glycolysis to plastidial glycolysis, providing mainly substrates for cell-energy metabolism (Fig. 4.3). Under these conditions, starch is rapidly broken down by the glycolytic pathway, the principal products being dihydroxyacetone phosphate and glycerate-3-P which are transported from the plastid by the phosphate translocator. The ATP required for the phosphofructokinase reaction may be provided by sub-

---

*The vacuole compartment is able to "buffer" the fluctuations of P_i concentrations occurring in the cytoplasm and therefore can exert a powerful control on cell metabolism (Foyer *et al.*, 1982; Rebeille *et al.*, 1983; Waterton *et al.*, 1983). Such a situation should prevent rapid phosphorolysis of starch, which is promoted by P_i (Preiss, 1984) insofar as phosphorylase contributes substantially to starch breakdown in plant cells.

strate level phosphorylation, by glyceraldehyde-3-P dehydrogenase, and by glycerate-3-P kinase (Fig. 4.3).

Bennoun (1982) demonstrated the existence of an electron transfer chain oxidizing NADPH at the expense of O_2 ("chlororespiration") in the thylakoid membranes of chloroplasts. This "chloroplast respiratory chain" shares at least one redox carrier, plastoquinone, with the photosynthetic electron chain. Likewise, Kow *et al.* (1982) have demonstrated that spinach chloroplasts have the ability to transfer electrons from NADPH (produced during the oxidation of glyceraldehyde-3-P to glycerate-3-P) to O_2 via ferredoxin and have termed this process "chloroplast respiration." It is possible, therefore, that these mechanisms ensure the recycling of NADPH generated by the glycolytic pathway converting starch into triose-P. Unfortunately, nothing is known on the recycling of NADPH in nongreen plastids.

In contrast to pea shoots, sucrose density gradient purified chloroplasts from *Vicia faba* and *Zea mays* (Murphy and Leech, 1978) and Percoll density gradient purified amyloplasts from cauliflower buds (Journet and Douce, 1984) have been reported to contain glycerate-3-P mutase. Consequently, as pointed out by Dennis and Miernyk (1982), the status of glycerate-3-P mutase is in dispute, and it is possible that the presence of this enzyme depends on the age and source of the plastids. Under these circumstances, it would be equally feasible for the pyruvate to be produced by plastidic glycolytic enzymes, the pyruvate then being imported into the cytoplasm. Proudlove and Thurman (1980) found that pyruvate readily enters pea chloroplasts by diffusion and that no special translocator is involved or required. However, excretion of pyruvate from starch by isolated plastids has never been reported. It is possible, therefore, that pyruvate formed inside the plastid from either starch or cytoplasmic triose-P is utilized for fatty acid synthesis instead of respiratory purposes. In support of this suggestion, it has been convincingly demonstrated that a pyruvate dehydrogenase complex is present in the plastids (for a review see Givan, 1983) and that a chloroplastic acetyl-CoA carboxylase is also present which converts acetyl-CoA into malonyl-CoA for use in the *de novo* synthesis of fatty acids. Furthermore, biosynthesis of palmitic, stearic, and oleic acids in plant tissues occurs solely in the plastids (Stumpf, 1980). Further experimental study of this important suggestion is clearly required.

B. Respiration and the Pentose Phosphate Pathway

It is convenient to distinguish between the two oxidative reactions of the pentose phosphate pathway that lead from glucose-6-P to ribulose-5-

P and the nonoxidative reactions that catalyze the interconversion of pentose, hexose, heptose, tetrose, and triose-P (ap Rees, 1974; see, however, J. F. Williams, 1980). This pathway does not end with the production of ribulose-5-P. It is a cyclic pathway that branches from, and returns to, glycolysis (Fig. 4.5). The hexose-to-pentose conversion is termed the oxidative portion of this "shunt," while the subsequent pentose-to-hexose conversion is nonoxidative. The oxidative portion produces NADPH. Glucose-6-P dehydrogenase (EC 1.1.1.49) and gluconate-6-P dehydrogenase (EC 1.1.1.44) have been demonstrated in many different plants (for a review see ap Rees, 1974, 1977). As pointed out by Turner and Turner (1980), the most effective point of control of the pentose phosphate pathway will be at the initial step i.e., the reaction in which glucose-6-P is utilized. In support of this suggestion, the glucose-6-P dehydrogenase plus lactonase step is far from equilibrium while gluconate-6-P dehydrogenase, ribulose-P epimerase, ribose-P isomerase, transketolase, and transaldolase are close to equilibrium (Ashihara and Komamine, 1974). Since the concentration of glucose-6-P in the cytosol is in the range of 2–5 mM (Martin *et al.*, 1982), the key to this control appears to be NADPH, and it is probable that glucose-6-P dehydrogenase activity is determined by the NADPH/NADP ratio. Furthermore, L. E. Anderson *et al.* (1974) have shown that the reduction of the enzyme by dithiothreitol leads to an inactivation of cytoplasmic glucose-6-P dehydrogenase.

It is theoretically possible to estimate the activity of the pentose phosphate pathway, relative to the activity of glycolysis, in the oxidation of glucose. Glycolysis in combination with the TCA cycle initially produces CO_2 from the C-1 and C-6 carbons of glucose in equivalent amounts; the shunt initially produces CO_2 derived only from C-1. The source of the CO_2 carbons may be determined by using isotopically labeled glucose. Thus, the ratio of C-1/C-6 derived CO_2 will be unity if the pentose phosphate pathway is not degrading glucose and will be greater than unity if it is. However, the ratio of the yields of CO_2 from carbons C-1 and C-6 of glucose is influenced by several variables and bears no simple relationship to the relative rates of catabolism by the pentose phosphate pathway or by the Krebs cycle (for a critical review see ap Rees, 1980). The patterns of $^{14}CO_2$ production and labeling of intermediates from specifically labeled glucose supplied to a wide range of plants strongly indicate that a proportion of hexose oxidation proceeds via glucose-6-P and gluconate-6-P dehydrogenases (initial C-6/C-1 ratios were significantly less than unity). The estimates obtained by these methods suggest that glycolysis predominates, however, and that only 5–20% of the glucose metabolism is catalyzed by the pentose phosphate pathway. For example, Agrawal and Canvin (1971) have demonstrated that 6–10% of

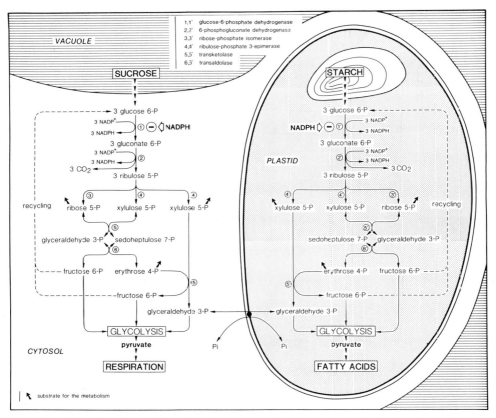

Fig. 4.5. Reactions of the oxidative pentose phosphate pathway in higher plant cells. This scheme indicates that in addition to the enzymes of the glycolytic pathway, the key enzymes of the oxidative pentose phosphate pathway have been detected in both the plastids and the cytoplasm. The ribulose-5-P formed in the pentose phosphate pathway is converted to fructose-6-P and glyceraldehyde-3-P, and the bulk of these products is then metabolized via glycolysis for fatty acid synthesis in the plastids or respiration in the cytoplasm. The reversibility of the nonoxidative reaction means that it could be used either to convert pentose phosphate to triose phosphate and hexose phosphate or to convert glycolytic intermediates to erythrose phosphate and pentose phosphate for use in the synthesis of aromatic compounds, nucleic acids, and cell wall constituents. The nonoxidative part of the pentose phosphate pathway should be considered as a pool of intermediates in "dynamic equilibrium that responds according to which compounds are removed" (ap Rees, 1980).

the glucose is metabolized via the pentose phosphate pathway in the developing endosperm of castor beans. There is also considerable evidence from measurement of the maximum catalytic activities of key enzymes that the activities of the two pathways change appreciably during development (ap Rees, 1977). Thus, studies using root segments (Fowler and ap Rees, 1970) or leaf tissues (Croxdale, 1983; Croxdale and Outlaw, 1983) have shown that glycolysis predominates in the most meristematic regions and that the contribution of the pentose phosphate pathway increases with cellular differentiation.

Theoretically, the pentose phosphate pathway can bring about the complete breakdown of glucose-6-P to CO_2, if fructose-6-P and glyceraldehyde-3-P are completely recycled (Fig. 4.5). The complete recycling of glyceraldehyde-3-P and fructose-6-P is most unlikely, however, because glucose-6-P enters the pentose phosphate pathway more readily than does fructose-6-P (ap Rees *et al.*, 1965). Furthermore, the restriction on recycling could be due to the partial inhibition of glucose-P isomerase by intermediates of the pentose phosphate pathway, such as phosphogluconate (Takeda *et al.*, 1967). Recycling of glyceraldehyde-3-P requires the presence of a cytosolic fructose 1,6-bisphosphate phosphatase. However, in plants that are neither gluconeogenic nor photosynthetic, the activities found were very low (S. M. Thomas and ap Rees, 1972). In addition, if present, this phosphatase should be strongly regulated, perhaps by fructose-2,6-P_2 (Stitt *et al.*, 1982b; Csèke *et al.*, 1982), in order to prevent a futile cycle between fructose-1,6-P_2 and fructose-6-P. To date, all the available data indicate that the glyceraldehyde-3-P formed in the pentose phosphate pathway is metabolized along with that formed in glycolysis (ap Rees, 1980).

It is classically considered that the primary function of the oxidative portion of the pentose phosphate pathway is to supply NADPH for use in reductive biosynthetic reactions in the cytoplasm (Osmond and ap Rees, 1969), whereas the important functions of the nonoxidative reactions of the pentose phosphate pathway are probably the provision of ribose-5-P for nucleotide and nucleic acid synthesis and the production of erythrose-4-P used in phenylpropanoid synthesis and aromatic amino acid synthesis. According to Croxdale (1983), the fact that meristematic regions have a greater capacity for glycolysis while older, maturing regions have a greater capacity for the pentose phosphate pathway is at odds with the advantages of the pentose phosphate pathway that could provide precursors needed by actively dividing young cells. It is possible, however, that the more meristematic regions receive metabolites and cofactors from others sources. The requirements of the cell for NADPH and the other products will vary, and so it is to be expected that mechanisms will exist to regulate the flux through the pathway. However, since plant mitochondria can couple the Ca^{2+}-dependent oxidation of NADPH to ATP synthesis via the respiratory chain, it can be argued that the physiological significance of the oxidative part of the pentose phosphate pathway has to be reconsidered. It is possible, therefore, that the rate of NADPH reoxidation by the mitochondrion could strongly increase the rate of the pentose phosphate pathway, because the most effective control of this pathway appears to be the concentration of NADPH. The rapid reoxidation of NADPH by the mitochondrion could play a significant role during active cell growth, because the supply of carbon skeletons (pentose-P, erythrose-4-P, etc.) for biosynthetic purposes may be more important than the supply of NADPH. Unfortunately, nothing is known about the control of NADPH oxidation by the mitochondrion *in vivo*. It is possible that cytoplasmic free-Ca^{2+} may play a part in the regulation of the mitochondrial NADPH oxidation and, therefore, of the pentose phosphate pathway functioning. The relative importance of the two modes of NADPH oxidation (biosynthetic purpose or mitochondrial oxidation) cannot be assessed at present.

The key enzymes of the oxidative pentose phosphate pathway, glucose-6-P dehydrogenase and gluconate-6-P dehydrogenase, have also been detected in plastids (chloroplasts and various plastids) (Schnarrenberger *et al.*, 1973; Miflin and Beevers, 1974; Emes and Fowler, 1983; Simcox *et al.*, 1977; Journet and Douce, 1984; for a review see Dennis and Miernyk, 1982; Table 4.1). The cytosolic and plastidial isoenzymes of these dehydrogenases have been separated and shown to have similar kinetic and physical properties, except for a difference in charge. A great deal of more direct work on a wider range of species is required to resolve the question of whether all plastids can catalyze the whole pentose phosphate pathway (Dennis and Miernyk, 1982). Very recently, however, Journet and Douce (1985) have shown that nongreen plastids isolated from cauliflower buds exhibit substantial transaldolase and transketolase activities. Examination of the relative activities of glucose-6-P dehydrogenase and gluconate-6-P dehydrogenase in chloroplasts and in leaves indicates that the relative activity of the oxidative pentose phosphate pathway may be higher in the chloroplasts than in the leaf as a whole (Stitt and ap Rees, 1979). The amount of these enzymes may vary greatly, however, in different types of plastids and in the same type of plastid at different stages of tissue development (Dennis and Miernyk, 1982). Since NADPH is copetitive inhibitor of the chloroplastic glucose-6-P dehydrogenase, with respect to NADP, the oxidative pentose phosphate pathway in plastids is also regulated by means of the NADPH/NADP ratio. In addition, the oxidative pentose phosphate pathway in the chloroplast is probably turned off in the light, because the glucose-6-P dehydrogenase is directly inhibited by a reductive process involving the electron transport chain of the thylakoids (for a review see Buchanan, 1980). This inhibition is prevented, in a competitive manner, by NADP (Lendzian, 1980) and by inorganic phosphate (Huber, 1979). Consequently, it has been suggested that the oxidative pentose phosphate pathway may contribute pentose phosphate intermediates during the inductive phase of photosynthesis (Huber, 1979). Stitt and Heldt, (1981b) have shown that both the oxidative pentose phosphate cycle and glycolysis make a significant contribution to chloroplastic dark metabolism in spinach. The distribution of flux through these pathways is determined by the concentration of ATP and the NADPH/NADP ratio as well as by other factors, such as external phosphate levels that promote transport of trioses from the plastids. They also suggest that when there is no cytosolic demand for phosphorylated intermediates, (i.e., when cytosolic phosphate concentration declines) phosphorolytic starch mobilization would continue at a much lower rate, providing mainly substrates for chloroplastic energy metabolism, with much of the hexose-P being metabolized in the oxidative pentose phosphate cycle. NADPH thus formed is utilized for biosynthetic activities, such as fatty acid synthesis. The surplus of NADPH can be oxidized either by malate dehydrogenase or a specific plastidial "respiratory chain" (Bennoun, 1982). Cooperation between the oxidative pentose phosphate pathway and nitrite assimilation in nongreen tissues, such as roots, could ensure a closely coupled supply of electrons via glucose-6-P dehydrogenase, as long as sufficient glucose-6-P is available (Emes and Fowler, 1979). In support of this hypothesis, a controlled flow of carbon through the oxidation pentose phosphate pathway, concurrent with the induction of nitrite assimilation, was recently reported (Emes and Fowler, 1983).

C. Mitochondrial Electron Transport Activity in the Light

In a number of plant species, the rate of photosynthesis, when plotted against light intensity, shows a break far below light saturation (Kok, 1949). This effect, known as the "Kok effect," has been attributed to a suppression of dark respiration at high light intensities (Hoch *et al.*, 1963). Perhaps the strongest direct evidence of light inhibition of mitochondrial respiration in leaves was obtained from *in vivo* experiments using $^{18}O_2$ (Canvin *et al.*, 1980). It was demonstrated that rates of $^{18}O_2$ uptake observed in darkness are completely suppressed in light. However, inhibition of dark respiration by light has repeatedly (Mangat *et al.*, 1974; Chevallier and Douce, 1976; Sawhney *et al.*, 1979), but not always, been observed, and the physiological evidence for operation of dark respiration in the light involving oxidation of sugars via glycolysis and the TCA cycle and the oxidative pentose phosphate pathway is conflicting (for an excellent review see Graham and Chapman, 1979). The question has some importance for the estimation of growth in plants, since apparent photosynthetic CO_2 fixation must be corrected for loss of CO_2 by respiratory activity. The most likely mediators of the interactions between respiration and photosynthesis are the adenine nucleotides (ATP and ADP), the pyridine nucleotides (NADP, NADPH, NAD, NADH), and perhaps inorganic phosphate.

The process of respiration in photosynthetic cells under light conditions is not easily described. Three reasons for this are: (a) the existence of photorespiration that introduces a set of O_2 uptake and CO_2 release reactions, (b) the presence of a second O_2 uptake reaction (the so-called Mehler reaction) in the chloroplast, and, (c) the lack of methods for direct measurement of the rate of respiration during photosynthesis. Indeed, O_2 uptake measurements on whole leaves of C_3 plants indicate that there is sufficient O_2 uptake via a Mehler-type reaction to be quantitatively important in energy balancing (Heber *et al.*, 1978; Badger and Canvin, 1981; Marsho *et al.*, 1979). Photorespiration is known to be of major quantitative significance in the light, but O_2 uptake by chloroplasts by other processes has been less carefully evaluated. Two groups have recently concluded that it is indeed a significant process in the light (Furbank *et al.*, 1982; Behrens *et al.*, 1982). Consequently, the components of light-dependent O_2 uptake are believed to include ribulose 1,5-bisphosphate oxygenase and glycolate oxidase, direct photoreduction of O_2 via a Mehler effect, and perhaps persistance of mitochondrial respiration during illumination (Gerbaud and André, 1980). In addition to "chloroplast respiration," photorespiration, and the normal mitochondrial respiration, one must add the cyanide-resistant respiration typical of plant mitochondria to obtain a complete list of the types of O_2 consuming reactions currently believed to occur in photosynthetic cells. Finally, in his interesting review article on the occurrence of respiration in illuminated green cells, Raven (1972a,b) criticized some of the conclusions that have been reached using gas exchange methods to prove that dark respiration is inhibited by light (see, for example, Mangat *et al.*, 1974). He suggested that reduced levels of CO_2 evolution in the light may be due to refixation rather than inhibition of dark respiration. Refixation was estimated by Raven (1972a,b) to comprise up to 17% of the total CO_2 fixation in the light.

Krause and Heber (1976) have suggested that the onset of photophosphorylation in the chloroplast in the light leads to a rapid increase in the phosphorylation potential in both the chloroplasts and the cytoplasm. Although the chloroplast is capable of directly transporting adenine nucleotides, particularly in young pea leaves (S. P. Robinson and Wiskich, 1977), indirect shuttle mechanisms (dihydroxyacetone phosphate–glycerate-3-P shuttle) across the chloroplastic envelope are believed to be sufficiently rapid to export phosphorylation energy. [The capacity of the chloroplastic malate—oxaloacetate shuttle is too low to mediate effective communication of reducing equivalents between the plastids and the cytosol (Giersch, 1982).] In this way, a high chloroplastic phosphorylation potential is thought to lead to elevated ATP/ADP ratios in the cytoplasm in the light and, therefore, slows down the rate of dark respiration. Earlier studies of subcellular compartmentation of adenine nucleotide levels in green leaves add support to this theory (Santarius and Heber, 1965; Heber and Santarius, 1970). Interestingly, Sawhney *et al.* (1978) have also suggested that in the light cytoplasmic ATP indirectly produced by photosynthesis inhibits mitochondrial electron transport. Therefore, NADH probably accumulates in the cytoplasm where it is utilized in nitrate reduction. Stitt *et al.* (1980) measured the level of ATP, ADP, and AMP in chloroplastic and extrachloroplastic fractions of spinach protoplasts. The extrachloroplastic ATP/ADP ratio increased from 10 in the dark to 20 after 30 sec of illumination. The ATP/ADP ratio then fell to almost the dark level after 4 min in the light. However, Stitt *et al.* (1980) measured the extrachloroplastic fraction of protoplasts which included both mitochondrial and cytosolic material, and this method is therefore too imprecise to allow for an accurate measurement of the true cytosolic ATP/ADP ratio. Hampp *et al.* (1982) have improved the technique used by Stitt *et al.* (1980) so that the mitochondria can also be separated from protoplasts by "rapid" centrifugation through two layers of silicone oil. They found that the level of ATP in the chloroplast increased in the light and reached a maximum level after 30 sec of illumination. This effect was mirrored in the cytoplasm, which showed an increase in the level of ATP after 1 min of illumination. Because of the immense technical problems involved in a separation of this type, the small spaces and, hence, rapid diffusability of metabolites within the mitochondrial matrix, these data cannot, in my opinion, be taken as representative of the *in vivo* state without further serious confirmation. Lilley *et al.* (1982) have developed a new method of fractionating wheat leaf protoplasts using membrane filtration. They criticized the silicone oil technique of Hampp *et al.* (1982) because 1 minute elapses between the disruption of the protoplasts and the quenching of the metabolism. The importance of the time factor is apparent from the fact that at a rate of 200 μmoles CO_2 reduction/mg chlorophyll/hr the chloroplastic pool of ATP turns over about six times per second (Heber *et al.*, 1982). Utilizing an ingenious arrangement of syringe, nylon net, and membrane filters, Lilley *et al.* (1982) were able to rupture the protoplasts, separate the released subcellular fractions, and quench them in $HClO_4$ in about one-tenth of a second, thereby ensuring that measured levels best reflected those present *in vivo* immediately prior to protoplast rupture. Using this method, Stitt *et al.* (1982a) measured changes in adenine nucleotide levels in chloroplasts, the cytosol, and mitochondria from wheat leaf protoplasts. The results obtained directly conflict with previous data. The main difference was found in changes in cytoplasmic ATP/ADP ratios. The cytosolic ATP/ADP ratio was lower in the light than in the dark (5.6 versus 7.5)! Such a result, therefore, does not support the widespread belief that mitochondrial oxidative phosphorylation is suppressed in the light by an increase of the cytosolic ATP/ADP quotient. The question thus remains, by what mechanism is mitochondrial respiration or electron transport switched off in leaf tissue in the light, if indeed, it is switched off? One speculative alternative would be the light-dependent formation of a specific endogenous inhibitor of mitochondrial respiration as postulated by Stitt *et al.* (1982a). Endogenous inhibitors of this type have, however, never been demonstrated. Alternatively, one might propose that mitochondrial respiration is in fact not inhibited in the light. Thus, Dry and Wiskich (1982) have questioned the proposal by Heber and Santarius (1970) that increases in the cytoplasmic ATP/ADP ratio of the leaf cells in the light leads to an inhibition of mitochondrial oxidative phosphorylation. Dry and Wiskich (1982) found that isolated pea leaf mitochondria oxidizing malate plus glutamate were only inhibited by extramitochondrial ATP/ADP ratios greater than 20. At ratios below 20, the rate of respiration was maximal and equivalent to the state 3 rate. Current data on the cytoplasmic levels of ATP and ADP in plant cells in the light suggest that the ATP/ADP ratios do not become sufficiently high to cause inhibition of mitochondrial respiration. In fact, Graham (1980) reviewed the literature and concluded that glycolysis and the TCA cycle can operate in illuminated green cells, although some modifications probably occur in relation to the dark pattern. When ^{14}C-labeled TCA cycle acids were fed to mung bean leaves in the light, they were incorporated into all of the components of the cycle (E. A. Chapman and Graham, 1974a,b). Likewise, a high CO_2 concentration under light conditions enhances the carbon traffic through the TCA cycle and related compounds, such as glutamate, glutamine, etc., presumably by increasing the supply of substrates for phosphoenolpyruvate carboxylase (Platt *et al.*, 1977). The results of E. A. Chapman and Graham (1974a,b), however, suggested that TCA cycle activity ceased within the first 3 min in the light and then returned to the dark rate after a longer period of illumination. These results provide a possible explanation for discrepancies in reports on the extent to which the TCA cycle is inhibited in the light. Short term labeling experiments would be misleading, because respiration appears to be inhibited only during the initial periods in the light. Alternative methods for reoxidation of mitochondrial NADH which are independent of electron transport have been proposed by Douce and Bonner (1972), Woo and Osmond (1976), and Journet et al. (1981). In these proposals, redox equivalents deriving either from NAD^+-linked TCA cycle dehydrogenases or glycine decarboxylase could be totally transferred into the extramitochondrial compartment by either the malate–oxaloacetate or the malate–aspartate shuttles, rather than being fed into the respiratory chain (see, for example, Fig. 3.18). Interestingly, the low K_m of the oxaloacetate transporter should allow it to compete successfully with matrix and cytosolic malate dehydrogenases (Oliver and Walker, 1984). This may help to explain why some authors have failed to find significant O_2 uptake by mitochondrial respiration in illuminated leaves. These results are consistent with the suggestion that glycolysis and the TCA cycle are modified in the light to allow a continuous anaplerotic carbon flow for supplying α-oxoacids that the chloroplast is unable to make (Larsson, 1979). The presence of $NADP^+$-isocitrate dehydrogenase in chloroplasts has been reported (Elias and Givan, 1977), and, hence, formation of some α-ketoglutarate within the plastids is possible, provided that isocitrate is available for the reaction. These compounds can be used for a variety of synthetic reactions, including amino acid (Miflin and Lea, 1980) and lipid formation (Stumpf, 1980). It is not known if the TCA cycle operates beyond succinate oxidation (if the TCA cycle is to be fully operational, then it is necessary to reoxidize the reduced flavoprotein produced at the succinic dehydrogenase step of the cycle, and this can only occur via the electron transport chain), and the operation of the mitochondrial electron chain in the light is a more uncertain aspect of the problem. It is, however, possible that under light conditions the major carbon flux occurs only through the part of the TCA cycle from pyruvate or malate to α-ketoglutarate.

The regulatory processes involved in the interaction of light with respiration are beginning to be understood, but obviously much remains to be done to fully describe the respiratory pathways of the green cell and the complex interactions of photosynthesis and dark respiration in the light. For example, nothing is known about the supply of TCA cycle intermediates to the chloroplast in the light for synthetic processes, especially the formation of amino acids. In addition, little is known about the cytoplasmic NAD redox states or about the mechanisms involved in the reoxidation of multiple cytosolic pools of NADH under light conditions. Thus, the redox state of NADH near the mitochondrion may be partially relevant in the control of external NADH oxidation by the mitochondrion. It is also obvious that measurement of the total amount of nicotinamide nucleotide is inadequate, because of problems of compartmentation. A more formidable problem is that of binding and compartmentation. Future research should be directed toward solving these problems that remain an important obstacle to studies on the effective collaboration between photosynthetic and oxidative phosphorylation.

D. Mitochondrial Electron Transport and β-Oxidation of Fatty Acids in Microbodies

During germination, fatty seeds such as castor bean endosperm convert their stored fat to sucrose, which is used to support the energy needs of the growing seedlings. At the time of maximal gluconeogenesis in castor bean endosperm, 1 gram of sucrose accumulates for each gram of fat metabolized. The fat is stored in spherosomes as triglycerides (Muto and Beevers, 1974). The conversion of fat to sucrose involves lipolysis, fatty acid activation, the β-oxidation spiral (Fig. 4.6), the glyoxylate cycle (condensing two acetyl-CoAs to form succinate), the conversion of succinate to phosphoenolpyruvate (via oxaloacetate), and the reversal of glycolysis (for a review see H. Beevers, 1980; Fig. 4.7). Until 1969, it had been assumed that for all plant tissues oxidation of long chain fatty acids to acetyl-CoA in β-oxidation occurred in the mitochondrion (see, however, D. R. Thomas and McNeil, 1976). Cooper and Beevers (1969) and Breidenbach and Beevers (1967) discovered a subcellular organelle that becomes abundant in the endosperm of castor bean at the time of germination and that carries out β-oxidation and the glyoxylate cycle. Because this organelle contains all of the enzymes catalyzing the glyoxylate cycle, namely citrate synthase, aconitase, isocitrate lyase, malate synthetase, and malate dehydrogenase, H. Beevers (1980) named them glyoxysomes (Fig. 4.7). Subsequently, he found that they also contained oxidases (in the first reaction of β-oxidation, catalyzed by the fatty acyl-CoA oxidase, two electrons are transferred from an acyl-CoA to molecular O_2, with the formation of an enoyl-CoA and H_2O_2) and catalases, making them a member of the microbody family (Tolbert, 1980). Interestingly, Gerhardt (1983) has reported that peroxisomes from spinach leaves, mung bean hypocotyls, and potato tubers catalyze a palmitoyl CoA–dependent KCN-insensitive O_2 uptake and that the activity of β-oxidation enzymes in the mitochondrial fraction is from contaminating peroxisomes rather than from mitochondrial constituents.

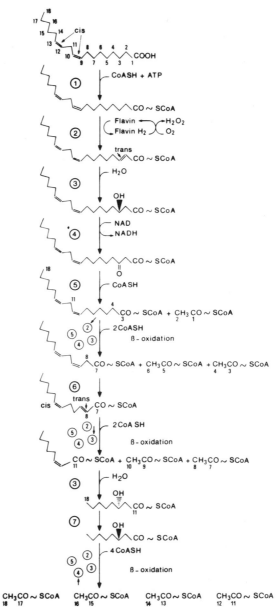

Fig. 4.6. Reaction sequence in the degradation of linoleic acid: oxidation (2), hydration (3), oxidation (4), and thiolisis (5). This scheme also indicates a reaction (6, isomerase) that shifts the position and configuration of the cis-Δ^9 double bond. Symbols: 1) fatty acyl-CoA synthetase; 2) fatty acyl-CoA oxidase; 3) enoyl-CoA hydratase; 4) L-3-hydroxy fatty acyl-CoA dehydrogenase; 5) 3-oxoacyl-CoA thiolase; 6) enoyl-CoA isomerase (converts the 3,4-cis-double bond to 2,3-trans); and 7) hydroxyacyl-CoA epimerase. Figure drawn by Marie-Claude Neuburger.

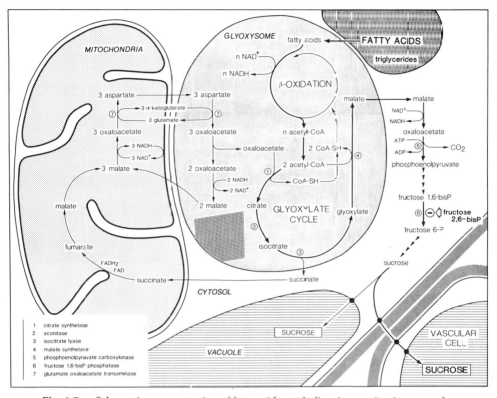

Fig. 4.7. Schematic representation of fatty acid metabolism in germinating castor bean endosperm. The glyoxysome from castor bean endosperm is now known to be the exclusive site of the enzymes involved in β-oxidation of fatty acids (see Fig. 4.6) and the glyoxylate cycle. The proposed malate–aspartate shuttle for the transfer of reducing equivalents between glyoxysomes and mitochondria, providing for *in situ* oxidation of NADH produced by glyoxysomal β-oxidation, is indicated (adapted from Mettler and Beevers, 1980). Finally, sucrose synthesized in the cytosol from malate via phosphoenolpyruvate and fructose-6-P (Nishimura and Beevers, 1981) is exported to the outside of the cell through the plasmalemma. Fructose-2,6-P_2 could function in the cytosol to regulate carbon flow (sucrose synthesis) through inhibition of an enzyme of gluconeogenesis (fructose-1,6-P_2 phosphatase).

Peroxisomal localization of β-oxidation enzymes has also been reported for Jerusalem artichoke tubers (Macey and Stumpf, 1982/1983).

Isolated glyoxysomes accumulate 1 mole of NADH for each mole of acetyl-CoA generated during β-oxidation of fatty acids (Fig. 4.7). The conversion of 2 moles of acetate to 1 mole of succinate in the glyoxylate cycle requires the oxidation of malate to oxaloacetate and thus, the production of another 1 mole of NADH (H. Beevers, 1980). Since the glyoxysome is not only incapable of oxidizing succinate but also lacks the cytochromes and other components of the electron transport system oxidizing NADH (Breidenbach and Beevers, 1967; see, however, Donaldson et al., 1981), it was concluded that NADH and succinate produced in the glyoxysomes are oxidized by the mitochondrion. Indeed, the only cellular site where the enzymes linking succinate and NADH oxidation to O_2 uptake occur, is the mitochondrion. The succinate seems to diffuse readily to the mitochondrion, where it is further metabolized to oxaloacetate (the glyoxysome is certainly a compartment for enzymes but it is not yet known whether it is a substrate compartment controlled by membrane transport systems; Schmitt and Edwards, 1983). The flow of carbon in the mitochondrion must be preferentially restricted to those reactions in which succinate is converted to oxaloacetate and perhaps pyruvate. In support of this suggestion, succinate and malate plus glutamate are the only substrates to be oxidized rapidly by purified castor bean mitochondria (Millhouse et al., 1983). Furthermore, external succinate does not exchange with internal malate in these mitochondria and ensures that the entering succinate is metabolized to oxaloacetate (Millhouse et al., 1983). However, oxaloacetate does not move out these mitochondria readily, and some other systems must be involved for rapid removal of oxaloacetate. According to Wiskich and Day (1979), malate oxidation by castor bean mitochondria is severely inhibited by oxaloacetate accumulation and glutamate is often an absolute requirement. This probably reflects the low level of malic enzyme in these mitochondria, but may also be due to restricted oxaloacetate efflux. Since internal α-ketoglutarate exchanged for malate and since aspartate was found to exchange with glutamate in a freely reversible way (Millhouse et al., 1983), it was suggested that during the course of succinate oxidation intramitochondrial oxaloacetate is preferentially transaminated with glutamate. The excess of α-ketoglutarate thus formed is excreted from the matrix via the α-ketoglutarate carrier (Fig. 4.7).

The problem of shuttling of NAD^+/NADH or corresponding reducing equivalents between glyoxysomes and mitochondria is of crucial importance in the conversion of fat to sucrose in fatty seedling tissues. The fate of the NADH formed in the glyoxysome is unknown at present. The first hypothesis regarding the manner in which NAD^+ is supplied to the glyoxysomal matrix would involve rapid entry of a pyridine nucleotide into the glyoxysome. The resulting NADH would return to the cytoplasm for subsequent oxidation. Under these circumstances, no involvement of glyoxysomal malate dehydrogenase or glutamate-oxaloacetate transaminase to generate NAD^+ within the glyoxysome is required. Movement across the glyoxysomal membrane would be either by diffusion or by carrier mediated exchange. However, this hypothesis is most unlikely because there is evidence against direct permeation of NADH through glyoxysomal membranes; e.g., latency of NADH-linked enzyme activities in microbody preparations has been observed (Donaldson *et al.*, 1981; Mettler and Beevers, 1980). Consequently, the possible confinement of NADH within the glyoxysomes reiterates the problem of oxidation of the NADH produced within the glyoxysomes by β-oxidation and the glyoxylate cycle (Lord and Beevers, 1972). The second mechanism would involve malate dehydrogenase (in the highly favored oxloacetate-to-malate direction) and glutamate-oxaloacetate transaminase in removing reducing equivalents from the glyoxysomes during the process of fatty acid oxidation. Several lines of evidence support this second mechanism. The required enzymes are present in both organelles and are highly active (Mettler and Beevers, 1980; Donaldson *et al.*, 1981; see, however, Schmitt and Edwards, 1983, in the case of leaf peroxisomes). Mettler and Beevers (1980) provided a glyoxysomal extract with NAD$^+$ and a substrate of β-oxidation (palmitoyl-CoA) and observed a reduction of NAD$^+$. Addition of aspartate and α-ketoglutarate resulted in a reoxidation of NADH which was sensitive to the transaminase inhibitor aminooxyacetate. The authors interpreted this as indicating the importance of a malate–aspartate shuttle, in this case to remove reducing equivalents from the glyoxysomal matrix. These sorts of experiments (using coupled enzymatic reactions within broken glyoxysomes) give some indication of the feasibility of shuttles. They must be interpreted with caution, however, since the shuttles in glyoxysomes, as currently envisioned, should be demonstrable in intact organelles using only endogenous pyridine nucleotides, as shown by Journet *et al.* (1980, 1982) with plant mitochondria. Figure 4.7 presents the overall stoichiometry of the hypothesized shuttle between glyoxysomes and mitochondria during gluconeogenesis metabolism in germinating castor bean (Mettler and Beevers, 1980). The overall process results in the formation of one unit of malate (to be used for sucrose synthesis in the cytosol) and the indirect transfer of NADH to the mitochondrion for oxidation. Considered in relation to the proposed shuttle, the failure to observe special transport systems in the glyoxysomes is disappointing (Mettler and Beevers, 1980). It is possible that experimental manipulations may have damaged the fragile glyoxysomal membrane and thus prevented detection of a transport system that operates *in vivo*. Nevertheless, it is clear that the various components of the shuttle can traverse the membrane of isolated glyoxysomes by diffusion (it is possible that porine is present in the membrane surrounding the glyoxysomes), and this may be adequate to sustain its operation *in vivo*. If the proposed interaction does occur, then mitochondria will produce massive amounts of ATP (nine molecules of ATP for each molecule of succinate formed) during the course of fatty acid oxidation. Since the endosperm is not growing at the time of fat breakdown, it is unlikely that much ATP is required for synthetic reactions (only part of the ATP formed is used for gluconeogenesis). There is no evidence of a cyanide-insensitive alternate respiration in these mitochondria, so the "fate" of this "excess" ATP needs to be considered. A good possibility suggested by Millhouse *et al.* (1983) is that the efficiency of oxidative phosphorylation is reduced, because fatty acids are known to uncouple mitochondria and some may be present in the endosperm during triglyceride hydrolysis. In support of this suggestion, Millhouse *et al.* (1983) have shown that nonpurified mitochondria contaminated by glyoxysomes were not coupled, although rates of O_2 consumption were high (purified mitochondria were coupled). This is probably due to fatty acids present in the crude particulate fraction and may reflect the *in vivo* condition so that "excess" ATP is not produced during respiration.

Although sucrose synthesis is a major metabolic process in the endosperm, our understanding of its control is limited. It is very likely that the production of fructose-6-P from fructose-1,6-P_2 is a regulated step in this pathway. Again fructose-2,6-P_2 could function in the endosperm cytosol to regulate carbon flow (i.e., sucrose synthesis and breakdown) through activation of a glycolytic enzyme (pyrophosphate-linked phosphofructokinase PP$_i$-PFK) and through inhibition of an enzyme of gluconeogenesis (fructose-1,6-P_2 phosphatase), as envisioned in the green leaf cytosol (Kruger and Beevers, 1984) (Fig. 4.7). Unfortunately, no specific inhibitors are as yet available for these enzymes. Finally, the kinetic properties of PP$_i$-PFK do not suggest an obvious physiological role for the enzyme. Castor bean PP$_i$-PFK is quite capable of catalyzing the reverse and forward reactions at almost equal maximum rates (Kombrink *et al.*, 1984).

Biogenesis of Plant Mitochondria

The last ten years have seen an explosive growth in the efforts devoted to studies of mammalian and yeast mitochondrial biogenesis, including that of the membrane lipids and their synthesis, the mitochondrial DNA and the identification of its gene products, the nature and characteristics of the mitochondrial synthesizing system, the integration (i.e., the accretion) of new material (newly synthesized lipids and proteins) into a preexisting structure, and finally, the interaction of mitochondrial genetic information with the classical nucleocytosolic system of the eukaryotic cell (for a recent review see Tzagoloff et al., 1979).

Unfortunately, very little information is available concerning plant mitochondrial biogenesis. The plant system is not ideal for such an approach because of the difficulty with which mitochondrial composition, structure, and function may be varied through the use of various inhibitors and through physiological and genetic manipulation of the cells. Thus, the unique ability of yeast to survive without functional mitochondria has allowed the study of mutations that interfere with mitochondrial biogenesis and function, leading to the impressive progress in mapping genes localized in mtDNA. In contrast, many mutations in plant mtDNA may be lethal, and, unlike many chloroplastic mutations, these mutations might not be characterized. Finally, it is extremely difficult to routinely prepare large amounts of plant mitochondria devoid of plastid contamination.

I. PHOSPHOLIPID AND UBIQUINONE SYNTHESIS

Although 20 years have passed since the general pathway of glycerolipid biosynthesis was described, largely through the work of Kennedy and his co-workers (Kennedy, 1961), the precise biosynthetic route for the production of phosphoglycerides in plant cells remains uncertain.

Enzymological studies of phospholipid-synthesizing enzymes in plant mitochondria have turned up no major differences from analogous enzymes in animal mitochondria. The subcellular location of the enzymes in the mitochondrion and endoplasmic reticulum is also similar to the situation of animal tissues (Mudd, 1980; T. S. Moore, 1982). In plant cells, the enzymes responsible for the synthesis of the major mitochondrial polar lipids (i.e., phosphatidylcholine, PC; phosphatidylethanolamine, PE; and perhaps phosphatidylinositol) occur predominantly in the ER. Thus, the major pathway for the synthesis of PC is via the CDP-choline–1,2-diacylglycerol choline phosphotransferase (EC 2.7.8.2.) reaction in plants. This enzyme has been described as being exclusively associated with the ER of castor bean endosperm (Lord *et al.*, 1972, 1973). Any activity associated with other subcellular fractions, commonly, has been attributed to contamination of that fraction by ER. Likewise, the biosynthesis of PC and PE by the nucleotide pathway has been conclusively demonstrated in spinach leaves (Devor and Mudd, 1971; Macher and Mudd, 1974; Marshall and Kates, 1973, 1974) and occurs almost exclusively in the "microsomal fraction" (i.e., a $100,000g$ pellet collected from a cell-free homogenate after removal of the larger cell organelles, such as plastids and mitochondria). T. S. Moore *et al.* (1973) measured the incorporation of inositol into lipids in castor bean endosperm organelles; again the enzymatic activity was almost exclusively in the ER. All of these results strongly suggest, but do not prove, that in plant cells the mitochondrial phospholipids may be synthesized by the ER system and then incorporated into mitochondrial membranes. More recently, however, Sparace and Moore (1981) convincingly demonstrated that CDP-choline–1,2-diacylglycerol choline phosphotransferase and CDP-ethanolamine–1,2-diacylglycerol ethanolamine phosphotransferase are normal components of plant mitochondria. In addition, there is evidence that plant mitochondria also have the capacity to synthesize PC via the methylation of PE (T. S. Moore, 1976).

It is not yet possible to say, however, if the mitochondrial synthetic activity *in vivo* is sufficient to fully support mitochondrial growth without the transfer of lipids from the ER, as in mammals. The question of the mitochondrial contribution to PC synthesis in higher plants warrants closer study.

If the major phospholipids of the mitochondrion are really synthesized at the level of the ER, we must imagine a direct transfer of PC and PE between the ER and the mitochondrial outer membrane (Fig. 5.1). Such a transfer can occur either by means of Wirtz and Zilversmit type proteins (Kader, 1977) or fusion (Crotty and Ledbetter, 1973) between the ER and the outer membrane. The chief difficulty arising in connection with exchange proteins is that, although they are extremely efficient in facilitating the exchange of phospholipids between membranes, they do not catalyze a net transfer of phospholipids. Nonetheless, K. W. A. Wirtz *et al.* (1980) demonstrated that the exchange proteins may very well be functioning in the capacity of phospholipid transferases, especially in an *in vivo* situation in which the membrane is growing and excess lipids are being produced (Fig. 5.1). Phospholipids released by the exchange protein and/or by lateral diffusion within a continuous membrane network into the outer layer of the outer envelope membrane could be transferred to the inner layer of the outer envelope membrane ("flip-flop" movement). It is possible that a specialized mechanism may enhance the rate of transfer of newly incorporated lipids across the bilayer, especially during membrane assembly (Rothman and Kennedy, 1977); then, at the point where the two membranes are in contact, undergo lateral fluid translocation into either the outer or inner membrane, depending on the lipid and its ultimate functional location (Ruigrok *et al.*, 1972). It is evident that the directed flow of lipids must necessarily be regulated to prevent all of the lipids from entering one compartment. Obviously, a better characterization of the mechanisms involved in the transport of phospholipids is needed. The details of the way in which these phospholipids become incorporated into the inner and outer mitochondrial membranes have yet to be determined.

Substantial evidence has accumulated regarding the ability of intact mitochondria to synthesize at least a portion of the phospholipids contituting their membranes. Among the phospholipids reportedly synthesized by plant mitochondria are phosphatidic acid, either via the sequential acylation of *sn*-glycerol-3-P by long chain fatty acyl-CoA thioesters or via phosphorylation of diacylglycerol (Bradbeer and Stumpf, 1960; Cheniae, 1965; Douce, 1971; Sparace and Moore, 1979),

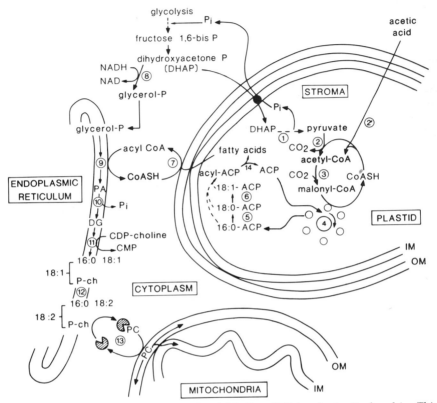

Fig. 5.1. Putative origin of phosphatidylcholine (PC) in plant mitochondria. This scheme emphasizes the proposed interrelationship between plastids and other cell compartments in the total synthesis of mitochondrial PC. Symbols and abbreviations: PA, phosphatidic acid; DG, diacylglycerol; ACP, acyl carrier protein; P-ch, phosphorylcholine; 1) plastid glycolysis; 2) pyruvate dehydrogenase; 2' acetyl-CoA synthetase; 3) acetyl-CoA carboxylase; 4) palmitoyl-ACP synthetase (a battery of soluble, nonassociated enzymes); 5) palmitoyl-ACP elongase; 6) stearoyl-ACP desaturase; 7) acyl-CoA synthetase; 8) sn-glycerol-3P dehydrogenase; 9) sn-glycerol-3P acyl-transferase; 10) phosphatidic acid phosphatase; 11) choline phosphotransferase; 12) desaturases; 13) phospholipid exchange protein; and 14) acyl-ACP thioesterase. (Courtesy of Marie-Claude Neuburger.)

CDP-diacylglycerol (Douce, 1968, 1971; Sumida and Mudd, 1968, 1970; Kleppinger-Sparace and Moore, 1985), phosphatidylglycerol (Douce, 1968; Douce and Dupont, 1969; T. S. Moore, 1974) and perhaps cardiolipin (Douce, 1968; Douce and Dupont, 1969; T. S. Moore, 1982). Cardiolipin is concentrated in the mitochondrial inner membrane in plants (Bligny and Douce, 1980), as in animals. The pathway of its biosynthesis proceeds by the sequence of enzyme-catalyzed reactions described in Fig. 5.2. The last reaction has not yet been satisfactorily demonstrated in plants. Douce and Dupont (1969) studied the incorporation of sn-[^{14}C]glycerol-3-P into lipids by purified cauliflower bud mitochondria. The mitochondria were allowed to generate CDP-diacylglycerol *in situ* before addition of the radioactive substrate, and trace amounts of cardiolipin were synthesized. In addition, they demonstrated that the enzyme that catalyzes the third reaction (Fig. 5.2) was inhibited by re-

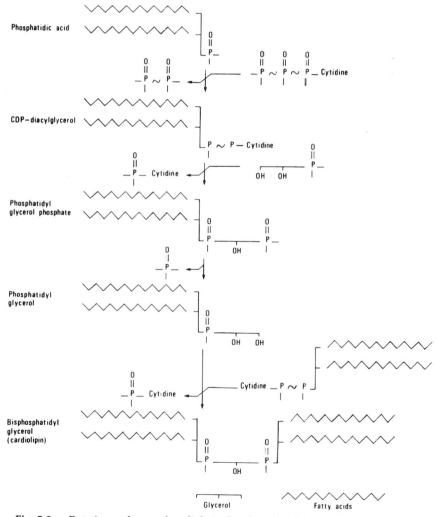

Fig. 5.2. Putative pathway of cardiolipin (bisphosphatidylglycerol) synthesis in higher plant mitochondria. (Courtesy of Marie-Claude Neuburger.)

agents that react with SH groups, such as $HgCl_2$. On the other hand, T. S. Moore (1974) did not detect cardiolipin when studying phosphatidylglycerol biosynthesis in castor bean endosperm mitochondria. Since the enzyme responsible for the conversion of phosphatidylglycerol to cardiolipin in the presence of CDP-diacylglycerol is firmly associated with the mitochondrial inner membrane isolated from animal tissues (McMurray and Jarvis, 1978), we believe that the plant mitochondria utilize endogenous phosphatidylglycerol and CDP-diacylglycerol as substrates for cardiolipin synthesis. Several attempts have been made to define the precise intramitochondrial sites of phospholipid synthesis which occur in mitochondria. Douce *et al.* (1973a) showed that CDP-diacylglycerol was synthesized exclusively on the inner membrane of mitochondria isolated from etiolated mung bean hypocotyls. Similar studies were made with mitochondria from the endosperm of castor beans and the results compared favorably with the mung bean system (Sparace and Moore, 1979). The synthesis of phosphatidylglycerol is localized in the inner membrane fraction, while the sequential acylation of *sn*-glycerol-3-P occurs in both the inner and outer membranes (Sparace and Moore, 1979). All of these results demonstrate that the mitochondrial inner membrane catalyzes the incorporation of *sn*-glycerol-3-P and fatty acids into phosphatidylglycerol and, perhaps, cardiolipin.

Questions about the synthesis of *sn*-glycerol-3-P and fatty acids, especially polyunsaturated fatty acids, remain. According to Leech and Murphy (1976) and Heinz (1977), *sn*-glycerol-3-P is synthesized in the cytoplasm. In support of this suggestion, and NAD^+-dependent glycerol-3-P dehydrogenase has been purified from spinach leaves. The enzyme appears to be in the cytoplasm and probably functions for production of *sn*-glycerol-3-P from dihydroxyacetone-P (Santora *et al.*, 1979). In recent years, increasing evidence has strongly suggested that, unlike prokaryotes and animals, higher plants have unique organelles that are specific sites of *de novo* fatty acid synthesis; in leaves, the chloroplast is the specific site and in nongreen tissues it is the proplastid (for a review, see Stumpf, 1980). Consequently, in plant cells plastids are the source of palmitic acid, stearic acid, and oleic acid for membranes of developing organelles such as mitochondria (for a review see Harwood, 1980; Fig. 5.1). Since the chloroplastic envelope contains an acyl-CoA synthetase that is associated with the outer membrane (Block *et al.*, 1983), it is very likely that fatty acids synthesized within the plastids and destined for further metabolism (phospholipid biosynthesis) in the cytosol are exported as acyl-CoA (Roughan *et al.*, 1979; Block *et al.*, 1983; Fig. 5.1). However, the mechanisms by which polyunsaturated fatty acids (e.g., linoleic acid) are synthesized are key problems in understanding lipid metabolism in plant mitochondria. It is possible that oleic acid synthesized within plastids is converted to an acyl-CoA form in the plastidial envelope and released into the cytosol where it is incorporated and desaturated into microsomal PC (Fig. 5.1). The latter may eventually be transferred to the mitochondrion (Fig. 5.1). The dual distribution in plant cells of several PC synthetic enzymes raises the possibility that different molecular species of a given phospholipid may be synthesized at different sites. Considering cardiolipin synthesis, it is possible that desaturation occurs after the formation of cardiolipin *in vivo* and that there is no need to derive linoleic acid present in cardiolipin from PC. The question concerning the specific positioning of unsaturated fatty acids in plant mitochondrial phospholipids remains unanswered, and further work has to be done in order to resolve this problem. It is certain that both the desaturases and acyltransferases involved in the acylation of *sn*-glycerol-3-P play an important role in fatty acyl-chain location.

Oxygen effects on the degree of unsaturation of mitochondrial membrane phospholipids have been investigated with sycamore cells in suspension cultures. By maintaining the O_2 concentration below 60 μM, the molar proportion of oleate increased dramatically, whereas that of the linoleate decreased. Under these conditions, the aeration of the culture medium ($[O_2] = 250$ μM) induced a rapid transformation (less than 10 hr at 25°C) of oleate to linoleate at the level of each polar lipid, such as PC, PE, and cardiolipin. These results demonstrated that the desaturases were controlled by the O_2 concentration in solution and that desaturation occurs *in vivo* after the formation of phospholipids (Rebeillé *et al.*, 1980). The implications of these findings are that proteins may be inserted into the membrane without concomitant synthesis and insertion of normal molecular species of a given phospholipid. Although these studies suggest a considerable degree of flexibility in the mechanism of membrane growth, it must nevertheless be taken for granted that in the normal situation the integration of genetically determined molecular species of phospholipids and proteins is coordinated. Membranes grow through the insertion of both proteins and phospholipids. The coordination of these two aspects of their growth, however, is poorly understood. In the mitochondrial membranes, this is a particularly intriguing question since its growth involves proteins and phospholipids from both the mitochondrion and the cytoplasm.

Mitrochondria isolated from various plant sources, including potato tubers, spinach leaves, and daffodil petals, readily incorporate isopentenyl diphosphate into prenyl chains (from C_{10} to the natural end product C_{50}) of ubiquinone intermediates. Likewise, isolated plant mitochondria form intermediates of the ubiquinone biosynthetic pathway from 4-

hydroxybenzoate. The biosynthetic steps in ubiquinone formation from the first prenylated intermediate 4-hydroxy-polyprenylbenzoic acid consist of a decarboxylation followed by hydroxylation and methylation (Lütke-Brinkhaus *et al.*, 1984). These results suggest that plant mitochondria have their own prenyltransferase and prenylation system similar to the plastid compartment (Lütke-Brinkhaus *et al.*, 1984).

II. PLANT MITOCHONDRIAL DNA

Mitochondrial genetics has its roots in the cytoplasmically inherited "petite" mutation of yeast described by Ephrussi *et al.* (1949a,b). One of the intriguing things about mitochondria is that they possess their own genetic system. For example, a marked feature of mitochondrial DNAs (mtDNAs) is their remarkably consistent buoyant densities in sucrose density gradients, which fall within the range, 1.700–1.706 g/cm^3. By contrast, nuclear DNAs show a wide range of values. Our detailed knowledge of the mitochondrial genetic system has been acquired mainly from studies with animal, *Saccharomyces cerevisiae*, and *Neurospora crassa* mitochondria (for a detailed review see Kroon and Saccone, 1980; Tzagoloff *et al.*, 1979; G. M. Gray, 1982). The mitochondrial genome of higher plants has only been intensively studied in recent years, but most of its physicochemical characteristics are now well established. The biogenesis of mitochondria involves the possession of a separate and distinctive genophore and associated enzymes for transcription and translation of the genetic information.

A. Structure of Higher Plant Mitochondrial DNA

Mitochondrial DNA from various higher plants exhibits a remarkably uniform buoyant density in the range 1.705–1.706 g/cm^3 (Suyama and Bonner, 1966; Table 5.1) corresponding to a G-C content of about 46% (Wells and Ingle, 1970; Table 5.1). It has been claimed that plant mtDNA consist of a homogeneous population of molecules (see for example, Kolodner and Tewari, 1972). These authors reported that the mtDNAs of pea as well as spinach, lettuce, and bean could be isolated in a circular conformation with a contour length of 30 μm. In fact, there is no general agreement in the literature on the size of the plant mitochondrial genome; different values have been obtained (for excellent reviews see Leaver and Gray, 1982; Lonsdale, 1984), depending on the method of analysis and plant species. Electron microscopy has been used to identify a heterogeneous population of circular molecules ranging from 0.5–30

TABLE 5.1

Physical Parameters of Plant Mitochondrial DNA[a]

Source	Density		Complexity (megadalton)[d]
	(g/cm^3)[b]	Percentage G + C[c]	
Muskmelon	1.7050	45.4	1600
Cucumber	1.7045	44.9	1000
Zucchini	1.7048	45.2	560
Watermelon	1.7057	46.1	220
Pea	1.7055	45.9	240
Corn	1.7046	45.0	320

[a] [From Ward *et al.* (1981). Copyright © 1981, M.I.T.]

[b] The number of determinations was 1 for watermelon, 5 for muskmelon, and 2 for the others. The mean is given.

[c] Calculated from the density, with 1.710 g/cm^3 as equivalent to 50.5% G + C and a change of 1 mg/cm^3 as equivalent to 0.98% G + C.

[d] The mitochondrial DNA values were obtained from Cot curve data (the complexity of a DNA can be approached by measuring the reassociation kinetics of the denatured molecules under standardized conditions).

μm in contour length, corresponding to a maximum molecular weight of 74×10^6, and linear molecules with molecular weights up to 120×10^6 (Vedel and Quetier, 1974; Levings *et al.*, 1979). DNA molecules of exceptional length have also been observed in citrus mtDNA preparations (Fontarnau and Hernandez-Yago, 1982). Additionally, a plethora of small discrete circular or linear DNAs have been identified in maize (Pring *et al.*, 1977; Kemble and Bedbrook, 1980), sugar beet (Powling, 1981), sorghum (Dixon and Leaver, 1982), and brassicacea (Boutry and Briquet, 1982; Palmer *et al.*, 1983) mtDNA. Yet on the basis of size alone, these small discrete circular DNA have dimensions that are much more akin to discrete elements known as plasmids or episomes, and some of them seem related to cytoplasmic male sterility. The proportion of circular molecules in total plant mtDNA has ranged from vanishingly small to substantial (Leaver and Gray, 1982; Fig. 5.3). The linear molecules may in fact have been the result of a break in the double helix of large circular molecules. However, the differently sized circular molecules were probably not produced during the isolation procedures, since the generation of differently sized circular molecules after lysis of the

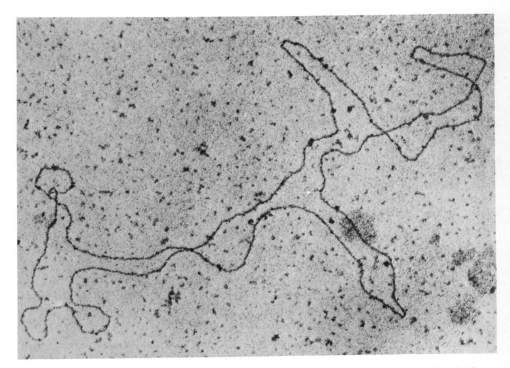

Fig. 5.3.   Electron micrograph of a 10-μm long circular molecule of mitochondrial DNA isolated from 10-day-old etiolated seedlings of wheat (*Triticum aestivum* var capitole). (Courtesy of M. Sevignac and F. Quetier, Laboratoire de Biologie Moléculaire Végétale, Orsay.)

mitochondriion would require ligation and conversion of the supercoiled conformation prior to dye buoyant density gradient centrifugation. Sparks and Dale (1980) have reported the isolation of intact supercoiled mtDNA from tobacco plant suspension culture cells (line W38). These supercoiled mtDNA size classes can be separated on agarose gels and discrete size classes (10–12) can be isolated.

The genome size of the plant mtDNA was also estimated kinetically from DNA reassociation rate measurements. Kinetic-complexity measurements of a range of plant mtDNAs give M_r values of 240×10^6 for pea, 320×10^6 for corn, and between 220×10^6 and 1660×10^6 in the cucurbitaceae family (Ward *et al.*, 1981). Another method that has been used for estimating the size of the plant mitochondrial genome is to sum the lengths of its restriction fragments (Fig. 5.4). Using this method Quetier and Vedel (1977) assigned a minimum genome size of 90×10^6–165×10^6 to the mitochondrial DNAs of potato, cucumber, wheat, and

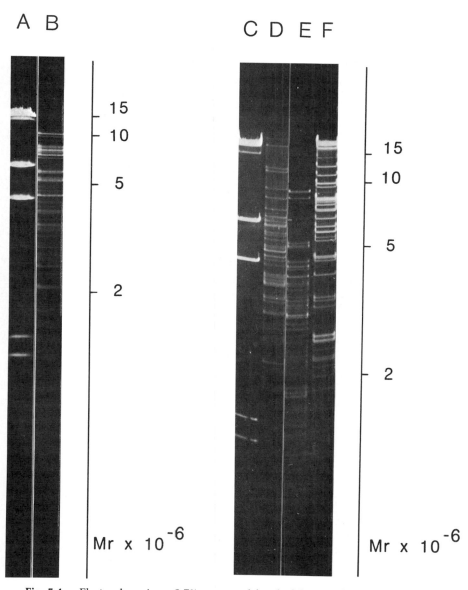

Fig. 5.4.   Electrophoresis on 0.7% agarose slab gel of digests of mtDNA from potato tuber hydrolyzed by: (B) *Bam*HI; (D) *Xho*I; (E) *Eco*RI; and (F) *Sal*I. The mtDNA has been extracted from mitochondria purified by self-generated Percoll gradients. Slots (A) and (C) show molecular weight markers of λ DNA digested with *Hind*III. (Courtesy of P. Lebacq and E. P. Journet.)

Virginia creeper. More recently, these workers (Belliard *et al.*, 1979) found that the total length of *Sal*I fragments of mitochondrial DNA from several types of tobacco ranges from 206 × 10⁶–290 × 10⁶. Bonen and Gray (1980) obtained as much as 230 × 10⁶ minimum complexity using *Xho*I restriction analysis of wheat mitochondrial DNA. In addition, restriction patterns are characterized by an unexpectedly large number of fragments with additive molecular weights generally severalfold higher than the estimated physical molecular weight. (Fig. 5.4). Furthermore, many restriction bands were anomalous in that they exhibited two- to threefold higher levels of fluorescence than did neighboring bands, as determined from peak heights on densitometric tracings. The comparison of the *Eco*RI restriction pattern given by highly purified potato mtDNA has been shown to be identical to that obtained for the mtDNA isolated from potato tubers prepared by conventional procedures. This eliminates the possibility, at least in this material, that the complex restriction patterns reported for higher plant mtDNA involve any artifacts in the organelle isolation, such as plastidial contaminations (P. Lebacq and E. Journet, unpublished data). On the contrary, with other mtDNA (mammalian, yeast, etc.) restriction endonuclease patterns are readily interpretable, and the fragment molecular weights sum to a value consistent in each case with the sequence complexity (determined by DNA–DNA reassociation kinetics) and the physical molecular weight (estimated by electron microscopy) (for a review see G. M. Gray, 1982). To rationalize this apparent discrepancy, it has been proposed that plant mtDNA may actually be heterogeneous, consisting of several types of physically "indistinguishable" molecules having different sequence arrangements of the same genetic information (Quetier and Vedel, 1977). In addition, the results of Bonen *et al.* (1980) make it unlikely that partial methylation of restriction sites could be contributing significantly to the observed complexity of the restriction digests of higher plant mtDNA. By using appropriate probes (e.g., cloned mtDNA fragments) to explore the basis of the unusual restriction patterns of plant mtDNA, it should be possible to gain additional insight into the organization of the mitochondrial genome in different plants. Obviously the size of the plant mitochondrial genome is large and variable (Belliard *et al.*, 1979; Bonen and Gray, 1980; Brennicke, 1980; Fortarnau and Hernandez-Yago, 1982; Lebacq and Vedel, 1981; Pring and Levings, 1978; Ward *et al.*, 1981; Table 5.1), and it is generally agreed that the mitochondrial genome of higher plants is the largest one identified to date. Most animal mtDNA have only some 15,000 nucleotides, but in yeast the mtDNA is five times as long and in higher plants it is as much as five times longer still. In addition, there is considerable mtDNA size heterogeneity within individual plant species as well as substantial differences in the range and frequency of mtDNA size classes among different plants. This situation is in marked contrast to the relative simplicity of fungal and animal mitochondrial genomes.

There are at least two explanations for the enormous sequence complexity of plant mitochondrial DNA, as compared with that for animals; (a) there are many more genes and/or regulatory sequences in plant mtDNA than in animal mtDNA; and (b) there is far more noncoding DNA (selfish or ignorant DNA) in the mitochondria of plants than in those of animals. If it is assumed that plant mtDNA contains one cistron for each of the 3 rRNAs and 20 tRNAs and taking into account the estimated number of molecular polypeptides synthesized by plant mitochondria (18 polypeptides, see later), it is possible to estimate the total required coding capacity (i.e., molecular weight of duplex DNA required for coding). The value obtained (16×10^6) is larger than the corresponding value calculated for animal and yeast mtDNA and allows one to suggest that plant mtDNA may contain a larger number of genes. However, even this estimate still leaves 80% of the potential coding capacity of plant mtDNA unaccounted for (Leaver and Forde, 1980). Interestingly, it has recently been shown by hydridization studies that the mtDNA size classes are directly related to one another and that most do share a degree of homology (Dale, 1981). For example, in the case of mtDNA from tobacco, the 28.8-kb molecule contains all of the restriction fragments found in the 10.1-kb molecule. Furthermore, digests of some of the higher-molecular-weight classes suggest that each successively larger molecule possesses most, if not all, of the sequences present in the smaller size classes in addition to new ones (Dale, 1981). From an evolutionary viewpoint, it is therefore possible that the plant mtDNA is derived from a single circular molecule, which, by a series of duplications, deletions, rearrangements, and other recombinational events, generates other size classes containing reiterated genes (see, however, Ward *et al.*, 1981) and noncoding DNA in varying proportions (Dale, 1981; Leaver and Gray, 1982). The reiterated genes may be analogous to amplified sequences, allowing plant mitochondria to increase the rate of synthesis or maintain higher concentrations of specific RNAs of their corresponding proteins. Unfortunately, the very attractive articles by Stern and Lonsdale (1982) and Lonsdale *et al.* (1983a) partially overturn this simple view by their demonstration that the mitochondrial DNA of maize plants contains a nucleotide sequence homologous to part of the chloroplastic DNA of the same and other species (i.e., chloroplastic DNA sequences have recombined with the mitochondrial genome). For example, the chloroplastic DNA sequence coding for the large subunit

polypeptide of ribulose 1,5-bisphosphate carboxylase has been identified in the mitochondrial DNA from maize (Lonsdale *et al.*, 1983a).
These chloroplastic genes are not transcribed, however, because mitochondria do not share common proteins with plastids (Hansmann and
Sitte, 1984). In fact, "DNA promiscuity" (Ellis, 1982) is a frequent event
in evolution. It could be that there are additional "regulation sequences"
in plant mtDNA that are involved in some form of communication with
the plastids (chloroplasts, amyloplasts, chromoplasts, etc.) and nuclear
genomes during the coordinated changes in cellular function associated
with the development and differentiation of the plant cell (Leaver and
Forde, 1980). Furthermore, in a very interesting article, J. D. Palmer and
Shields (1984) reported the first complete restriction map for a higher
plant mitochondrial genome–that of chinese cabbage (*Brassica camp-
estris*). This genome is organized as three physically distinct circular
molecules. The largest circle, 218 kb in size, bears the entire sequence
complexity of the genome (including two copies of an approximately 2-
kb element present as direct repeats separated by 135 and 83 kb) which
constitutes a "master" chromosome while two smaller circles contain
distinct, 135- and 83-kb subsets of the master molecule and one copy
each of the 2-kb repeat element. These three chromosomes appear to
interconvert via reciprocal recombination across the major repeat element (2 kb) present on all three molecules (see also Chetrit *et al.*, 1984;
Falconet *et al.*, 1984; Lonsdale *et al.*, 1984). Likewise in maize, the structure of the mitochondrial genome has been predicted from restriction
mapping data to exist in one of two states, either as a single large circular
molecule of approximately 1200 kb or three circular molecules, two of
which arise by intramolecular recombination between paired direct repeats from the larger of the three circles (Lonsdale *et al.*, 1983b; Fig. 5.5).
The remarkable similarity of these two unrelated mitochondrial genomes explains many of the puzzling features of plant mitochondrial
DNA. Under these circumstances, these very large circular molecules
would be extremely unlikely to survive mtDNA isolation procedures.
Therefore, it is not unexpected that isolated mtDNA consist of a series of
fragmented linear molecules, with a low proportion of circular DNA
species (Lonsdale, 1984).

Fig. 5.5.   Possible arrangements for repeated sequences in the mitochondrial genome.
(A) Single pair of inverted repeats, recombination leads to sequence inversion, "flip-flop."
(B) Single pair of direct repeats, recombination leads to the generation of two circular
products, "loop-out." (C) A single pair of inverted and direct repeats which are nested.
This gives four possible genomic configurations. The larger the number of direct and
inverted repeats, the more complex the genome organization becomes. Site-specific recombination would tend to limit the number of different circles that could be generated
from a master chromosome containing many repeats. [From Lonsdale (1984).]

A

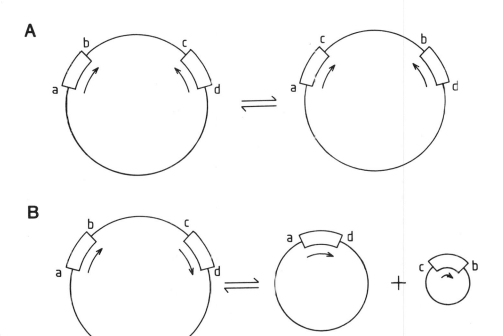

B

C

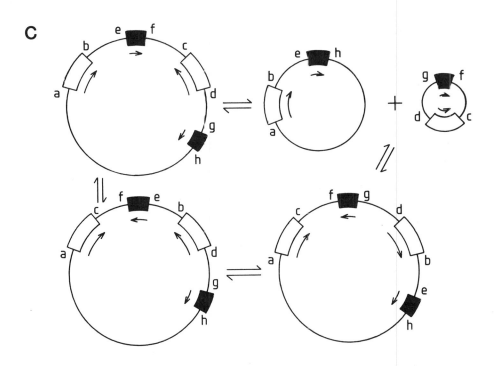

In summary, both linear and circular mtDNA molecules in varying proportions and length distributions have been observed by different investigators in different plant species. While some of these discrepancies undoubtedly have their origin in preparative artifacts and differences in the analytical techniques, the possibility remains that real differences in the organization of the mitochondrial genome may exist in different plants. This raises the important question of how such heterogeneity is transmitted from generation to generation with regard to the biological aspects of plant mtDNA replication. It is possible, however, as suggested by Lonsdale (1984) that in the specialized tissues where cell division is occurring, mtDNA recombination is suppressed, such that the only replicating DNA species would be the master chromosome. There are a number of interesting questions for which we have no answers: whether mtDNA replication is in any way coordinated with the actual duplication of the organelle itself is unknown, as is the site of synthesis within the organelle. Although the biosynthesis of mtDNA has been well documented in animals and fungi (Clayton, 1982), very little is known about the replication of mtDNA in higher plants. Such a situation is probably related to the much greater structural complexity of plant mtDNA. However, isolated mitochondria from wheat embryos are able to synthesize relatively long stretches of deoxynucleotides over the entire mitochondrial genome (Richard *et al.*, 1983). Furthermore, a DNA polymerase resistant to aphidicolin (an inhibitor of nuclear DNA replication) has been characterized in wheat embryo mitochondria, and this enzyme differs in some of its properties (especially at the level of the template recognition) from the animal mitochondrial DNA polymerase γ (Litvak *et al.*, 1983; Christophe *et al.*, 1981). In fact, four DNA polymerases have been purified from wheat embryo. Two of these polymerases (nuclear?) are of the α-type, one is a γ-like DNA polymerase (plastid?), and the last is the mitochondrial polymerase (Castroviejo *et al.*, 1982). Finally, the nature and pattern of integration of plastidial DNA sequences within plant mitochondrial DNA deserves special attention (see Lonsdale *et al.*, 1983a; Stern and Lonsdale, 1982). The frequency of these recombination events, the mechanisms by which DNA transposition has occurred, and the times during evolution when they took place are unclear. The ability of the plant mitochondrion to take up foreign DNA and incorporate it into its genome may partly explain the large size of plant mitochondrial genomes, relative to the mitochondrial genome sizes of fungi and mammals (Ward *et al.*, 1981).

B. Organization of Higher Plant Mitochondrial DNA

Recent methodological advances centered on restriction endonuclease analysis, DNA cloning, and nucleotide sequence analysis have contributed to an understanding of mtDNA organization in yeast and mammals. Indeed, the entire nucleotide sequence of human mtDNA [16,569 bp (base pairs)] has recently been determined (S. Anderson *et al.*, 1981), and this provides the most complete molecular description currently availabe for any organelle genome. However, most of our current knowledge of mitochondrial genes has come from studies of *Saccharomyces cerevisiae* which has proven to be especially suitable for genetic analysis (Tzagoloff *et al.*, 1979). A major conclusion to emerge from comparative studies is that mtDNA is functionally conservative, encoding basically the same genes in all of the yeast and mammals examined to date. Prominent among these are genes for the ribosomal and transfer RNA component (22–24 tRNA species) of a distinctive mitochondrial protein–synthesizing system, the function of which is to translate a limited number of mtDNA-encoded messenger RNAs that specify essentially the same set of polypeptides in all mitochondria (Tzagoloff *et al.*, 1979). These include subunit 6 of the ATP synthase (oligomycin-sensitive ATPase) complex, specified in yeast by the *oli-2/pho-1* gene, the three largest subunits of cytochrome oxidase (COI, COII, and COIII products, respectively, of the *oxi-3*, *oxi-1*, and *oxi-2* loci in yeast), and the apoprotein of cytochrome *b* (CYT B, encoded by the *cob/box* gene). The only well-documented exception to a universal coding function for mtDNA is the ATPase subunit 9 (the dicyclohexylcarbodiimide binding protein) for which the active gene (the *oli-1/pho-2* locus) is mitochondrially located in *Saccaromyces cerevisiae* (Tzagoloff *et al.*, 1979) and plants (Leaver and Gray, 1982) but nuclear in *Neurospora crassa* and probably also in humans (Tzagoloff *et al.*, 1979). Mitochondrial gene products, therefore, play a key role in oxidative phosphorylation and coupled oxidative phosphorylation, even though they make only a modest quantitative contribution to the overall biogenesis of mitochondria (Schatz and Mason, 1974). However, the striking feature to emerge also from these studies is that the structure and gene arrangement vary tremendously among mitochondrial DNAs. For example, the mammalian mitochondrial genome is extraordinarily tightly packed and shows an extreme economy in gene arrangement. In this case, coding sequences for rRNAs, tRNAs, and proteins immediately abut one another, with few or no noncoding nucleotides separating individual genes. Intervening sequences are absent and there even appears to be a small amount of gene overlap, in which the same stretch of DNA encodes parts of two different proteins in different reading frames (S. Anderson *et al.*, 1981). Furthermore, it appears there is only one major promotor, or initiation site, for RNA synthesis. The primary product of transcription is, therefore, a full-length RNA copy of each strand (Attardi, 1981a,b). The copy is then cleaved to yield the final RNAs (rRNA, tRNA, and mRNA). On the other hand, in *Saccharomyces cerevisiae,* mitochondrial genome organization is strikingly different. Thus, several genes are split and contain a number of intervening sequences. The *oxi-3* gene, for example, is about 10 kbp long and consists of seven to eight exons that account for only 16% of the gene sequence (Bonitz *et al.,* 1980a,b). The CYT B coding sequence in yeast mtDNA consists of six exons ranging in size from 14 to 415 kbp and separated by five introns, between 733 and 1670 bp long (Lazowska *et al.,* 1980). Thus, the CYT-B gene in yeast mitochondria is six- to seventimes as long as the same gene in human mitochondria. What purpose is served by having introns present is still obscure, because they are "spliced out" during the transcription of the message. It would seem, therefore, that mammalian mtDNA has evolved to a condition like present day bacteria in which the genes are efficiently packed without introns (except perhaps in some "archaebacterium") and with a minimum of "spacer" DNA between genes. In summary, a comparison of genome organization in the two best studies of mtDNAs, those of mammalian and yeast, does strongly suggest that much of the difference in size between small and large mtDNA can be accounted for by the presence of intra- and intergenic sequences in the latter that are absent in the former (Borst and Grivell, 1981). In support of this suggestion, the distance between the two rRNA genes [in the direction of transcription, from small subunit (16 S) gene to large (23 S)] roughly correlates with the size of the mitochondrial genome (G. M. Gray, 1982).

The major problems in undertaking a detailed analysis of plant mtDNA are the presence of both linear and circular mtDNA molecules in varying proportions and length distributions in different plant species, the lack of well-defined genetic markers (the extreme difficulty in isolating plant mitochondrial mutants has made a more classical genetic approach infeasible), and the inability to isolate discrete species of plant mitochondrial mRNAs as probes for transcriptional mapping. The first plant mitochondrial genes to be directly identified were those for the structural RNAs of the ribosome. Bonen and Gray (1980) used hybridization techniques to demonstrate that the 26 S and 18 S mitochondrial rRNAs and the mixed population of mitochondrial tRNAs are encoded in *Triticum aestivum* (wheat) mtDNA. In addition, they demonstrated that genes for the large subunit (26 S) and the small subunit (18 S) mitochondrial rRNAs are very far apart on the mtDNA, as in *Saccharomyces cerevisiae* but in contrast to the mammalian. For example, in maize the 26 S rRNA gene is separated from the 18 S by 16,000 bp of DNA (Stern *et al.,* 1982). Likewise, the genes coding for 18 S and 26 S are far apart on the cauliflower mitochondrial genome (Chetrit *et al.,* 1984). Interestingly, the wheat mitochondrial 26 S ribosomal RNA gene has no intron and is present in multiple copies arising by recombination (a repeated sequence that includes the rRNA gene flanked by 4 different sequences) (Falconet *et al.*, 1985). It appears, therefore, that the recombination process could be responsible for most of the complexity of restriction profiles observed in wheat mtDNA. It is noteworthy that plant mitochondria (but not those of other eukaryotes) contain a distinctive 5 S RNA species (Leaver and Harmey, 1976; Cunningham *et al.*, 1976) the gene for which is closely linked with that of the 18 S RNA on wheat, *Secale cereale* (Bonen and Gray, 1980), and maize (Stern *et al.*, 1982) mtDNA. This novel localization of the 5 S RNA gene contrasts with that found in prokaryotic and chloroplastic genomes, in which the 5 S rRNA genes are closely linked with those of the 23 S not the 16 S rRNA. In *Triticum aestivum*, tRNA genes appear to be broadly distributed throughout the mitochondrial genome, with some apparent clustering in regions containing 18 S and 5 S rRNA genes (Bonen and Gray, 1980). For example, the 3' end of the tRNA$_{fMet}$ gene is separated by only one base pair from the 5' end of the 18 S ribosomal RNA in wheat mitochondrial DNA (M. W. Gray and Spencer, 1983). This is, in fact, a rather surprising finding, given the "spaciousness" of plant mtDNA. The only other gene that has been clearly identified as being encoded in plant mtDNA is that for subunit II of cytochrome *c* oxidase (Fox and Leaver, 1981). This was done by using yeast mtDNA from the *oxi-1* gene (coding for subunit II of cytochrome *c* oxidase) as a probe for cross-hybridization to the homologous gene in maize. The gene was localized to a single 2.4-kbp maize mtDNA *Eco*RI fragment that was cloned in the bacterial plasmid pBR322 and sequenced. The maize gene comprises two coding regions separated by a single, centrally located, intervening sequence of approximately 794 bp (Fig. 5.6). The presence of an intron in the gene for subunit II of cytochrome *c* oxidase in maize was unexpected in that the corresponding gene in yeast and mammals is not interrupted (G. M. Gray, 1982). As in maize, the wheat COII gene is split; however, the wheat intron is 1.5 times longer because it contains an insert relative to maize (Bonen *et al.*, 1984). Leaver and his colleagues (see Hack and Leaver, 1983) have also identified and sequenced the maize mitochondrial genes for cytochrome *c* oxidase subunit I and apocytochrome *b*. Clearly, an examination of plant mitochondrial genes as well as the proteins they encode will prove very enlightening.

III.   THE MITOCHONDRIAL TRANSLATION SYSTEM

The experiments of McLean *et al.* (1958) first showed the uptake of amino acids and their incorporation into proteins through peptide link-

-63 5'-CACCCAATCCTCGATCTGAATATTGGTGAGTACTATGTCTCATTCACAAATCTATCCTTGTCT

┌──────── Possible initiation codons ────────┐

MET LEU LEU THR LEU GLY LEU VAL LEU LEU LEU VAL LEU PRO MET ILE LEU ARG SER LEU GLU CYS ARG PHE LEU
1   ATG CTA CTC ACT CTC GGT TTG GTC CTA CTT CTG GTG CTG CCA ATG ATT CTT CGT TCA TTA GAA TGT CGA TTC CTC
                                                      HinfI   TaqI HinfI

THR ILE ALA LEU CYS ASP ALA ALA GLU PRO TRP GLN LEU GLY SER GLN ASP ALA ALA THR PRO MET MET GLN GLY
76 ACA ATC GCT CTT TGT GAT GCT GCG GAA CCA TGG CAA TTA GGA TCT CAA GAC GCA GCA ACA CCT ATG ATG CAA GGA
                                                      Sau3A   HinfI

ILE ILE ASP LEU HIS HIS ASP ILE PHE PHE PHE LEU ILE LEU ILE LEU VAL PHE VAL SER ARG MET LEU VAL ARG
151 ATC ATT GAC TTA CAT CAC GAT ATC TTT TTC TTC CTC ATT CTG ATT TTG GTT TTC GTA TCA CGG ATG TTG GTT CGC
                                                                                    ⁄

ALA LEU TRP HIS PHE ASN GLU GLN THR ASN PRO ILE PRO GLN ARG ILE VAL HIS GLY THR THR ILE GLU ILE ILE
226 GCT TTA TGG CAT TTC AAC GAG CAA ACT AAT CCA ATC CCG CAA AGG ATT GTT CAT GGA ACT ACT ATC GAA ATT ATT
                                                                              TaqI

ARG THR ILE PHE PRO SER VAL ILE PRO LEU PHE ILE ALA ILE PRO SER PHE ALA LEU LEU TYR SER MET ASP GLY
301 CGG ACC ATT TTT CCT AGT GTC ATT CCA TTG TTC ATT GCT ATA CCA TCG TTT GCT CTG TTA TAC TCA ATG GAC GGG
    ⁄

VAL LEU VAL ASP PRO ALA ILE THR ILE LYS ALA ILE GLY HIS GLN TRP TYR ARG SER
376 GTA TTA GTA GAT CCA GCC ATT ACT ATC AAA GCT ATT GGA CAT CAA TGG TAT CGG AGT GCGCCTCTTAACGAGGGTGATTT
                Sau3A   ⁄

456 AAGTGCAACGAAATGTACCGGTGGTTCGCGAAGCATCTGGCTTACCGGTCATCTCCCATTCCCGTCGTCGAGAGACTAAAAGAACTATAGCATGCCAGA
                                                                      TaqI

555 AACGGGGAGTTGAGGTGGTTAGACCTATACCCCGAAATGCTCCCAGCATAGGAGCCTATGGTTCCATTCTTGTTATTGCTGGAGGTACACATACCTCTT

654 CTCGGTGTGGTGGAGCGATATACGAAAAATAGATGCTAAGCCCGCAATGTCCGATAACGGGGCTTCAGTAGTGAATCTATCGGCACCACAGCAGTGGCA
                                                                              HinfI

753 TACAACTTTGGACCTAAGGGCCGGCCCCGTTACCTTTCGGAATGGGGGATCCCCGTTGGCAACAACCACGGTAGTAGTTGCGGAACTACTGGGCCAAGA
                                        BamHI

852 GAGGACAACCTGTTGTTCCTGCTCCTCCTTCTTCGCTTCGGGGACGGAGGTCCTACGGTAGGTAAGAGCACGCACAAGCACTTGGCCGAAGGGGACCAG

951 CGCTTCTACTCCTCCACCGAGGAGCCGTTCTTGCGGAGAAGCAAGGGATGTCGTGAACGGTGGGAGGTCAAAGAAAGAGAATTGACCTCTGAATACAGTG

1050 ATCCTATGATCTAGATAGACTCCGTCCTTTTTTTTTTAGATAAGGGTGACTCAAGAGGGGGGAGAACTACCTAACTAAAGAAGAATAGCGCTCTTTAAAA
     Sau3A   HinfI   HinfI

                                                                              TYR GLU TYR SER ASP
1149 ATAAGAGTAGGCGTGGAGAGCTTTTTGCGGGGAAACTTGCAAGTCAAGTTTGGGGGGAGGCGGGCGTCGACCCAACCT TAT GAG TAT TCG GAC
                                                                  TaqI

TYR ASN SER SER ASP GLU GLN SER LEU THR PHE ASP SER TYR THR ILE PRO GLU ASP ASP PRO GLU LEU GLY GLN
1242 TAT AAC AGT TCC GAT GAA CAG TCA CTC ACT TTT GAC AGT TAT ACG ATT CCA GAA GAT GAT CCA GAA TTG GGT CAA
                                                        HinfI   Sau3A

SER ARG LEU LEU GLU VAL ASP ASN ARG VAL VAL VAL PRO ALA LYS THR HIS LEU ARG MET ILE VAL THR PRO ALA
1317 TCA CGT TTA TTA GAA GTT GAC AAT AGA GTG GTT GTA CCA GCC AAA ACT CAT CTA CGT ATG ATT GTA ACA CCC GCT

ASP VAL PRO HIS SER TRP ALA VAL PRO SER SER GLY VAL LYS CYS ASP ALA VAL PRO GLY ARG SER ASN LEU THR
1392 GAT GTA CCT CAT AGT TGG GCT GTA CCT TCC TCA GGT GTC AAA TGT GAT GCT GTA CCT GGT CGT TCA AAT CTT ACC

SER ILE SER VAL GLN ARG GLU GLY VAL TYR TYR GLY GLN CYS SER GLU ILE CYS GLY THR ASN HIS ALA PHE THR
1467 TCC ATC TCG GTA CAA CGA GAA GGA GTT TAC TAT GGT CAG TGC AGT GAG ATT TGT GGA ACT AAT CAT GCC TTT ACG

PRO ILE VAL VAL GLU ALA VAL THR LEU LYS ASP TYR ALA ASP TRP VAL SER ASN GLN LEU ILE LEU GLN THR ASN
1542 CCT ATC GTC GTA GAA GCA GTG ACT TTG AAA GAT TAT GCG GAT TGG GTA TCC AAT CAA TTA ATC CTC CAA ACC AAC

1617 TAA ACCGGGGAAGCTGAAGCGGAAATGCAATTCTCGGGTGAGGGAAGGCTTCGCTCGCTCGCTCAAAAAGCTCTAACGCTCGTTTACGAGTGGAGTGC

1715 ATAAGCCCTTATTGAAGTAG-3'

Fig. 5.6.   DNA sequence of the maize gene for cytochrome oxidase subunit II. Predicted coding regions (exons) have been written in triplets and translated according to the standard genetic code. Presumed nontranslated regions, including the central intervening sequence (intron), have been written without interruptions. Slanted arrows indicate the three CGG codons that occur in this gene. [From Fox and Leaver (1981). Copyright © 1981, M.I.T.]

ages by muscle and liver mitochondria. In a relatively short time, mito-chondrial protein (mt protein) synthesis and organelle biogenesis in yeast and mammals have become extremely active fields of research (for review see Buetow and Wood, 1978).

A. Structure

1. Plant Mitochondrial Ribosomes and Ribosomal RNA

Mitoribosomes (mt ribosomes) differ considerably from both their cytoplasmic and prokaryotic homologues and, furthermore, are indi-vidually variable in structure (Buetow and Wood, 1978). Sedimentation values for mt ribosomes range from 55 S, in the case of mammals, to 73 S, in the case of fungi. In general, the sedimentation values of mt ribosomes are lower than those of the corresponding cytoplasmic ribosomes. It is intriguing that the constituent rRNAs of the two un-equally sized subunits of mitoribosomes vary in size over at least a threefold range, parallelling the sizes of the mitochondrial genomes en-coding them (G. M. Gray, 1982). It is clear, therefore, that this size differential poses a number of questions about the evolution of mitoribo-somes, because their protein components (proteins of mt ribosomes from different sources show a wide range of molecular weights and vary widely from one another) are almost exclusively nuclear gene products (Boynton *et al.*, 1980). (No direct role in the translation process has been unambiguously assigned to any ribosomal protein.) In summary, during evolution, mitochondrial ribosomes have diverged to very different structures, having different sedimentation coefficients, different sizes of rRNAs, and different sets of ribosomal proteins.

Mitochondrial ribosomal proteins from higher plants have not been the subject of such studies up to now. Heavy contamination by pro-plastids during the isolation of mitochondria and the scarcity of mito-chondrial ribosomes (Fig. 1.9) have made these kinds of analyses haz-ardous. Leaver and Harmey (1973) indicated that a variety of higher plant mitochontria contain distinctive ribosomes and rRNAs, which can be distinguished from their cytosol counterparts by sedimentation in sucrose gradients and gel electrophoresis (see also Pollard *et al.*, 1966; Pring, 1974; Quetier and Vedel, 1974). Depending on the species, plant mitoribosomes sediment at 77–78 S, close to the value of 80 S for plant cytoplasmic ribosomes, and contain rRNA components having apparent molecular weights of 0.69×10^6–0.78×10^6 (small subunit) and 1.12×10^6–1.3×10^6 (large subunit). When determined by gel electrophoresis, however, the apparent molecular weights of mitochondrial rRNAs of different species can vary and depend on the condition used (ionic strength, temperature, etc.). Thus, mobilities of these RNAs appear to depend on secondary structure as well as on molecular weight. For example, Pring and Thornbury (1975) found that under nondenaturating conditions molecular weights of maize mitochondrial rRNAs are higher than the corresponding cytoplasmic RNAs. Consequently, the values obtained for molecular weights of mitochondrial rRNAs by electophoresis, and especially under nondenaturating conditions, should not be viewed as final. As mentioned previously, plant mitochondria (but not those of other eukaryotes) contain a distinctive 5 S rRNA species (Leaver and Harmey, 1976; Cunningham et al., 1976). The wheat mitochondrial 5 S can be distinguished from the cytosolic 5.8 S rRNA on the basis of its T1 RNase and pancreatic RNase oligonucleotide "fingerprints" (Cunningham et al., 1976). In addition, the complete nucleotide sequence of the 5 S rRNA encoded by wheat mtDNA has been determined. Somewhat surprisingly, this molecule is not typically prokaryotic (eubacterial) in structure (Spencer et al., 1981). The role of the 5 S rRNA appears to be in the joining of the large ribosomal subunit to the small subunit mRNA/tRNA complex (Buetow and Wood, 1978). As analyzed by one-dimensional polyacrylamide gel electrophoresis, numerous differences in electrophoretic patterns appear between the proteins of mitochondrial and cytoplasmic ribosomes of higher plants (Vasconcelos and Bogorad, 1971; Leaver and Harmey, 1973). The protein complement of the potato tuber mitochondrial ribosomes has been analyzed by two-dimensional electrophoresis in polyacrylamide gels to determine the number and molecular weights of the ribosomal proteins (C. Pinel, R. Douce, and R. Mache, unpublished data). The 33 S small subunit contained 35 protein species ranging in molecular weight from 8000 to 60,000. The 50 S large subunit contained 33 protein species ranging in molecular weight from 12,000 to 46,000 (Fig. 5.7). In addition, Pinel et al. (1984) reported that none of the chloroplastic 30 S or 50 S ribosomal proteins comigrates with plant mitochondrial 33 S or 50 S ribosomal proteins, respectively. In short, plant mt ribosomes differ in a number of respects from all of the mitoribosomes characterized to date.

2. Plant Mitochondrial Messenger RNA

Attardi (1981a,b) have clearly identified a ca. 18 poly(A)-containing presumptive mRNA molecule in HeLa mitochondria. The poly(A) sequence is added posttranscriptionally to the 3' end of mtDNA-coded RNA species. The length of this poly(A) track is about 55 nucleotides. It is interesting to note that the presence of poly(A) at the 3' end of many mitochondrial RNA species represents a definite eukaryotic trait. HeLa mitochondrial mRNA lack one eukaryotic attribute, however, namely the presence of a "cap" structure at their 5' ends (Attardi, 1981a,b). It has not been possible to detect any polyadenylated RNA species in yeast mitochondrial mRNA, however, yeast mitochondrial mRNA possess a 3' noncoding sequence of 50 nucleotides (Borst and Grivell, 1981).

The isolation of plant mitochondrial mRNA is proving to be extremely difficult, in good part because mitochondrial mRNA represents such a small percentage of the total cellular RNA and, therefore, large amounts of starting material are required. Thus, the isolation is not always practical. Another difficulty is the lengthy preparation time needed to ensure mitochondrial purity (i.e., without plastidial or cytoplasmic materials), a requirement that increases the chances of mRNA degradation even if the media contains RNase inhibitors. Consequently, very little is known concerning mRNA in higher plant mitochondria. Evidence is lacking for an identified plant mitochondrial mRNA that is translated *in vitro* by a subcellular system into an identifiable mitochondrial translation product. According to Leaver and Gray (1982), plant mitochondrial mRNAs lack a poly(A) tail. Hybridization of labeled DNA probes, from exon and intron regions of the maize gene for subunit II of cytochrome *c* oxidase, to total maize mitchondrial RNA, fractionated by agarose gel electrophoresis has shown that: (a) the gene is transcribed as a higher-molecular-weight precursor and (b) the precursors are spliced to give a major stable RNA species with a molecular size of ca. 2.6 kbp, which presumably functions as the mature mRNA (Fox and Leaver, 1981). As yet there is no indication of whether or not this maize mitochondrial intron plays an active role in RNA processing.

3. Plant Mitochondrial Transfer RNAs

Mitochondrial transfer RNAs (tRNAs), especially those of mammals, display many deviations from the general structural pattern followed by both bacterial and eukaryotic cytoplasmic tRNAs (M. W. Gray and Doolittle, 1982; G. M. Gray, 1982). Yeast mitochondrial tRNAs have a particularly low G+C content, although most of the stem and loop structures of the standard clover leaf model are of normal size. Mammalian mitochondrial tRNAs are more atypical and are on the average somewhat smaller than normal, due to small loops I and IV. In certain cases, these structural pecularities have been correlated with an altered and/or expanded codon recognition pattern, so that mitochondria have provided the first exceptions to the "universal" genetic code (see, for example, Bonitz *et al.*, 1980a,b). The mitochondrial genome not only employs some different "code words" but also deciphers them more economically (Bonitz *et al.*, 1980a,b). The code is embodied in transfer RNA mole-

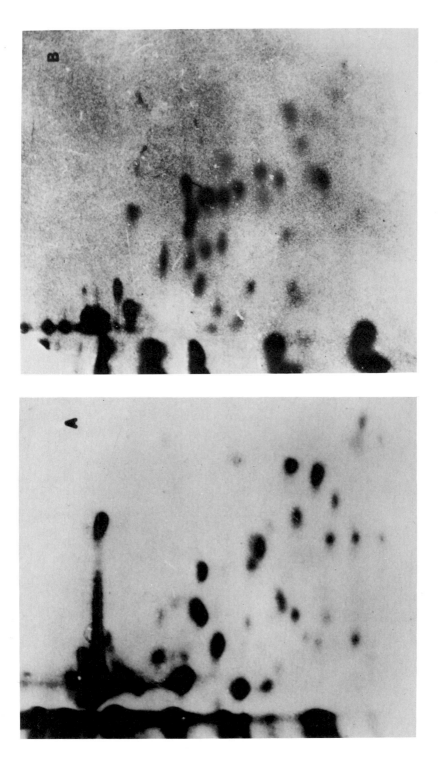

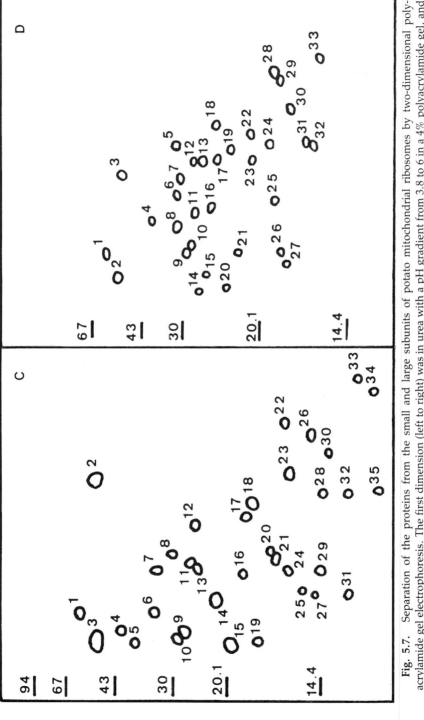
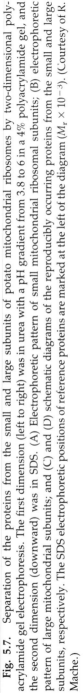

Fig. 5.7. Separation of the proteins from the small and large subunits of potato mitochondrial ribosomes by two-dimensional polyacrylamide gel electrophoresis. The first dimension (left to right) was in urea with a pH gradient from 3.8 to 6 in a 4% polyacrylamide gel, and the second dimension (downward) was in SDS. (A) Electrophoretic pattern of small mitochondrial ribosomal subunits; (B) electrophoretic pattern of large mitochondrial subunits; and (C) and (D) schematic diagrams of the reproducibly occurring proteins from the small and large subunits, respectively. The SDS electrophoretic positions of reference proteins are marked at the left of the diagram ($M_r \times 10^{-3}$). (Courtesy of K. Mache.)

cules, each specific for an amino acid and containing a sequence of three nucleotides (an anticodon) that pairs in a reverse direction with three complementary nucleotides (a codon in a messenger RNA strand). One of the most striking differences is seen in the codon UGA, which in all other genetic systems (including chloroplastic DNA) is a stop codon, whereas in all mitochondria investigated to date, it codes for the amino acid tryptophan. The economy arises as a result of using a smaller amount of tRNA species without importing any from the cytosol. Only about 24 tRNA species are employed in the mitochondrial system, whereas at least 32 tRNAs are needed by nonmitochondrial protein synthesizing systems (there are 4^3 codons, but organisms make do with fewer transfer RNAs because a single transfer RNA can read more than one codon). In fact, the 24 anticodons in mammalian mitochondria pair with only 60 codons because there are four unpaired "stop" codons for polypeptide chain termination in the mammalian mitochondrial code (Jukes, 1983). Each of the 24 tRNAs in mammalian mitochondria appears to be able to read a family of either two or four synonymous codons. For example, anticodon UAG pairs with CUU, CUC, CUA, and CUG (*leucine*). The first anticodon base, U, pairs with U, C, A, or G. There even appears to be some small deviation in the code used by different mitochondria. Thus, AUA is read as methionine rather isoleucine in mammalian mitochondria, whereas CUA is read as threonine rather leucine in yeast mitochondria. There is insufficient evidence at present to permit a decision as to whether the mitochondrial genetic code is more primitive or more highly evolved than has been described for other prokaryotic and eukaryotic systems. According to Bonitz *et al.* (1980a,b), mitochondria represent an evolutionary simplification in which a minimum of tRNAs have been conserved without compromising functional efficiency. It is possible that the narrow role of mitochondrial protein synthesis in the production of a select class of constitutive proteins may have been a crucial factor that led to the reduced number of tRNAs found in present day mitochondria (Bonitz *et al.*, 1980a,b). The divergent sequence of fungal and animal mitochondrial tRNA have so far made it impossible to draw any firm conclusions about their evolutionary origin (Cedergren, 1982).

The presence of organelle-specific tRNAs and their associated amino acyl tRNA synthetases in plant mitochondria is now well documented (Guillemaut and Weil, 1975; Guillemaut *et al.*, 1972, 1975; Jeannin *et al.*, 1976; Sinclair and Pillay, 1981; Swamy and Pillay, 1982), and there seems to be little similarity in the chromatographic properties of isoaccepting tRNAs from the chloroplasts and mitochondria of the same plant species. In addition, the presence of a formylatable initiator tRNA$_{Met}$ in plant mitochondria (formylmethionine is found at the N-terminal amino acid of polypeptides) has been first demonstrated by Weil and his colleagues (Weil, 1979). The presence of a formylatable tRNA$_{Met}$ was also shown in the chloroplasts of various plants (Weil, 1979). In the plant cytoplasm, on the contrary, the initiator tRNA species is a tRNA$_{Met}$ that cannot be formylated. Unfortunately, the structural properties of almost no plant mitochondrial tRNAs have been determined so far, probably because of the great difficulties encountered in obtaining sufficient amounts of tRNA from intact purified mitochondria. Future success in this area will depend on a reliable method for isolating large amounts of intact plant mitochondria from fast growing tissues, devoid of plastid contamination. Such a situation is in marked contrast with yeast, for which at present sequences have been determined for 23 of the estimated 24 mitochondrial tRNA genes and virtually all of their anticodons present in loop II are now known. M. W. Gray and Spencer (1983) discussed the structural features of an initiator methionine tRNA (tRNA$_{fMet}$) from the mitochondria of wheat, as deduced from the sequence of its gene. This is the first mitochondrial tRNA sequence to be reported for higher plants. In contrast to mitochondrial tRNA sequences from animal and fungal mitochondria, it displays strong structural affinity with eubacterial–chloroplastic tRNA$_{fMet}$ sequences. Consequently, based on the secondary structure for wheat mitochondrial tRNA$_{fMet}$ as well as that of a tRNA$_{Pro}$, it appears that plant mitochondrial tRNAs will turn out to be considerably more conservative in structure than other mitochondrial tRNAs studied to date, although there is some evidence of altered codon recognition in plant mitochondria (Fox and Leaver, 1981). Obviously, it will be necessary to examine other mitochondrial tRNAs from higher plants as well as their genes to draw any firm conclusions about their evolutionary origin. It is possible that plant mitochondrial tRNAs may provide an important link between tRNAs of eubacterial and fungal mitochondria in the construction of a mitochondrial phylogenetic tree. Finally, we cannot tell whether the decoding mechanism in mitochondria represents a highly evolved form that has emerged in response to selective pressures on the mitochondrion, or whether it represents the remnant of a primitive system (Rosamond, 1982).

B.   Products of Plant Mitochondrial Protein Synthesis

Protein synthesis by intact mitochondria isolated from several species has been studied to a considerable extent and was reviewed recently (Kroon and Saccone, 1980). Much of the work has had the goal of identifying which of the membrane proteins of the mitochondrion are synthesized within the organelle on its own 70 S ribosome. The most direct means of identifying the chloramphenicol-sensitive protein biosynthesizing activities of mitochondria is to study their ability to incorporate labeled amino acid precursors *in vitro* (Schatz and Mason, 1974). This approach, of course, assumes that cytoplasmic mRNAs are not imported into the mitochondrion and translated on mitochondrial ribosomes. Among the mitochondrial translation products from yeast cells, seven hydrophobic polypeptides of the inner mitochondrial membrane have been identified. Yeast and *Neurospora* mitochondria also synthesize a polypeptide of molecular weight ca. 52,000 which copurifies with one of the polypeptides (S-5) of the small subunit of the mitochondrial ribosome (Lambowitz *et al.*, 1976; LaPolla and Lambowitz, 1981). Attardi (1981a,b) has identified by two-dimensional electrophoresis through an SDS–polyacrylamide gradient slab gel and an SDS–8 M urea–polyacrylamide gradient slab gel ca. 25 discrete components of the M_r range 3500–51,000 in HeLa cell mitochondria. Unfortunately, there is still very little information on the nature of mitochondrial translation products in animal cells. Among these, only the three subunits of cytochrome *c* oxidase have been identified with any degree of confidence.

Leaver and co-workers (1982) have analyzed the proteins synthesized *in vitro* by intact sterile plant mitochondria. The validity of data concerning protein synthesis *in organelle* greatly depends on the purity and intactness of the mitochondria. Mitochondria have been isolated from a number of plant tissues and "optimal" conditions established for radioactive amino acid incorporation into the proteins (Forde *et al.*, 1978). Incorporation is dependent upon the mitochondria being intact and possessing coupled oxidative phosphorylation (or an ATP generating system) and displays characteristic sensitivity toward inhibitors of respiration and organellar protein synthesis (Leaver and Forde, 1980). Amino acid incorporation was insensitive to cycloheximide but completely inhibited by chloramphenicol, confirming that protein synthesis was not occurring on the cytoplasmic ribosomes. In addition, the resistance to ribonuclease indicates that incorporation was taking place within the protecting boundaries of intact mitochondria. Mitochondria from maize seedlings and all of the other higher plants examined thus far synthesize an essentially similar spectrum of some 18–20 major polypeptides and several minor polypeptides ranging in the molecular weight from 8000–54,000 (Fig. 5.8) (Leaver *et al.*, 1983; Boutry *et al.*, 1984). Interestingly, the number of mitochondrial translation products is the same among plants with widely different genome sizes. The two highest-molecular-weight

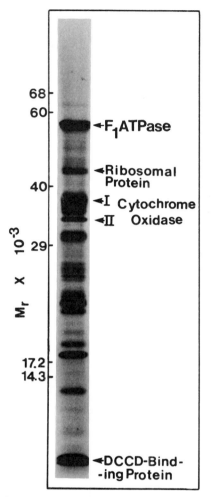

Fig. 5.8. Polypeptides synthesized by isolated maize mitochondria. Mitochondrial translation products were labeled by incubating mitochondria from five-day-old dark-grown maize shoots, in a medium containing [^{35}S]methionine for 90 min. The proteins were solubilized in SDS, electrophoresed in a 15% (w/v) polyacrylamide slab gel, and labeled polypeptides were detected by autoradiography of the dried gel. The SDS electrophoretic positions of reference proteins are marked at the left of the electrophoresis. [From Leaver *et al.* (1982).]

polypeptides are easily solubilized from broken mitochondria whereas the remainder are membrane bound (Leaver *et al.*, 1982). The soluble polypeptide, with an apparent molecular weight of 44,000, copurifies with mitochondrial ribosome proteins and is probably equivalent to the var 1 protein described in yeast mitochondria. By immunoprecipitation with monospecific antibodies prepared against yeast cytochrome oxidase subunits, two of the products have been tentatively identified as subunits I and II of cytochrome c oxidase (Leaver and Forde, 1980). The smallest plant mitochondrial translation produce has an estimated molecular weight of about 8000, is soluble in butanol or in a chloroform–methanol mixture, and specifically binds [^{14}C]DCCD. These characteristics suggest that it is equivalent to the proteolipid subunit of the mitochondrial translocating ATPase (subunit 9), which is encoded in mtDNA in yeast but in the nucleus of animals and *Neurospora*. The nature of the remaining plant mitochondrial translation products can only be a matter of conjecture at present. It appears, however, that plant mitochondria synthesize and, by extrapolation, probably code for additional polypeptides not synthesized by mitochondria from fungi and animals. In support of this suggestion, the α-subunit of the ATPase F_1 complex comigrates on one- and two-dimensional isoelectric focusing SDS–polyacrylamide gels, with the major polypeptide synthesized by isolated plant mitochondria (mitochondrial translation products were labeled with [^{35}S]methionine; Fig. 5.9). These results suggest that the α-subunit but not the β-subunit of mitochondrial ATPase F_1 is synthesized by the mitochondrion (Boutry *et al.*, 1983; Hack and Leaver, 1983). Again, this contrasts with the situation in animal and fungal cells in which the α-subunits of ATPase F_1 are encoded by the nuclear genome and synthesized on cytosolic ribosomes (Hack and Leaver, 1983). In addition, the observations that 20 major polypeptides are synthesized by plant mitochondria may in fact be an underestimate, as it is likely that a number of labeled mitochondrial polypeptides may have the same electrophoretic mobility. In fact, SDS–gel electrophoresis poorly resolves polypeptides with an $M_r < 14,000$. Better resolution of these small polypeptides will be achieved in a 12–18% gradient gel containing 8 M urea (Chua, 1980). The use of appropriate two-dimensional gels, in conjunction with a high rate of protein synthesis, should help to establish more clearly the number and identity of polypeptides synthesized by plant mitochondria. It is, however, possible that some of the labeled bands represent premature termination products of a defective translation system. Interestingly, Tassi *et al.* (1983) have shown that erythromycin inhibited protein synthesis in chloroplasts but not in mitochondria of *Nicotiana sylvestris*. They concluded, therefore, that this antibiotic is a good tool to discriminate *in vivo* between chloroplastic and mitochondrial protein synthesis in the control of gene expression in plant cells.

Considering the large size of the plant mitochondrial genome, the analysis of the nature of the unidentified mitochondrial translation

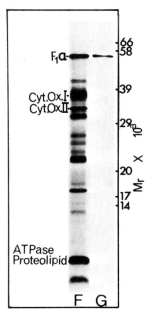

Fig. 5.9. Immunoprecipitation of the α-subunit of ATPase F$_1$ from *in vitro* labeled maize mitochondrial translation products. [^{35}S]Methionine-labeled mitochondrial translation products were immunoprecipitated, analyzed on a 15% (w/v) SDS–acrylamide gel, and autoradiographed. Abbreviations: F, total mitochondrial translation products and G, immunoprecipitation with antibody raised against α-subunit of yeast ATPase F$_1$. [From Hack and Leaver (1983).]

products in plant cells and of their relationships represents a formidable challenge for the future and may reveal unexpected facets of the expression and regulation of the mitochondrial genome in the plant cells. Further studies of the regulatory processes on the levels of transcription and translation seem desirable, but our limited knowledge of these reactions imposes severe experimental limitations. Finally, nothing is known about regulatory sequences in mtDNAs.

C. Variation in Mitochondrial Translation Products Associated with Male-Sterile Cytoplasm in Plants

It is well established that the cytoplasm is a frequent source of heritable variability in eukaryotic cells (Sager, 1972). In higher plants, one of the best examples of this type of inheritance is cytoplasmic male sterility (cms), a trait that causes pollen to abort in the anthers. The cms phenotype that is clearly inherited in a non-Mendelian manner is widely used in the commercial production of F_1 hybrid seed varieties by pre-venting self-pollination of the seed parent. At least 84 sources of cms have been described in maize (Duvick, 1965), but the Texas (or T) source has been extensively utilized because of the ease of fertility restoration and the absence of anther exsertion. Subsequent genetic analysis has shown that there are at least three types of male sterility, designated T, C, (charrua), and S (USDA), each of which is suppressed by different nuclear genes. In fact, certain plant lines carry specific nuclear genes known as restorer genes (Rf genes) which restore fertility by suppress-ing the male-sterile phenotype. However, the unusual nature of the S cytoplasm is evident in the frequency of its spontaneous reversion to a male-fertile condition. Using these revertants to male fertility in subse-quent crossing programs, it was determined that some plants reverted due to nuclear mutations, whereas in others the mutations were cyto-plasmic (Laughnan *et al.*, 1981). It was estimated that 85% of U.S. hybrid corn was produced with cms-T in 1970. In that year, a new race (desig-nated race T) of *Helminthosporium maydis* caused an epidemic of southern corn leaf blight. Race T of *Helminthosporium maydis* displayed increased virulence on lines carrying T male-sterile cytoplasm but has little effect on those with normal fertile (N) and other male-sterile types (cms-C; cms-S).

Several distinct lines of evidence suggest that mitochondria are carriers of genetic determinants that condition cms in plants (J. D. Palmer *et al.*, 1983; Levings, 1983). *Helminthosporium maydis* produces host-specific pathotoxins (polyketo–polyalcohol structure; Kono *et al.*, 1980). The tox-ins preferentially affect mitochondria of lines with T male-sterile cyto-plasm (Miller and Koeppe, 1971). The toxin increases the permeability of the inner membrane of T cytoplasm but not N cytoplasm, mitochondria to NAD^+, and various other cofactors, such as CoA, and, therefore, inhibits respiration dependent on NAD^+-linked substrates. Added NAD^+ partially or fully restored toxin-inhibited electron transport in T cytoplasmic mitochondria (Matthews *et al.*, 1979; Berville *et al.*, 1984). In addition, toxin treatment of T mitochondria uncouples oxidative phos-phorylation (Bednarski *et al.*, 1977) and induces swelling in either ionic or nonionic media, as observed by a decrease in absorbance of the mitochon-drial suspension (Gengenbach *et al.*, 1973; Miller and Koeppe, 1971) or by electron microscopy (Gengenbach *et al.*, 1973). Nuclear genes that restore fertility to T male-sterile cytoplasm modify the response of mitochondria to the *Helminthosporium maydis* toxin, suggesting an alteration of the T cytoplasm mitochondria. The striking heterogeneity of mtDNA among cytoplasms (normal-fertile and male-sterile), with apparently less varia-tion in ctDNA, provides additional and compelling evidence that mitochondria may be involved in the male sterility and disease susceptibility traits in maize (Pring and Levings, 1978; Kemble and Bedbrook, 1980), sugarbeet (Powling, 1981), and sorghum (Pring *et al.*, 1982). Forde *et al.* (1978), Leaver and Forde (1980), and Forde and Leaver (1980) have clearly shown that there are discrete qualitative differences among the polypeptides synthesized by normal, cms-T, cms-S, and cms-C mitochondria. For example, mitochondrial synthesis of a 13,000-M_r variant polypeptide is a cytoplasmic trait that is associated with the T form of cms in maize (Fig. 5.10). The nature of the mutation responsible for synthesis of this polypeptide is not known. In addition to being controlled by mitochondrial genes, the synthesis of the 13,000-M_r polypeptides is under the specific control of certain nuclear genes. When alleles that restore fertility are introduced into the "nuclear background" of lines carrying the T cytoplasm, there is a specific reduction in the rate of synthesis of the 13,000-M_r polypeptides by isolated mitochondria. Leaver and Forde (1980) suggested, therefore, that the 13,000-M_r polypeptide synthesized by cms-T mitochondria could be responsible for the altered membrane structure leading to the formation of a binding site for the toxin. Likewise, an analysis of the polypeptides synthesized by isolated mitochondria has enabled fertile (Kafir) cytoplasm to be distinguished from cms-milo cytoplasm in sorghum (Dixon and Leaver, 1982). Isolated mitochondria from *Faba* beans carrying two different determinants of cytoplasmic male sterility have also been compared to fertile lines. In these cases, [^{35}S]methionine labeling *in situ* of the mitochondrion revealed an additional polypeptide ($M_r = 25,000$), detected only in the two male-sterile cytoplasms (Boutry and Briquet, 1982). The function of these polypeptides is entirely unknown, and there is no direct evidence implicating the variant polypeptides in the expression of the male-sterile phenotype. Interestingly, electrophoretic comparisons of mitochondrial polypeptides from maize lines susceptible and resistant to *Helminthosporium maydis* race T made by Diano (1982), strongly suggests that there is no qualitative difference between the mitochondrial polypeptides of the two inbreds. It is possible, therefore, that the methods used by Diano (1982) were not sensitive enough to detect differences. It is also possible that the synthetic patterns observed by Forde *et al.* (1978) were generated by differential responses to *in vitro* labeling. Further experiments are therefore necessary before any conclusions on the role of these labeled polypeptides can be reached.

The analysis of the mitochondrial DNAs from fertile and male-sterile cytoplasms of maize also provides a means of identifying and distinguishing these cytoplasms (Kemble and Bedbrook, 1980; Pring and Leving, 1978; Pring *et al.*, 1977, 1982; Dixon and Leaver, 1982). For exam-

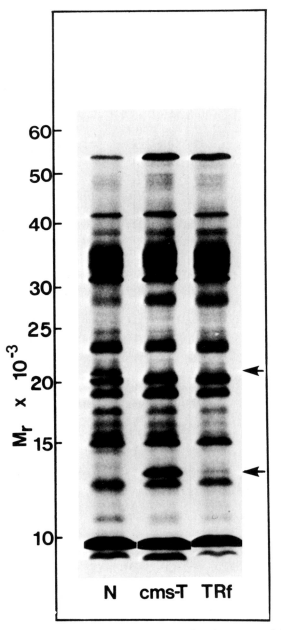

Fig. 5.10. Polypeptides synthesized by mitochondria from normal (N), male-sterile (cmS-T), and restoral male-fertile (TRf) cytoplasms of maize. Mitochondria were isolated and incubated in a medium containing [^{35}S]methionine and an energy generating system. After 90 min of incorporation, mitochondrial polypeptides were fractionated by SDS–polyacrylamide gel electrophoresis and autoradiographed. The level of incorporation into

ple, the male-sterile S cytoplasms are characterized by two linear episomes S1 (6.2 kbp) and S2 (5.4 kbp). The S1 and S2 molecules are structurally related in that they have characteristic inverted terminal repeated sequences. (When the isolated molecules were denatured and analyzed by electron microscopy, a large percentage of the molecules were found in the form of single stranded circles.) The 5' termini are probably covalently linked to a protein, probably involved in the initiation of DNA replication and analogous to adenovirus. They resemble, therefore, transposable elements by possessing terminal inverted repeats and additional regions of sequence homology (Kim *et al.*, 1982). S-Episomes can recombine with homologous sequences present on the circular chromosomes to convert a high proportion of the mtDNA into linear molecules with copies of S1 and S2 residing at the ends of the linear molecules (Schardl *et al.*, 1984). Interestingly, the replication or maintenance of the S1 and S2 episomes is under nuclear control and the most obvious change in the fertile revertants is the loss of free and replicating S1 and S2 episomes (Laughnan *et al.*, 1981). It is also possible, as suggested by Pring *et al.* (1982), that cms in other species may also be related to small discrete, perhaps mobile, genetic elements. In support of this attractive suggestion, Kemble *et al.* (1983) suggested the presence of common DNA sequences in the maize nuclear and mitochondrial genomes. These episomal DNAs (virus-like DNAs?) may provide the mechanism whereby sequences are transported from the mitochondrion to the nucleus. Obviously, further work will be required to establish whether one or several small mtDNA species are responsible for cms. Besides their possible role in the male-sterile phenotype, these small DNA species may be interesting in other ways and require a more detailed study. It is possible that these small DNA species represent a nuclear DNA minicircle that has been sequestered and maintained by the mitochondrion. The identification of their self-replication would be important and might lead one to consider these DNA as potential vectors for plant transformation.

IV. TRANSPORT OF NEWLY SYNTHESIZED PROTEINS INTO MITOCHONDRIA

Mitochondria from a variety of organisms have been shown to synthesize a relatively small number of polypeptides. Approximately 10% of

mitochondrial translation products was similar in all three types of mitochondria, and each gel track was loaded with the same amount of trichloracetic acid–insoluble radioactivity. The arrow indicates those polypeptides synthesized only by mitochondria from male-sterile cytoplasms. (Courtesy of C. J. Leaver.)

the mitochondrial proteins in a given cell are synthesized within the mitochondrion (Schatz and Mason, 1974). These polypeptides are mostly hydrophobic subunits of the complexes of the respiratory chain associated with the inner mitochondrial membrane. Consequently, most mitochondrial proteins are coded for by nuclear genes. From the results of studies on protein synthesis by isolated mitochondria and studies on protein synthesis by cells in the presence of chloramphenicol (which inhibits protein synthesis on mitochondrial ribosomes) or in the presence of cycloheximide (which inhibits protein synthesis on cytoplasmic ribosomes), it is well established that those mitochondrial proteins coded for by nuclear genes are synthesized on cytoplasmic, non-mitochondrial ribosomes (Schatz and Mason, 1974). Thus, most mitochondrial proteins are synthesized outside the organelle.

Globular proteins cannot in general penetrate biological membranes. The large number of charged groups on their surfaces force them to stay in an aqueous environment. There must, therefore, be mechanisms on both mitochondrial membranes which control the entry of cytoplasmic proteins into their destined mitochondrial location (Fig. 5.11). The mechanism involved in transferring proteins from the cytoplasm into the mitochondrion plays a fundamental role in the interactions that necessarily exist between the nuclear and mitochondrial genomes. How do these polypeptides get into the mitochondrion? So far two mechanisms have been proposed to explain the passage of several proteins through the mitochondrial membrane: direct injection and specific protein carriers (Zimmerman *et al.*, 1981). Indeed, there are cytologically only two compartments involved in protein synthesis and expression of nuclear genes, namely soluble and membrane-bound polysomes. Yet there is a much larger number of membrane choices and, therefore, there must be an additional mechanism for directing these proteins to the appropriate membrane, e.g., the outer mitochondrial membrane. It is necessary to remember that the mitochondrial proteins encoded by the nuclear genome have to pass through two membranes. In addition, nothing is known about a possible passage of mRNAs, in one direction or the other, through the mitochondrial membranes.

A. Direct Injection

Direct injection easily explains the passage of protein through the outer mitochondrial membrane. It does not explain the passage of protein through the inner membrane, unless the passage occurs at the point where both membranes are in contact. According to this mechanism, proteins are synthesized on membrane-bound ribosomes and then

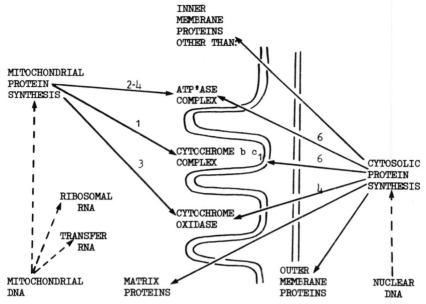

Fig. 5.11. The contribution of nuclear and mitochondrial genetic systems to mitochondrial biogenesis in yeast and *Neurospora*. The number of polypeptide subunits of the
particular enzyme complex of the inner mitochondrial membrane synthesized by each
genetic system is indicated. [From Leaver and Gray (1982) and reproduced, with permission, from the *Annu. Rev. Plant Physiol.* **33**, 375. Copyright © 1982 by Annual Reviews Inc.]

threaded through the membrane as they are being synthesized, in a
manner similar to the transmembrane vectorial transport of secretary
proteins demonstrated to occur in animal cells (Blobel and Sabatini,
1971) and extended by Rothman and Lenard (1977) to membrane
proteins.

The validity of this mechanism for passage through the outer membrane was tested by electron microscopic studies. Growing spheroplasts
prepared from *Saccaromyces cerevisiae* revealed ribosome-like particles
aligned along the outer mitochondrial membrane (Kellems and Butow,
1972). These electron micrographs show a striking similarity between
the association of cytoplasmic ribosomes with yeast mitochondria and
the association of cytoplasmic ribosomes with the endoplasmic reticulum in secretory cells of higher eukaryotes. However, Chepko *et al.*
(1979) were not able to demonstrate associations of cytoplasmic ribosomes with the envelope of the mitochondrion in *Phaseolus aureus*. In
addition, in our laboratory we were unable to observe any association of
ribosomes with the outer mitochondrial membrane, although the density of 80 S ribosomes in the young spinach leaf (greening leaves, 2 mm in

height) cells studied was considerable (Fig. 1.8). The biological significance of these yeast mitochondrially bound polysomes remains unknown. A number of ultrastructural observations have indicated the
existence of occasional continuities between the outer mitochondrial
membrane and the endoplasmic reticulum in cells from plant tissues
(see Chapter 1). Biochemical analyses have also indicated similarities
between these membranes. For example, in plants an NADH-depen-
dent cytochrome *c* reductase activity similar to the "microsomal"
NADH-cytochrome b_5 reductase is associated with the outer mitochondrial membrane (Douce *et al.*, 1972) and the phospholipid composition
of the outer mitochondrial membrane is closer to that of the "microsomal" membrane than that of the inner mitochondrial membrane (Bligny and Douce, 1978). Based on these observations, it has been suggested
that the association between the outer mitochondrial membrane and the
endoplasmic reticulum may facilitate the translocation of newly synthesized mitochondrial proteins from the lumen of the endoplasmic reticulum to the intermembrane space of the mitochondrion. Under these
conditions, the endoplasmic reticulum may be regarded as a "passive
corridor." In support of this suggestion, several laboratories reported
that in animals injected with various labeled amino acids (pulse labeling), mitochondrial proteins, such as the apoprotein of cytochrome *c*,
were synthesized on rough "microsomes" (reviewed by Gonzalez-Cadavid, 1974). However, the redistribution of newly synthesized mitochondrial polypeptides, such as cytochrome *c*, during tissue homogenization following the pulse labeling was not fully assessed and could
have been the source of artifacts (Ades, 1982). This consideration severely limits the conclusions one may draw from the studies on the *in
vivo* pulse labeling of mitochondrial proteins.

While it is reasonably clear that the membranes of the endoplasmic
reticulum may be in some cases morphologically associated with outer
mitochondrial membranes, there is as yet no convincing evidence that
this association is related to the biosynthesis of mitochondrial proteins.
In fact, several independent laboratories determined the subcellular
sites of synthesis of mitochondrial proteins using a different approach
than the *in vivo* pulse-labeling method. In these studies, polysomes
either free or extracted from microsomes were translated in cell-free
translation systems (wheat germ and rabbit reticulocyte systems), and
the products synthesized on these polysomes were then screened im-
munochemically for specific mitochondrial proteins. Invariably, synthesis appeared to take place on the free ribosomes. In addition, all of
these elegant reconstitution experiments *in vitro* demonstrated convincingly that protein synthesis and transport are independent events, in

contrast with the cotranslational transfer of secretory proteins across the endoplasmic reticulum, and that mitochondrial proteins synthesized in heterologous translation systems are always released into the postribosomal supernatant fraction. The mitochondrial proteins discharged into the supernatant are accessible to added protease or antibody. Thus, mitochondrial proteins synthesized *in vitro* are apparently neither enclosed in membrane vesicles nor deeply buried in a lipid phase. For instance, using specific antibodies Korb and Neupert (1978) were able to demonstrate that apocytochrome *c* was released from the free ribosomes into the cytosol, was translocated in the mitochondrion, and became associated with the hemin moiety in the organelles. Such a result contrasts with the early studies, suggesting that the apoprotein of cytochrome *c* was synthesized on rough microsomes (Gonzalez-Cadavid, 1974). It is interesting to note that chloroplasts are remarkably similar to mitochondria, both in structure and in the fact that several of the chloroplasts' proteins have to be synthesized on cytoplasmic ribosomes and transferred into the organelles (Chua and Schmidt, 1979).

B. Envelope Carriers

Mitochondrial proteins synthesized on free polysomes (reviewed by Ades, 1982; and Hay *et al.*, 1984) appear to be released in the cytosol to find their way to the mitochondrion. The incorporation of these polypeptides is, therefore, a posttranslational event, the specificity of which must depend on the recognition by specific membrane components of structural features that are expressed in the newly synthesized polypeptides when the polypeptides are released in the cytosol. Most mitochondrial, chloroplastic, and probably glyoxysomal and peroxysomal proteins utilize posttranslational mechanisms of this kind to reach their functional destination. This implies that a special class of proteins exists in the outer mitochondrial membrane which catalyze the unidirectional influx of all these proteins (coiled or uncoiled) that are made on cytoplasmic ribosomes but which are destined to function in the mitochondrion (Neupert and Schatz, 1981). Furthermore, mitochondrial proteins appear to be synthesized generally as precursors (the molecular weights of the extra segments present on cytoplasmic precursors vary from 2000 to 10,000) which become processed to their mature molecular weights at some point during their translocation into the mitochondrion (Neupert and Schatz, 1981; Fig. 5.12; Table 5.2). The preteolytic conversion of these precursors to their mature size involves a neutral matrix-located protease that has been purified 100-fold from yeast mitochondria (Böhni *et al.*, 1983; Cerletti *et al.*, 1983). EDTA and micromolar concentra-

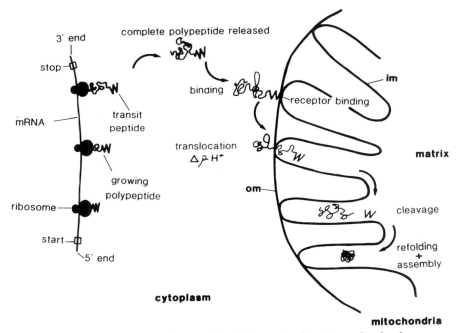

Fig. 5.12.   Steps in import of a mitochondrial protein. Step 1: a molecule of precursor is translated on a cytoplasmic polysome and released into the cytosol. Step 2: the precursor binds, by virtue of its NH_2-terminal recognition sequence, to a receptor (receptor binding) on the mitochondrial outer membrane (OM). Step 3: permanent or transient contact between the mitochondrial membranes allows the precursor to utilize the transmembrane potential ($\Delta\bar{\mu} H^+$) of the inner membrane for translocation. Step 4: the NH_2-terminal recognition sequence is removed in the matrix by a specific protease. Step 5: the processed precursor is assembled into functional forms within the mitochondria. [Adapted from Hay et al. (1984).]

tions of the hydrophobic chelating agent O-phenanthroline (1,10-phenanthroline) inhibit the processing of precursors (Zwizinski and Neupert, 1983). The protease cleaves the precursors of several imported proteins of the matrix and the mitochondrial inner membrane but is inactive against all mature mitochondrial proteins. For example, cytochrome c_1 in yeast is transported and assembled in a two step processing event that seems to involve a "detour" of the precursor to the matrix before it reaches its functional site at the outer face of the inner membrane (Gasser et al., 1982). Proteolytic processing is not necessary for the translocation of precursor proteins across mitochondrial membranes but rather occurs subsequent to this event (Zwizinski and Neupert, 1983).

Import of cytoplasmically synthesized mitochondrial proteins into or across the mitochondrial inner membrane requires an electrochemical

TABLE 5.2

Cytoplasmically Synthesized Mitochondrial Proteins[a]

Protein location	Protein	Organism	Apparent molecular size (Kilodaltons)		Import observed	
			Mature	Precursor	In vivo	In vitro
Matrix						
	ATPase F_1					
	α-Subunit	Yeast	58	64	+	+
	β-Subunit	Yeast	54	56	+	+
	γ-Subunit	Yeast	34	40	+	+
	Citrate synthase	Yeast	47	50		+
		N. crassa	45	47	+	
	RNA polymerase subunit	Yeast	45	47	+	
	Mn^{2+} superoxide dismutase	Yeast	24	26	+	
Inner membrane	Adenine nucleotide translocater	N. crassa	32	32	+	+
	ATPase F_1F_0					
	Subunit IX (= DCCD-binding protein)	N. crassa	8.2	14	+	+
	Cytochrome c oxidase					
	Subunit IV	Yeast	14	17		
	Subunit V	Yeast	12.5	15	+	+
	Subunit VI	Yeast	12.6	17–20		
	Subunit VII	Yeast	5–7.5	5–7.5		

(continued)

TABLE 5.2 (*Continued*)

Protein location	Protein	Organism	Apparent molecular size (Kilodaltons)		Import observed	
			Mature	Precursor	*In vivo*	*In vitro*
	Ubiquinol-cytochrome c reductase (= bc_1 complex)					
	Subunit I	Yeast	44	44.5	+	+
	Subunit II	Yeast	40	40.5	+	+
	Subunit V	Yeast	25	27	+	+
	Subunit VI	Yeast	17	25	+	
	Subunit VII	Yeast	14	14		
	Subunit VIII	Yeast	11	11		
	Cytochrome c_1	Yeast	31	37	+	+
	Subunit I	N. crassa	50	51.5		+
	Subunit II	N. crassa	45	47.5		+
	Subunit V	N. crassa	25	28		+
	Subunit VI	N. crassa	14	14		+
	Subunit VII	N. crassa	11.5	12		+
	Subunit VIII	N. crassa	11.2	11.6		+
	Cytochrome c_1	N. crassa	31	38		+
Intermembrane space	Cytochrome c	N. crassa	12	12	+	+
Outer membrane	Porin	N. crassa	31	31	+	+

[a] [Adapted from Hay *et al.* (1984).]

potential across the inner membrane (Neupert and Schatz, 1981; Schleyer *et al.*, 1982; Fig. 5.12). The first step in the rather complex transport reaction may be binding to a complementary structure [mitochondrial import receptor(s) on the surface of the mitochondrion (Fig. 5.12)]. Evidence for such a receptor has been presented in the case of apocytochrome *c* and the ADP/ATP translocator (Zimmermann and Neupert, 1980; Hay *et al.*, 1984). If the mitochondrial transmembrane potential is abolished, import does not occur but the precursor binds to the mitochondrial surface. Upon reestablishment of the membrane potential, the bound precursor is imported. The binding observed very likely represents an interaction with receptor sites and, thus, is an early step in the import pathway (Zwizinski *et al.*, 1983). It is possible that different receptors of the mitochondrial surface mediate the specific recognition of precursor proteins by the mitochondrion as a first step in the transport process. In addition, the insertion of all major proteins of the outer mitochondrial membrane, such as porine, differ in at least three respects from the pathways by which many proteins are imported into the matrix and the inner membrane; import is not dependent on an energized inner membrane, is not accompanied by NH_2-terminal cleavage, and does not require a trypsin-sensitive component on the mitochondrial surface (Freitag *et al.*, 1982; Gasser and Schatz, 1983).

The physiological significance of the presence of a small peptide called the transit peptide (Chua and Schmidt, 1979) in most of the mitochondrial polypeptides that are synthesized by free cytoplasmic ribosomes is still obscure (Table 5.2). Thus, it is not clear whether the extensions present on the precursors have any special features. This question will obviously be resolved with the accumulation of amino acid sequence data. We can, however, formulate the following hypothesis; (a) the transit peptide is specifically involved in the passage of the polypeptide through both mitochondrial membranes. Thus, in yeast, subunit 9 of the mitochondrial ATPase is coded for by mitochondrial DNA and is translated on mitochondrial ribosomes (Tzagoloff and Meagher, 1972). It is inserted into the inner membrane from the matrix side. No cleavage of a presequence is involved in this process: in fact, the initiating formylmethionine is retained in the mature protein (Macino and Tzagoloff, 1979). In contrast, in *Neurospora* and higher organisms, subunit 9 is coded for by a nuclear gene, is synthesized by cytoplasmic polysomes as a larger precursor, and is posttranslationally transferred into the inner membrane with concomitant proteolytic processing (Zimmermann *et al.*, 1981). However, since at least some soluble (matrix proteins) and some membrane proteins (cytochrome oxidase subunits) seem to enter in the same way, the mechanism responsible for their subsequent divergence (presumably triggered by a second, later, additional peptide segment) remains to be determined. This raises the possibility that proteolytic cleavage at the NH_2 terminus may not be the only posttranslational modification that accompanies uptake of precursor polypeptides into the mitochondrion. In support of this suggestion, cytochrome c_1 undergoes two proteolytic events before it is assembled into complex III (for review see, Hay *et al.*, 1984); (b) the transit peptide might prevent aggregation of hydrophobic proteins synthesized in the hydrophilic medium of the protein synthetic machinery. It is possible, however, that some hydrophobic proteins exist in a water soluble form as precursors and upon integration into the membrane undergo a conformational change and, therefore, become hydrophobic. Thus, a few precursors, such as those of the pore forming protein of the outer membrane and the adenine nucleotide translocator of the inner membrane, lack transient NH_2-terminal extensions, but nonetheless seem to differ from the mature proteins in conformation; and (c) the transit peptide could play a fundamental role either in the assembly of whole protein molecules or their incorporation into the membrane lipid bilayer. For example, according to Wickner (1979), the NH_2-terminal additional peptide is thought to activate the protein for membrane assembly by altering the folding pathway. In other words, the presence of the transit peptide may introduce a basic structural feature into the precursors which would simply allow the protein to diffuse through the mitochondrial membrane. In order to understand the exact function of the transit peptide(s), i.e., to decide which hypothesis is correct, it is absolutely necessary to determine and compare its(their) complete amino acid sequence and to study the localization and biochemical properties of the very specific endoproteases (processing enzymes) involved in its(their) removal. Unfortunately, amino acid sequence data on the terminal ends of the precursor are yet to be obtained. Furthermore, the mechanism of action, submitochondrial distribution, and substrate specificities of the processing enzymes are yet to be established.

It is possible that besides enzymes, some regulatory proteins must also cross the mitochondrial membranes in order to trigger nucleic acid and protein synthesis, either in the matrix or the cytoplasm to ensure a balanced synthesis of the subunits made outside and inside the mitochondrion. Thus, Maeshima and Asahi (1981) and Hattori and Asahi (1982) have shown that in sweet potato root tissue a control is exerted on the synthesis of mitochondrial proteins on the cytoplasmic ribosomes by products, probably proteins, synthesized in the mitochondrion. Clearly, there is still considerable work to be done to provide an understanding of the integration of the mitochondrion and nucleocytoplasmic systems.

We believe, however, that the balance observed between cellular and mitochondrial growth may be achieved by mitochondrial products exerting influence over the expression of nuclear genes that code for mitochondrial proteins.

What is the future of research on plant mitochondrial biogenesis? In the next 10 years most or all of the nucleotide sequences of representative plant mitochondrial genomes will be known. The next steps would then be to learn more about the expression of mitochondrial genes. Present knowledge in this area is surprisingly scant. Can different genes on a mitochondrial DNA be transcribed at different rates, and if yes, how is this controlled? Do RNA splicing events have a regulatory function? A major unsolved problem is the interaction between nuclear, plastidial, and mitochondrial genes. The isolation of nuclear and perhaps plastidial genes for mitochondrial proteins will help in understanding these interactions. Finally, much work still remains to be done in order to fully understand the structure, range in size, and mitochondrial defects which exist in plant mtDNA. It is possible, as an example, that cytoplasmic male sterility in higher plants can be linked to alterations in the mitochondrial genome.

V. ORIGIN OF PLANT MITOCHONDRIA

The structure and function of mitochondria appear to have remained virtually invariant throughout evolution. Two general theories have been proposed to explain the origin of mitochondria. According to the nonsymbiotic theory, mitochondria evolved by compartmentalization of existing genetic material within the ancestral protoeukaryote. Assumptions vary concerning the mechanism by which compartmentalization of segments of the prokaryotic genome within membrane-limited vesicles occurred (Allsopp, 1969; Raff and Mahler, 1972; Uzzell and Spolsky, 1974). The second, or endosymbiotic theory, assumes that mitochondria arose by successive symbiotic events. According to Margulis (1975), the mitochondria in eukaryotic cells originated as free-living aerobic eubacterial species that found shelter within primitive eukaryotic cells (fermentative anaerobe?) and then became permanent symbiotic elements within them. Possibly, the mitochondria of animals, plants, and fungi originated from a single symbiosis preceding the divergences of these kingdoms (Dayhoff and Schwartz, 1981). This is especially apparent when one contrasts the large variety of different electron transport and energy coupling pathways in different bacteria with the single pathway present in all eukaryotic mitochondria. Although there are some recognizable differences between plant and animal mitochondria, these differences are not sufficient at present to indicate multiple origins for mitochondrfia in different groups of organisms (unlike plastids). In both theories of organelle evolution, the properties of organelle genomes and protein-synthesizing systems have assumed a central role. Several differences in the protein-synthesizing machinery of the two cellular systems suggest a different evolutionary origin for each (Mahler, 1981). For example, mitochondrial ribosomes are sensitive to erythromycin and chloramphenicol and are insensitive to cycloheximide, like bacterial ribosomes and unlike cytoplasmic ribosomes. Similarly, mitochondrial protein synthesis is initiated by formylmethionyl tRNA, as in bacterial protein synthesis, while cytoplasmic protein synthesis is initiated by methionyl tRNA. M. W. Gray and Doolittle (1982) argued that the strongest evidence for the proposed prokaryotic nature of mtDNA was derived from T_1-oligonucleotide comparisons of wheat mitochondrial DNA with eubacterial 16 S and other small subunit rRNAs. In support of this suggestion, Schnare and Gray (1982) have sequenced the last 100 nucleotides of the 18 S rRNA of wheat mitochondria and have directly demonstrated a high degree of homology with eubacterial 16 S. Likewise, the nucleotide sequence has been determined for a 664-bp region of maize mtDNA which contains the 3' end of the 18 S RNA gene. Specific regions of the 18 S rRNA gene show a striking homology with the corresponding gene in *E. coli* (Chao *et al.*, 1983). The large degree of homology found between the mitochondrial sequence and the bacterial sequence provides further support for the view that plant mitochondria have an endosymbiotic eubacterial origin (G. M. Gray, 1982; M. W. Gray and Doolittle, 1982). A simple conclusion might be that mitochondria did have a prokaryotic origin, however, this is really an unwarranted conclusion. The detailed analysis of the yeast and perhaps plant mitochondrial genomes has shown a number of surprising features. Mitochondrial genes are organized like nuclear genes in higher eukaryotes and not like bacterial genes. They contain intervening sequences and are surrounded by relatively nontranscribed spacers. However, the identification of introns and RNA splicing in mitochondria present the same problems as finding these features in the nucleus. Doolittle (1978) and Darnell (1978) have argued that split genes were present in the genome of the common ancestor of both eukaryotes and prokaryotes. The fact that introns and RNA splicing have not been found in *E. coli* or blue-green algae says little about the huge variety of unstudied bacterial genera. For instance, prokaryotes comprise two phylogenetically distinct groups, the eubacteria and the "archaebacteria" (the best characterized members of which are the halobacteria and the methanogenic bacteria) (Woese and Fox, 1977). Kaine *et al.* (1983) have cloned a number of tRNA genes and gene clusters from several (distinctly related) species of archaebacteria to look for introns in that group and have found them!

Differing evolutionary pressures on the "free-living" and "intracellular progenitors" of present day eubacteria and mitochondria, respectively, could have resulted in the subsequent elimination of these "extra sequences" at different rates and to different extents, leading in the case of mammals to a mitochondrial genome having little or no noncoding sequences (G. M. Gray, 1982). Under these conditions, repetitive intergenic sequences and introns, both in some larger mtDNAs (e.g., plants) and in nuclear DNA, become retained primitive traits, vestiges of genomic organization in the "universal progenote." If so, plant mtDNA in particular may hold additional important clues to mitochondrial evolution, because it is possible that the plant mitochondrial genome is evolving less rapidly than its animal and fungal counterparts.

John and Whatley (1977), on the basis of an incisive comparison of respiratory chain and respiratory control properties, demonstrated the great similarities of the electron transport chains of *Paracoccus denitrificans* and mitochondria. These include NADH dehydrogenase and succinate dehydrogenase, the proton translocating ATPase, and the enzymes of the TCA cycle. Furthermore, cytochrome aa_3 is the terminal oxidase in *Paracoccus*, as well as in all mitochondria, and the amino acid sequences of cytochrome *c* from a variety of eukaryotes bear a close resemblance to the cytochrome c_2 from a group of bacteria which includes *Paracoccus* (Table 5.3). According to Whatley (1981), we have, therefore, in *Paracoccus* and a few other bacteria, such as *Rhodopseudomonas sphaeroides*, a present day model that contains a great many of the mitochondrial features to be expected of the ancestral bacterium that supposedly gave rise to the mitochondria. This elegant and attractive hypothesis is strengthened by the fact that several workers have reported the occurrence of endosymbiotic bacteria within the multinucleate amoeba (a primitive eukaryote with a highly vacuolated cytoplasm that lacks mitochondria) throughout its entire life cycle (Whatley, 1981). In this case, the bacteria has taken on the characteristics of a functioning cell organelle and very often, its behavior is regulated largely by the activity of the host. This demonstrates the ease with which symbiotic prokaryotes may have been incorporated into the cytoplasm of their host's cells and has provided a wealth of information on the evolutionary potential of such a symbiotic system. Nevertheless, as indicated by Cavalier-Smith (1975), the strongest criticism of the symbiosis theory is that it fails to explain how the eukaryotic condition itself

TABLE 5.3

Mitochondrial Features of Restricted Distribution among Aerobic Bacteria[a]

Mitochondria and *Paracoccus*	Other aerobic bacteria
Respiratory chain	
Two *b*-type and two *c*-type cytochromes readily distinguishable by difference spectrophotometry	One or two *b*-type cytochromes; zero, one, or two *c*-type cytochromes
Cytochrome aa_3 as oxidase	Cytochrome aa_3, a_1, a_2, or *o* as oxidase
Ubiquinone 10 as sole quinone	Ubiquinone 10 or 8 in Gram-negatives; naphthoquinone in Gram-positives
Sensitive to low concentrations of antimycin	Largely insensitive or require higher concentrations of antimycin
NADH respiration inhibited by rotenone	Generally insensitive
Oxidative phosphorylation	
H^+/O Ratio of 6 for the NADH $\rightarrow O_2$ span	H^+/O Ratios of 2, 4, or 6 observed, under similar conditions
Respiratory control released by ADP or by uncouplers	Respiratory control rarely observed
Membrane phospholipid	
Phosphatidylcholine as a major constituent	Phosphatidylcholine rarely observed
All fatty acids, straight-chain saturated and unsaturated	Branched-chain and cyclopropane fatty acids in many bacteria

[a] [From Whatley *et al.* (1979).]

(that is the "cell with nucleus") evolved. Furthermore, this theory does not explain why nuclear genes control the numerous steps of mitochondrial growth and development. It is at least awkward to base arguments for the xenogenous origin of organellar genomes on traits not now encoded in these genomes. It is highly likely that in the course of evolution the initial bacterium, which was supposed to have been taken up endosymbiotically by the "protoeukaryote" host to yield the mitochondrion, became adapted to a symbiotic existence and lost various functions that had been partly taken over by the host nucleus. The shifting of genes from one genome to another within a eukaryotic cell by a gene transfer mechanism seems a reasonable possibility. For example, mitochondrial cytochrome *c* is coded for by the nuclear DNA. Obviously cytochrome *c* in free-living bacteria, such as *Paracoccus*, is coded for by the bacterial DNA. Likewise the α-subunit of ATPase F_1 is a mitochondrial translation product in plants but not in those of mammals and fungi, in which the α-subunit is synthesised on cytoplasmic ribosomes. How could such transfers of DNA have taken place, and are they still happening? Unfortunately, there has not yet been any conclusive demonstration of genetic transfer from a prokaryotic cell to an eukaryotic nucleus. Several possibilities can be entertained. Mitochondria may occasionally lyse and release their DNA into the cytoplasm, which could enter other organelles, such as the nucleus, and become partially integrated into their genomes. It is well established that DNA injected into the cytoplasm of animal cells can be integrated into the nucleus and expressed (W. F. Anderson and Diacumakos, 1981). The discovery by Stern and Lonsdale (1982) that mitochondrial and chloroplastic DNAs share a common sequence strongly supports this view. In addition, experiments have shown that the prokaryote *Agrobacterium tumefaciens* contains a plasmid that can be transferred to a plant cell (Schell *et al.*, 1979). Following transfer of the plasmid, the plant cell develops into a crown gall. If the plasmid containing the information for this morphogenetic change could be shown to join the nuclear genome, then we would have a demonstration of genetic transfer from a prokaryotic cell to an eukaryotic nucleus, the type of transfer required by the evolutionary theory of endosymbiosis (Whatley, 1981). Finally, Kemble *et al.* (1983) reported the presence of common DNA sequences in the maize nuclear and mitochondrial genomes. These episomal DNAs, therefore, may provide the mechanism whereby sequences are transported from the mitochondrion to the nucleus.

With time, the proliferation of the "promitochondria" became synchronized with the cell division of its host and its limiting membrane, the prokaryotic plasma membrane, acquired the appropriate carriers (the adenine nucleotide carrier) necessary for its metabolism to be integrated with the metabolism of its host. It is possible that the specific carriers, including the outer membrane protein(s), responsible for recognition or transport of cytoplasmically synthesized proteins are the only entirely new components necessary in the evolutionary transition from the "promitochondria" to the "mitochondria."

Lastly, while the symbiosis theory readily explains the origin of the inner membrane of the mitochondrion (this membrane derives from the plasma membrane of the wall-free prokaryotic ancestor), it does not explain the origin of the outer membrane of the mitochondrion. Mitochondrial membranes may have existed in their present form at least since upper Silurian times (Uzzel and Spolsky, 1974). It is possible that in the course of evolution this membrane may have arisen either from the inner membrane of the mitochondrion or from the endoplasmic reticulum. It is also possible that the outer membrane could represent a boundary membrane of an endocytotic vacuole formed by a mitochondria-free protoeukaryote. It is clear, but not absolutely certain, that the original function of this enveloping membrane was either to isolate the "invading" prokaryote or to protect the prokaryote against the cell environment. Interestingly, Gram-negative bacteria also contain a membrane external to the cytoplasmic membrane (DiRienzo *et al.,* 1978). Functionally, this outer membrane is quite distinct: it acts as a diffusion barrier against various compounds and contains nonspecific pore proteins analogous to the outer mitochondrial porines that allow the diffusion of low-molecular-weight substrates. Consequently, it is possible that the outer membrane of the mitochondrion is homologous with the outer membrane of the prokaryotic symbiont and not with the vacuolar membrane of the host cell.

It is clear that a great deal of research is needed before the notion that mitochondria are derived from true symbiotic organelles can be asserted with any degree of confidence. The presence of introns and intergenic spacers in the yeast and plant mitochondrial genome does little to clarify the origin of the mitochondrial genome in eukaryotes. Finally, the reason mitochondria possess genomes and protein-synthesizing systems, even in the most highly evolved living creatures, is less clear now than it was several years ago.

Bibliography

Abou-Khalil, S., and Hanson, J. B. (1977). *Arch. Biochem. Biophys.* **183,** 581–587.

Abou-Khalil, S., and Hanson, J. B. (1979a). *Plant Physiol.* **64,** 276–280.

Abou-Khalil, S., and Hanson, J. B. (1979b). *Plant Physiol.* **64,** 281–284.

Ackerman, J. J. H., Grove, T. H., Wong, G. G., Gadian, D. G., and Radda, G. K. (1980). *Nature (London)* **283,** 167–170.

Ackrell, B. A. C., Kearney, E. B., and Coles, C. J. (1977). *J. Biol. Chem.* **252,** 6963–6965.

Ades, I. Z. (1982). *Mol. Cell. Biochem.* **43,** 113–127.

Agrawal, P. K., and Canvin, D. T. (1971). *Can. J. Bot.* **49,** 267–272.

Akerboom, T. P. M., Bookelman, H., Zuurendonk, P. F., Van der Meer, R., and Tager, J. M. (1978). *Eur. J. Biochem.* **84,** 413–420.

Åkerman, K. E. O., and Moore, A. L. (1983). *Biochem. Biophys. Res. Commun.* **114,** 1176–1181.

Åkerman, K. E. O., and Wikström, M. K. F. (1976). *FEBS Lett.* **68,** 191–197.

Albertsson, P.-Å. (1971). "Partition of Cell Particles and Macromolecules," 2nd ed. Wiley, New York.

Alexandre, A., Reynafarje, B., and Lehninger, A. L. (1978). *Proc. Natl. Acad. Sci. U.S.A.* **75,** 5296–5300.

Alexandre, A., Galliazzo, F., and Lehninger, A. L. (1980). *J. Biol. Chem.* **255,** 10721–10728.

Allsopp, A. (1969). *New Phytol.* **68,** 591–600.

Altmann, R. (1890). "Die Elementarorganismen und ihre Beziehungen zu den Zellen," Veit, Leipzig.

Amzel, L. M. (1981). *J. Bioenerg. Biomembr.* **13**, 109–120.
Anderegg, G., L'Eplattenier, F., and Schwarzenbach, G. (1963). *Helv. Chim. Acta* **46**, 1390–1400.
Anderson, J. M. (1983). *Plant Physiol.* **71**, 333–340.
Anderson, J. W., and Done, J. (1977). *Plant Physiol.* **60**, 504–580.
Anderson, J. W., and Walker, D. A. (1983). *Planta* **159**, 77–83.
Anderson, L. E., Ng, T. C. L., and Park, K. E. Y. (1974). *Plant Physiol.* **53**, 835–839.
Anderson, S., Bankier, A. T., Barrell, B. G., de Bruijn, M. H. L., Coulson, A. R., Drouin, J., Eperon, I. C., Nierlich, D. P., Roe, B. A., Sanger, F., Schreier, P. H., Smith, A. J. H., Staden, R., and Young, I. G. (1981). *Nature (London)* **290**, 457–465.
Anderson, W. F., and Diacumakos, E. G. (1981). *Sci. Am.* **245**, 60–65.
ap Rees, T. (1974). *Int. Rev. Biochem.* **11**, 89–127.
ap Rees, T. (1977). *Symp. Soc. Exp. Biol.* **31**, 7–32.
ap Rees, T. (1980). *In* "The Biochemistry of Plants," Vol. 2: Metabolism and Respiration (D. D. Davies, ed.), pp. 1–29. Academic Press, New York.
ap Rees, T., Blanch, E., Graham, D., and Davies, D. D. (1965). *Plant Physiol.* **40**, 910–914.
ap Rees, T., Fuller, W. A., and Wright, B. W. (1976). *Biochim. Biophys. Acta* **437**, 22–35.
ap Rees, T., Wright, B. W., and Fuller, W. A. (1977). *Planta* **134**, 53–56.
ap Rees, T., Bryce, J. H., Wilson, P. M., and Green, J. H. (1983). *Arch. Biochem. Biophys.* **227**, 511–521.
Arron, G. P., and Edwards, G. E. (1979). *Can. J. Biochem.* **57**, 1392–1399.
Arron, G. P., and Edwards, G. E. (1980a). *Plant Physiol.* **65**, 591–594.
Arron, G. P., and Edwards, G. E. (1980b). *Plant Sci. Lett.* **18**, 229–235.
Arron, G. P., Spalding, M. H., and Edwards, G. E. (1979a). *Biochem. J.* **184**, 457–460.
Arron, G. P., Spalding, M. H., and Edwards, G. E. (1979b). *Plant Physiol.* **64**, 182–186.
Asami, K., Juntti, K., and Ernster, L. (1970). *Biochem. Biophys. Acta* **205**, 307–311.
Ashihara, H., and Komamine, A. (1974). *Plant Sci. Lett.* **2**, 331–337. 6, 86–89
Attardi, G. (1981a). *Trends Biochem. Sci. (Pers. Ed.).* **6**, 86–89.
Attardi, G. (1981b). *Trends Biochem. Sci. (Pers. Ed.)* **6**, 100–103.
Avron (Abramsky), M., and Biale, J. B. (1957). *J. Biol. Chem.* **225**, 699–708.
Awasthi, Y. C., Chuang, T. F., Keenan, T. W., and Crane, F. L. (1971). *Biochim. Biophys. Acta* **226**, 42–52.
Axelrod, B., and Beevers, H. (1972). *Biochim. Biophys. Acta* **256**, 175–178.
Azcón-Bieto, J., and Osmond, C. B. (1983). *Plant Physiol.* **71**, 574–581.
Azcón-Bieto, J., Lambers, H., and Day, D. A. (1983a). *Plant Physiol.* **72**, 598–603.
Azcón-Bieto, J., Day, D. A., and Lambers, H. (1983b). *Plant Sci. Lett.* **32**, 313–320.
Azzi, A., Chappell, J. B., and Robinson, B. H. (1967). *Biochem. Biophys. Res. Commun.* **29**, 148–157.
Badger, M. R., and Canvin, D. T. (1981). *In* "Photosyntheis," Vol. 4; Regulation of Carbon Metabolism (G. Akoyunoglou, ed.) pp. 151–161, Balaban Int. Sci. Ser., Philadelphia, Pennsylvania.
Bagshaw, V., Brown, R. and Yeoman, M. M. (1969). *Ann. Bot. (London)* **33**, 35–44.
Bahr, J. T., and Bonner, W. D., Jr. (1973a). *J. Biol. Chem.* **248**, 3441–3445.
Bahr, J. T., and Bonner, W. D., Jr. (1973a). *J. Biol. Chem.* **248**, 3446–3450.
Baird, B. A., and Hammes, G. G. (1979). *Biochim. Biophys. Acta* **549**, 31–53.
Baker, J. E., and Lieberman, M. (1962). *Plant Physiol.* **37**, 90–97.
Baker, J. E., Elfvin, L.-G., Biale, J. B., and Honda, S. I. (1968). *Plant Physiol.* **43**, 2001–2022.
Baldus, B., Kelly, G. J., and Latzko, E. (1981). *Phytochemistry* **20**, 1811–1814.
Balogh, A., Wong, J. H., Wötzel, C., Söll, J., Cséke, C., and Buchanan, B. B. (1984). *FEBS Lett.* **169**, 287–292.

Bandlow, W. (1972). *Biochim. Biophys. Acta* **282**, 105–122.
Barber, D. J., and Thurman, D. A. (1978). *Plant, Cell Environ.* **1**, 297–302.
Barker, J., Khan, M. A. A., and Solomos, T. (1967). *New Phytol.* **66**, 577–582.
Beattie, D. S., Clejan, L., Chen, Y-S., Lin, C-IP., and Sidhy, A. C. (1981). *J. Bioenerg. Biomemb.* **13**, 357–373.
Beconi, M. T., Sanchez, R. A., and Boveris, A. (1983). *Plant Sci. Lett.* **32**, 125–132.
Bednarski, M. A., Izawa, S., and Scheffer, R. P. (1977). *Plant Physiol.* **59**, 540–545.
Beeckmans, S., and Kanarek, L. (1981). *Eur. J. Biochem.* **117**, 527–535.
Beevers, H. (1961). "Respiratory Metabolism in Plants." Harpers, New York.
Beevers, H. (1980). *In* "The Biochemistry of Plants," Vol. 4; Lipids: Structure and Function (P. K. Stumpf, ed.), pp. 117–130. Academic Press, New York.
Beevers, H., Stiller, M. L., and Butt, V. S. (1966). *In* "Plant Physiology: A treatise," (F. C. Steward, ed.), Vol. 4, B, pp. 119–262. Academic Press, New York.
Beevers, L., and Hageman, R. H. (1980). *In* "The Biochemistry of Plants," Vol. 5: Amino Acids and Derivatives (B. J. Miflin, ed.), pp. 115–168. Academic Press, New York.
Behrens, P. W., Marsho, T. V., and Radmer, R. J. (1982). *Plant Physiol.* **70**, 179–185.
Beinert, H., (1977). *In* "Iron–Sulfur Proteins" (W. Lovenberg, ed.), Vol. 3, pp. 61–100. Academic Press, New York.
Beinert, H., and Thompson, A. J. (1983). *Arch. Biochem. Biophys.* **222**, 333–361.
Beinert, H., Ackrell, B. A. C., Vinogradov, A. D., Kearney, E. B., and Singer, T. P. (1977). *Arch. Biochem. Biophys.* **182**, 95–106.
Belliard, G., Vedel, F., and Pelletier, G. (1979) *Nature (London)* **281**, 401–403.
Ben Abdelkader, A., and Mazliak, P. (1970). *Eur. J. Biochem.* **15**, 250–262.
Bendall, D. S., and Bonner, W. D., Jr. (1966). *In* "Hemes and Hemoproteins" (B. Chance, R. W. Estabrook, and T. Yonetani, eds.), pp. 485–488. Academic Press, New York.
Bendall, D. S., and Bonner, W. D., Jr. (1971). *Plant Physiol.* **47**, 236–245.
Bendall, D. S., and Hill, R. (1956). *New Phytol.* **55**, 206–212.
Bennoun, P. (1982). *Proc. Natl. Acad. Sci. U.S.A.* **79**, 4352–4356.
Benson, A. A., and Calvin, M. (1950). *J. Exp. Bot.* **1**, 63–68.
Bergman, A. (1983). Ph.D. thesis. Univ. of Umeå, Umeå, Sweden.
Bergman, A., Gardeström, P., and Ericson, I. (1980). *Plant Physiol.* **66**, 442–445.
Bergman, A., Gardeström, P., and Ericson, I. (1981). *Physiol. Plant.* **53**, 528–532.
Bernardi, P., and Azzone, G. F. (1981). *J. Biol. Chem.* **256**, 7187–7192.
Bervillé, A., Ghazi, A., Charbonner, M., and Bonavent, J. F., (1984). *Plant Physiol.* **76**, 508–517.
Bertagnolli, B. L., and Hanson, J. B. (1973). *Plant Physiol.* **52**, 431–435.
Beyer, R. E., Peters, G. A., and Ikuma, H. (1968). *Plant Physiol.* **43**, 1395–1400.
Bieleski, R. L. (1964). *Anal. Biochem.* **9**, 431–442.
Bird, I. F., Cornelius, M. J., Keys, A. J., and Whittingham, C. P. (1972a). *Biochem. J.* **128**, 191–192.
Bird, I. F., Cornelius, M. J., Keys, A. J., and Whittingham, C. P. (1972b). *Phytochemistry* **11**, 1587–1594.
Bird, I. F., Cornelius, M. J., Dyer, T. A., and Keys, A. J. (1973). *J. Exp. Bot.* **24**, 211–215.
Bird, I. F., Cornelius, M. J., Keys, M. T., and Whittingham, C. P. (1974). *Phytochemistry* **13**, 59–64.
Birnberg, P. R., Jayroe, D. L., and Hanson, J. B. (1982). *Plant Physiol.* **70**, 511–516.
Black, C. C., Goldstein, L. D., Ray, T. B., Kestler, D. P., and Mayne, B. C. (1976). *In* "CO_2 Metabolism and Plant Productivity" (R. H. Burns and C. C. Black, eds.), pp. 113–139. Univ. Park Press, Baltimore, Maryland.
Bligny, R., and Douce, R. (1977). *Plant Physiol.* **60**, 675–679.

Bligny, R., and Douce, R. (1978). *Biochim. Biophys. Acta* **529,** 419–428.
Bligny, R., and Douce, R. (1980). *Biochim. Biophys. Acta* **617,** 254–263.
Blobel, G., and Sabatini, D. (1971). *In* "Biomembranes," (L. A. Manson, ed.), Vol. 2, pp. 193–195. Plenum, New York.
Block, M. A., Dorne, A. J., Joyard, J., and Douce, R. (1983). *FEBS Lett.* **153,** 377–381.
Boggess, S. F., Koeppe, D. E., and Stewart, C. R. (1978). *Plant Physiol.* **62,** 22–25.
Böhni, P. C., Daum, G., and Schatz, G. (1983). *J. Biol. Chem.* **258,** 4937–4943.
Boller, T., and Kende, H. (1979). *Plant Physiol.* **63,** 1123–1132.
Bomsel, J. L., and Pradet, A. (1968). *Biochim. Biophys. Acta.* **162,** 230–242.
Bonen, L., and Gray, M. W. (1980). *Nucleic Acids Res.* **8,** 319–335.
Bonen, L., Huh, T. Y., and Gray, M. W. (1980). *FEBS Lett.* **111,** 340–346.
Bonen, L., Boer, P. H., and Gray, M. W. (1984). *EMBO J.* **3,** 2531–2536.
Bonitz, S. G., Berlani, R., Coruzzi, G., Li, M., Macino, G., Nobrega, F. G., Nobrega, M. P., Thalenfeld, B. E., and Tzagoloff, A. (1980a). *Proc. Natl. Acad. Sci. U.S.A.* **77,** 3167–3170.
Bonitz, S. G., Corruzzi, G., Thalenfeld, B. E., Tzagoloff, A., and Macino, G. (1980b). *J. Biol. Chem.* **255,** 11,927–11,941.
Bonner, W. D., Jr. (1961). *In* "Haematin Enzymes" (J. E. Falk, R. Lemberg, and R. K. Morton, eds.), pp. 479–497. Pergamon, Oxford.
Bonner, W. D., Jr. (1965). *In* "Plant Biochemistry" (J. Bonner and J. E. Varner, eds.), pp. 89–123. Academic Press, New York.
Bonner, W. D., Jr. (1967). *In* "Methods in Enzymology" (R. W. Estabrook and M. E. Pullman, eds.) Vol. 10, 126–133. Academic Press, New York.
Bonner, W. D., Jr. (1973). *In* "Phytochemistry," Vol. 3; Inorganic Elements and Special Groups of Compounds (L. P. Miller, ed.), pp. 221–261. Van Nostrand-Reinhold, Princeton, New Jersey.
Bonner, W. D., Jr. and Plesnicar, M. (1967). *Nature (London)* **214,** 616–617.
Bonner, W. D., Jr., and Prince, R. C. (1984). *FEBS Lett.* **177,** 47–50.
Bonner, W. D., Jr., and Rich, P. R. (1978). *In* "Plant Mitochondria" (G. Ducet and C. Lance, eds.), pp. 241–247. Elsevier, Amsterdam.
Bonner, W. D., Jr., and Rich, P. R. (1983). *Plant Physiol.* **72,** 19 (Suppl.) Abst. no. 103.
Bonner, W. D., Jr., and Voss, D. O. (1961). *Nature, (London)* **191,** 682–684.
Bonner, W. D., Jr., Clark, S. D., and Rich, P. R. (1985). *J. Biol. Chem.* (in press).
Borst, P., and Grivell, L. A. (1981). *Nature (London)* **290,** 443–444.
Boutry, M., and Briquet, M. (1982). *Eur. J. Biochem.* **127,** 129–135.
Boutry, M., Briquet, M., and Goffeau, A. (1983). *J. Biol. Chem.* **258,** 8524–8526.
Boutry, M., Faber, A. M., Charbonnier, M., and Briquet, M. (1984). *Plant Mol. Biol.* **3,** 445–452.
Bowman, E. J., and Ikuma, H. (1976). *Plant Physiol.* **58,** 433–437.
Bowman, E. J., Ikuma, H., and Stein, H. J. (1976). *Plant Physiol.* **58,** 426–432.
Boyer, P. D., Chance, B., Ernster, L., Mitchell, P., Racker, E., and Slater, E. C. (1977). *Annu. Rev. Biochem.* **46,** 955–1026.
Boynton, J. E., Gillham, N. W., and Lambowitz, A. M. (1980). *In* "Ribosomes, Structure, Function and Genetics" (G. Chambliss, G. R. Craven, J. Davies, K. Davis, L. Kahan, and M. Nomura, eds.), pp. 903–950. Univ. Park Press, Baltimore, Maryland.
Bradbeer, C., and Stumpf, P. K. (1960). *J. Lipid Res.* **1,** 214–220.
Branton, D. (1969). *Annu. Rev. Plant Physiol.* **20,** 209–238.
Breidenbach, R. W., and Beevers, H. (1967). *Biochem. Biophys. Res. Commun.* **27,** 462–469.
Brennicke, A. (1980). *Plant Physiol.* **65,** 1207–1210.

Brown, R. H., Rigsby, L. L., and Akin, D. E. (1983). *Plant Physiol.* **71**, 437–439.

Brunton, C. J., and Palmer, J. M. (1973). *Eur. J. Biochem.* **39**, 283–291.

Bryce, J. H. (1983). Ph. D. thesis. Univ. of Cambridge, Cambridge, England.

Buchanan, B. B. (1980). *Annu. Rev. Plant Physiol.* **31**, 341–374.

Buetow, D. E., and Wood, W. M. (1978). *Subcell. Biochem.* **5**, 1–85.

Burke, J. J., Siedow, J. N., and Moreland, D. E. (1982). *Plant Physiol.* **70**, 1577–1585.

Camargo, I. J. B., Kitajima, E. W., and Costa, A. S. (1969). *Phytopathol. Z.* **64**, 282–288.

Cammack, R., and Palmer, J. M. (1973). *Ann. N.Y. Acad. Sci.* **222**, 816–823.

Cammack, R., and Palmer, J. M. (1977). *Biochem. J.* **166**, 347–355.

Canvin, D. T. (1979). *In* "Encyclopedia of Plant Physiology," Vol. 6; Photosynthesis II. Photosynthetic Carbon Metabolism and Related Processes (M. Gibbs and E. Latzko, eds.), pp. 368–396. Springer-Verlag, Berlin and New York.

Canvin, D. T., Berry, J. A., Badger, M. R., Fock, H., and Osmond, C. B. (1980). *Plant Physiol.* **66**, 302–307.

Carde, J. P., Joyard, J., and Douce, R. (1982). *Biol. Cell.* **44**, 315–324.

Carmelli, C., and Biale, J. B. (1970). *Plant Cell Physiol.* **11**, 65–81.

Carnal, N. W. and Black, C. C. (1979). *Biochem. Biophys. Res. Commun.* **86**, 20–26.

Carnal, N. W. and Black, C. C. (1983). *Plant Physiol.* **71**, 150–155.

Castroviejo, M., Fournier, M., Gatius, M., Gandar, J. C., Labouesse, B., and Litvak, S. (1982). *Biochem. Biophys. Res. Commun.* **107**, 294–301.

Cataldo, D. A. (1979). *Plant Physiol.* **5**, 912–917.

Cavalier-Smith, T. (1975). *Nature (London)* **256**, 463–468.

Cavalieri, A. J., and Huang, A. H. C. (1980). *Plant Physiol.* **66**, 588–591.

Cedergren, R. J. (1982). *Can. J. Biochem.* **60**, 475–479.

Cerletti, N., Böhni, P. C., and Suda, K. (1983). *J. Biol. Chem.* **258**, 4944–4949.

Chance, B. (1972). *In* "Methods in Enzymology" (A. San Pietro, ed.), Vol. 24, 322–336. Academic Press, New York.

Chance, B., and Hackett, D. P. (1959). *Plant Physiol.* **34**, 33–49.

Chance, B., and Thorell, B. (1959). *J. Biol. Chem.* **234**, 3044–3050.

Chance, B., and Williams, G. R. (1955). *J. Biol. Chem.* **217**, 383–393.

Chance, B., and Williams, G. R. (1956). *Adv. Enzymol. Relat. Areas Mol. Biol.* **17**, 65–134.

Chance, B., Ernster, L., Garland, P. B., Lee, C. P., Light, P. A., Ohnishi, T., Ragan, C. I., and Wong, D. (1967). *Proc. Natl. Acad. Sci. U.S.A.* **57**, 1498–1505.

Chance, B., Saronio, C. and Leigh, J. S., Jr. (1979). *Biochem. J.* **177**, 931–961.

Chao, S., Sederoff, R. R., and Levings, C. S., III (1983). *Plant Physiol.* **71**, 190–193.

Chapman, E. A., and Graham, T. (1974a). *Plant Physiol.* **53**, 879–885.

Chapman, E. A., and Graham, D. (1974b). *Plant Physiol.* **53**, 886–892.

Chapman, E. A., Bain, J. M., and Gove, D. W. (1975). *Aust. J. Plant Physiol.* **2**, 207–212.

Chapman, E., Wright, L. C., and Raison, J. K. (1979). *Plant Physiol.* **63**, 363–366.

Chapman, K. S. R., and Hatch, M. D. (1977). *Arch. Biochem. Biophys.* **184**, 298–306.

Chappell, J. B. (1968). *Br. Med. Bull.* **24**, 150–157.

Chappell, J. B., and Crofts, A. R. (1966). *In* "Regulation of Metabolic Processes in Mitochondria" (J. M. Tager, S. Papa, E. Quagliariello, and E. Slater, eds.), pp. 293–316. Elsevier, Amsterdam.

Chappell, K. J., and Haarhoff, K. M. (1967). *In* "Biochemistry of Mitochondria" (E. C. Slater, Z. Kaniuga, and L. Wojtczak, eds.), pp. 75–90. Academic Press, New York.

Chauveau, M., and Lance, C. (1982). *Bull. Soc. Bot. Fr.* **129**, 123–134.

Chauveau, M., Dizengremel, P., and Lance, C. (1978). *Physiol. Plant,* **42**, 214–220.

Chen, C. H., and Lehninger, A. L. (1973). *Arch. Biochem. Biophys.* **157**, 183–196.

Chen, R. F. (1967). *J. Biol. Chem.* **242,** 173–181.

Chen, J. and Heldt, H. W. (1983). *Proc. Int. Congr. Photosynth., 6th, 1982* **1,** (Abstr.) no. 112, 4.

Cheniae, G. M. (1965). *Plant Physiol.* **40,** 235–243.

Chepko, G., Weistrop, J. S., and Margulies, M. M. (1979). *Protoplasma* **100,** 385–392.

Chesnoy, L. (1974). *C. R. Hebd. Seances Acad. Sci. Ser: D* **278,** 727–730.

Chetrit, P., Mathieu, C., Muller, J. P., and Vedel, F. (1984). *Curr. Genet.* **8,** 413–421.

Chevallier, D., and Douce, R. (1976). *Plant Physiol.* **57,** 400–402.

Chollet, R. (1974). *Arch. Biochem. Biophys.* **163,** 521–529.

Chollet, R. (1977). *Trends Biochem. Sci. (Pers. Ed.)* **2,** 155–159.

Chrispeels, M. J., Vatter, A. E., and Hanson, J. B. (1966). *J. Microsc. (Oxford)* **85,** 29–44.

Christophe, L., Tarrago-Litvak, L., Castroviejo, M., and Litvak, S. (1981). *Plant Sci. Lett.* **21,** 181–192.

Chua, N. H. (1980). *Methods Enzymol.* **69,** 434–446.

Chua, N. H., and Schmidt, G. W. (1979). *J. Cell Biol.* **81,** 461–483.

Clandinin, M. T., and Cossins, E. A. (1972). *Biochem. J.* **128,** 29–40.

Clandinin, M. T., and Cossins, E. A. (1974). *Phytochemistry* **13,** 585–591.

Clandinin, M. T., and Cossins, E. A. (1975). *Phytochemistry* **14,** 387–391.

Clayton, D. A. (1982). *Cell* **28,** 693–705.

Clermont, H., and Douce, R. (1970). *FEBS Lett.* **9,** 284–286.

Clore, G. M., Andréasson, L. E., Karlsson, B., Aasa, R., and Malmström, B. G. (1980). *Biochem. J.* **185,** 139–154.

Clowes, F. A. L., and Juniper, B. E. (1968). "Plant Cells," Blackwell, Glasgow.

Coleman, J. O. D., and Palmer, J. M. (1971). *FEB Lett.* **17,** 203–208.

Coleman, J. O. D., and Palmer, J. M. (1972). *Eur. J. Biochem.* **26,** 499–509.

Collot, M., Wattiaux-De Coninck, S., and Wattiaux, R. (1975). *Eur. J. Biochem.* **51,** 603–608.

Cook, N. D., and Cammack, R. (1984). *Eur. J. Biochem.* **141,** 573–577.

Cooper, T. G., and Beevers, H. (1969). *J. Biol. Chem.* **244,** 3507–3513.

Cossins, E. A. (1980). *In* "The Biochemistry of Plants." Vol. 2: Metabolism and Respiration (D. D. Davies, ed.), pp. 365–418. Academic Press, New York.

Cottingham, I. R., and Moore, A. L. (1983). *Biochim. Biophys. Acta* **724,** 191–200.

Cottingham, I. R., and Moore, A. L. (1984). *Biochim. J.* **224,** 171–179.

Coulon-Morelec, M. J., and Douce, R. (1968). *Bull. Soc. Chim. Biol.* **50,** 1547–1560.

Cox, G. F. (1969). *Methods Enzymol.* **13,** 47–51.

Cox, G. F., and Davies, D. D. (1967). *Biochem. J.* **105,** 729–734.

Cox, G. F., and Davies, D. D. (1969). *Biochem. J.* **113,** 813–827.

Cox, G. F., and Davies, D. D. (1970). *Biochem. J.* **116,** 819–831.

Cowley, R. C., and Palmer, J. M. (1980). *J. Exp. Bot.* **31,** 199–207.

Craig, D. W., and Wedding, R. T. (1980a). *J. Biol. Chem.* **255,** 5763–5768.

Craig, D. W., and Wedding, R. T. (1980b). *J. Biol. Chem.* **255,** 5769–5775.

Crane, F. L. (1959). *Plant Physiol.* **34,** 128–131.

Crane, F. L. (1965). *In* "Biochemistry of Quinones" (R. A. Morton, ed.), pp. 183–206. Academic Press, London and New York.

Crane, F. L., Lester, R. L., Widmer, C., and Hatefi, Y. (1959). *Biochim. Biophys. Acta* **32,** 73–79.

Cross, R. L. (1981). *Annu. Rev. Biochem.* **50,** 681–714.

Crotty, W. J., and Ledbetter, M. C. (1973). *Science* **182,** 839–840.

Croxdale, J. G. (1983). *Plant Physiol.* **73,** 66–70.

Croxdale, J. G. and Outlaw, W. H., Jr. (1983). *Planta* **157,** 289–297.

Cséke, C., and Buchanan, B. B. (1983). *FEB Lett.* **155,** 139–142.
Cséke, C., Weeden, N. F., Buchanan, B. B., and Uyeda, K. (1982). *Proc. Natl. Acad. Sci. U.S.A.* **79,** 4322–4326.
Cséke, C., Stitt, M., Balogh, A., and Buchanan, B. B. (1983). *FEBS Lett.* **162,** 103–106.
Cséke, C., Balogh, A., Wong, J. H., and Buchanan, B. B. (1985). *Physiol. Veg.* (in press).
Cunningham, R. S., Bonen, L., Doolittle, W. F., and Gray, M. W. (1976). *FEBS Lett.* **69,** 116–122.
Dahlhelm, H., Schober, H., and Ficker, K. (1982). *Biochem. Physiol. Pflanz.* **177,** 156–166.
Dale, R. M. K. (1981). *Proc. Natl. Acad. Sci. U.S.A.* **78,** 4453–4457.
Dalgarno, L. and Birt, L. M. (1963). *Biochem. J.* **87,** 586–596.
Darnell, J. E., Jr. (1978). *Science* **202,** 1257–1260.
Davies, D. D. (1953). *J. Exp. Bot.* **4,** 173–183.
Davies, D. D. (1980). *In* "The Biochemistry of Plants," Vol. 2: Metabolism and Respiration" (D. D. Davies, ed.), pp. 581–611. Academic Press, New York.
Davies, D. D., and Teixeira, A. N. (1975). *Phytochemistry* **14,** 647–656.
Davis, E. J., and Davis-Van Thienen, W. I. A. (1978). *Biochem. Biophys. Res. Commun.* **83,** 1260–1266.
Davis, K. A., and Hatefi, Y. (1971). *Biochemistry* **10,** 2509–2516.
Day, D. A. (1980). *Plant Physiol.* **65,** 675–679.
Day, D. A., and Hanson, J. B. (1977a). *Plant Sci. Lett.* **11,** 99–104; 127.
Day, D. A., and Hanson, J. B. (1977b). *Plant Physiol.* **59,** 139–144.
Day, D. A., and Hanson, J. B. (1977c). *Plant Physiol.* **59,** 630–635.
Day, D. A., and Lambers, H. (1983). *Physiol. Plant.* **58,** 155–160.
Day, D. A., and Wiskich, J. T. (1974a). *Plant Physiol.* **53,** 104–109.
Day, D. A., and Wiskich, J. T. (1974b). *Plant Physiol.* **54,** 360–363.
Day, D. A., and Wiskich, J. T. (1975). *Arch Biochem. Biophys.* **171,** 117–123.
Day, D. A., and Wiskich, J. T. (1977a). *Phytochemistry* **16,** 1449–1502.
Day, D. A., and Wiskich, J. T. (1977b). *Plant Sci. Lett.* **9,** 33–36.
Day, D. A., and Wiskich, J. T. (1978). *Biochim. Biophys. Acta* **501,** 396–404.
Day, D. A., and Wiskich, J. T. (1980). *FEBS. Lett.* **112,** 191–194.
Day, D. A., and Wiskich, J. T. (1981a). *Arch. Biochem. Biophys.* **211,** 100–107.
Day, D. A., and Wiskich, J. T. (1981b). *Plant Physiol.* **68,** 425–429.
Day, D. A., and Wiskich, J. T. (1984). *Physiol. Veg.* **22,** 241–261.
Day, D. A., Rayner, J. R., and Wiskich, J. T. (1976). *Plant Physiol.* **58,** 38–42.
Day, D. A., Arron, G. P., Christoffersen, R. E., and Laties, G. G. (1978a). *Plant Physiol.* **62,** 820–825.
Day, D. A., Bertagnolli, B. L., and Hanson, J. B. (1978b). *Biochim. Biophys. Acta* **502,** 289–297.
Day, D. A., Arron, G. P., and Laties, G. G. (1979). *J. Exp. Bot.* **30,** 539–549.
Day, D. A., Arron, G. P., and Laties, G. G. (1980). *In* "The Biochemistry of Plants," Vol. 2: Metabolism and Respiration (D. D. Davies, ed.), pp. 197–241. Academic Press, New York.
Day, D. A., Neuburger, M., Douce, R., and Wiskich, J. T. (1983). *Plant Physiol.* **73,** 1024–1027.
Day, D. A., Neuburger, M., and Douce, R. (1984). *Arch Biochem. Biophys.* **231,** 233–242.
Day, D. A., Neuburger, M., and Douce, R. (1985). *Aust. J. Plant Physiol.* (in press).
Dayhoff, M. D. (1972). *In* "Atlas of Protein Sequence and Structure," Vol. 5, Suppl. 1 (1973), 2 (1976) and 3 (1978). Nat. Biomedical Res. Foundation, Washington, D.C.
Dayhoff, M. D., and Schwartz, R. M. (1981). *Ann. N. Y. Acad. Sci.* **361,** 92–104.

De Fonseka, K., and Chance, B. (1982). *Biophys. J.* **37,** 402–422.

Deléage, G., Penin, F., Godinot, C., and Gautheron, D. C. (1983). *Biochim. Biophys. Acta* **725,** 464–471.

Delrot, S., and Bonnemain, J. L. (1981). *Plant Physiol.* **67,** 560–564.

Denis, M. (1981). *Biochim. Biophys. Acta.* **634,** 30–40.

Denis, M., and Bonner, W. D., Jr. (1978). *In* "Plant Mitochondria" (G. Ducet and C. Lance, eds.), pp. 35–42. Elsevier/North-Holland, Amsterdam.

Denis, M., and Clore, G. M. (1981). *Plant Physiol.* **68,** 229–235.

Denis, M., and Ducet, G. (1975). *Physiol. Veg.* **13,** 709–720.

Denis, M., and Richaud, P. (1982). *Biochem. J.* **206,** 379–385.

Dennis, D. T., and Coultade, T. P. (1967). *Biochem. Biophys. Acta* **146,** 129–137.

Dennis, D. T., and Miernyk, J. A. (1982). *Annu. Rev. Plant Physiol.* **33,** 27–50.

De Santis, A., Borraccino, G., Arrigoni, O., and Palmieri, F. (1975). *Plant Cell Physiol.* **16,** 911–923.

De Santis, A., Arrigoni, O., and Palmieri, F. (1976). *Plant Cell Physiol.* **16,** 911–923.

Devor, K. A., and Mudd, J. B. (1971). *J. Lipid Res.* **12,** 403–411.

Diano, M. (1982). *Plant Physiol.* **69,** 1217–1221.

Diano, M., and Ducet, G. (1971). *C. R. Hebd. Seances Acad. Sci., Ser. D* **273,** 943–945.

Dickerson, R. E. (1972). *Sci. Am.* **226,** 58–68.

Dickinson, H. G., and Potter, U. (1978). *J. Cell Sci.* **29,** 147–169.

Dieter, P., and Marmé, D. (1980). *Planta* **150,** 1–8.

Dieter, P., and Marmé, D. (1984). *J. Biol. Chem.* **259,** 184–189.

Dilley, R. A., and Crane, F. L. (1963). *Plant Physiol.* **38,** 452.

Diolez, P., and Moreau, F. (1983). *Physiol. Plant.* **59,** 177–182.

Diolez, P., and Moreau, F. (1985). *Biochim. Biophys. Acta* **806,** 56–63.

DiRienzo, J. M., Nakamura, K., and Inouye, M. (1978). *Annu. Rev. Biochem.* **47,** 481–532.

Dittrich, P. (1976). *Plant Physiol.* **57,** 310–314.

Di Virgilio, F., and Azzone, G. F. (1982). *J. Biol. Chem.* **257,** 4106–4113.

Dixon, L. K., and Leaver, C. J. (1982). *Plant Mol. Biol.* **1,** 89–102.

Dizengremel, P. (1977). *Plant Sci. Lett.* **8,** 283–289.

Dizengremel, P. (1983). *Physiol. Veg.* **21,** 743–752.

Dizengremel, P., and Kader, J. C. (1980). *Phytochemistry* **19,** 211–214.

Dizengremel, P., and Lance, C. (1976). *Plant Physiol.* **58,** 147–151.

Dizengremel, P., and Lance, C. (1982). *Bull. Soc. Bot. Fr.* **129,** 19–36.

Dizengremel, P., Chauveau, M., and Lance, C. (1973). *C. R. Hebd. Seances Acad. Sci. Ser. D* **277,** 239–242.

Donaldson, R. P., and Beevers, H. (1977). *Plant Physiol.* **59,** 259–263.

Donaldson, R. P., Tully R. E., Young, O. A., and Beevers, H. (1981). *Plant Physiol.* **67,** 21–25.

Doolittle, W. F. (1978). *Nature (London)* **272,** 581–582.

Douce, R. (1965). *C. R. Hebd. Seances Acad. Sci. Ser. D* **260,** 4067–4070.

Douce, R. (1968). *C. R. Hebd. Seances Acad. Sci. Ser. D* **267,** 534–537.

Douce, R. (1970). Ph. D. Thesis. Univ. of Paris, Paris, France.

Douce, R. (1971). *C. R. Hebd. Seances Acad. Sci. Ser. D* **272,** 3146–3149.

Douce, R., and Bonner, W. D., Jr. (1972). *Biochem. Biophys. Res. Commun.* **47,** 619–624.

Douce, R. and Dupont, J. (1969). *C. R. Hebd. Seances Acad. Sci. Ser. C* **268,** 1657–1660.

Douce, R. and Joyard, J. (1979). *Adv. Bot. Res.* **7,** 1–116.

Douce, R., and Joyard, J. (1980). *In* "The Biochemistry of Plants," Vol. 4: Lipids Structure and Function (P. K. Stumpf, ed.), pp. 321–362. Academic Press, New York.

Douce, R., and Lance, C. (1972). *Physiol. Veg.* **10,** 181–198.

Douce, R., Guillot-Salomon, T., Lance, C., and Signol, M. (1968). *Bull. Soc. Fr. Physiol. Veg.* **14,** 351–373.

Douce, R., Christensen, E. L., and Bonner, W. D. (1972). *Biochim. Biophys. Acta* **275,** 148–160.

Douce, R., Mannella, C. A., and Bonner, W. D., Jr. (1973a). *Biochem. Biophys. Res. Commun.* **49,** 1504–1509.

Douce, R., Mannella, C. A., and Bonner, W. D., Jr. (1973b). *Biochim. Biophys. Acta* **292,** 105–116.

Douce, R., Moore, A. L., and Neuburger, M. (1977). *Plant Physiol.* **60,** 625–628.

Dry, I. B. (1984). Ph. D. Thesis. Univ. of Adelaide, Adelaide, Australia.

Dry, I. B., and Wiskich, J. T. (1982). *Arch. Biochem. Biophys.* **217,** 72–79.

Dry, I. B., and Wiskich, J. T. (1983). *FEBS Lett.* **151,** 31–35.

Dry, I. B., Day, D. A., and Wiskich, J. T. (1983a). *FEBS Lett.* **158,** 154–158.

Dry, I. B., Nash, D., and Wiskich, J. T. (1983b). *Planta* **158,** 152–156.

Ducet, G. (1960). *Ann. Physiol. Veg.* **1,** 19–28.

Ducet, G. (1978). *Physiol. Veg.* **16,** 753–772.

Ducet, G. (1979). *Planta* **147,** 122–126.

Ducet, G. (1980). *Physiol. Plant.* **50,** 241–250.

Ducet, G. (1981). *Physiol. Plant.* **52,** 161–166.

Ducet, G., and Diano, M. (1978a). *Plant Sci. Lett.* **11,** 217–226.

Ducet, G., and Rosenberg, A. J. (1962). *Annu. Rev. Plant Physiol.* **13,** 171–200.

Ducet, G., Diano, M., and Denis, M. (1970). *C. R. Hebd. Seances Acad. Sci. Ser.* D **270,** 2288–2291.

Ducet, G., Gidrol, X., and Richaud, P. (1983). *Physiol. Veg.* **21,** 385–394.

Duggleby, R. G., and Dennis, D. T. (1970a). *J. Biol. Chem.* **245,** 3745–3750.

Duggleby, R. G., and Dennis, D. T. (1970b). *J. Biol. Chem.* **245,** 3751–3754.

Dupéron, R., Meance, J., Lartillot, S., and Dupéron, P. (1975). *Physiol. Veg.* **13,** 539–548.

Dupont, J. (1981). *Physiol. Plant.* **52,** 225–232.

Dupont, J. (1983). Ph. D. Thesis. Univ. Pierre et Marie Curie, Paris, France.

Dupont, J., Rustin, P., and Lance, C. (1982). *Plant Physiol.* **69,** 1308–1314.

Dutton, P. L. (1971). *Biochim. Biophys. Acta* **226,** 63–80.

Dutton, P. L. (1978). *In* "Methods in Enzymology" (S. Fleisher and L. Packer, eds.), Vol. 54, 411–435. Academic Press, New York.

Dutton, P. L., and Storey, B. T. (1971). *Plant Physiol.* **47,** 282–288.

Dutton, P. L., and Wilson, D. F. (1974). *Biochim. Biophys. Acta* **346,** 165–212.

Dutton, P. L., Wilson, D. F., and Lee, C. P. (1970). *Biochemistry* **9,** 5077–5082.

Duvick, D. N. (1965). *Adv. Genet.* **13,** 1–56.

Early, F. G. P., and Ragan, C. I. (1980). *Biochem. J.* **191,** 429–436.

Earnshaw, M. J. (1977). *Phytochemistry* **16,** 181–184.

Edwards, D. L., Rosenberg, F., and Maroney, P. A. (1974). *J. Biol. Chem.* **249,** 3551–3556.

Edwards, G., and Walker, D. A. (1983). "C₃, C₄: Mechanisms, and Cellular and Environmental Regulation, of Photosynthesis." Blackwell, Oxford.

Edwards, G. E., Robinson, S. P., Tyler, N. J. C., and Walker, D. A. (1978). *Plant Physiol.* **162,** 313–319.

Edwards, G. E., Foster, J. G., and Winter, K. (1982). *In* "Crassulacean Acid Metabolism" (I. P. Ting and M. Gibbs, eds.), pp. 92–111. Am. Soc. Plant Physiologists, Rockville, Maryland.

Edwards, J., Reese, T., Wilson, P. M., and Morrell, S. (1984). *Planta* **62,** 188–191.

Elias, B. A., and Givan, C. V. (1977). *Plant Physiol.* **59,** 738.

Ellis, J. (1982). *Nature (London)* **299,** 678–679.

Elthon, T. E., and Stewart, C. R. (1982). *Plant Physiol.* **70,** 567–572.

Emes, M. J., and Fowler, M. W. (1979). *Planta* **145,** 287–292.

Emes, M. J., and Fowler, M. W. (1983). *Planta* **158,** 97–102.

Engel, W. D., Schägger, H., and Von Jagow, G. (1980). *Biochim. Biophys. Acta* **592,** 211–222.

Ephrussi, B., Hottinguer, H., and Chimenes, A. M. (1949a). *Ann. Inst. Pasteur, Paris* **76,** 351–367.

Ephrussi, B., Hottinguer, H., and Tavlitzki, J. (1949b). *Ann. Inst. Pasteur, Paris* **76,** 419–442.

Erecinska, M., Wilson, D. F., Mukai, Y. and Chance, B. (1970). *Biochem. Biophys. Res. Commun.* **41,** 386–392.

Erecinska, M., Stubbs, M., Miyata, Y., Ditre, C. M., and Wilson, D. F. (1977). *Biochim. Biophys. Acta* **462,** 20–35.

Ernster, L., and Lee, C. P. (1967). *In* "Methods in Enzymology" (R. W. Estabrook and M. E. Pullman eds.) Vol. 10, pp. 729–744. Academic Press, New York.

Ericson, I., Sahlström, S., Bergman, A., and Gardeström, P. (1983). *Proc. Int. Congr. Photosynth. 6th, 1982* (Abstr.). **2,** 404–411.

Estabrook, R. W. (1966). *In* "Hemes and Hemoproteins" (B. Chance, R. W. Estabrook, and T. Yonetani, eds.), pp. 405–409. Academic Press, New York.

Faiz-ur-Rahman, A. T. M., Trewavas, A. J., and Davies, D. D. (1974). *Planta* **118,** 195–210.

Falconet, D., Lejeune, D., Quétier, F., and Gray, M. W. (1984). *Embo J.* **3,** 297–302.

Falconet, D., Delorme, S., Lejeune, B., Sevignac, M., Delcher, E., Bazetoux, S., and Quétier, F. (1985). *Curr. Genetics* (in press).

Fernández-Morán, H., Oda, T., Blair, P. V., and Green, D. E. (1964). *J. Cell Biol.* **22,** 63–100.

Fillingame, R. H. (1980). *Annu. Rev. Biochem.* **49,** 1079–1113.

Flemming, W. (1882). "Zellsubstanz, Kern-und Zellteilung." Vogel, Leipzig.

Fontarnau, A., and Hernandez-Yago, J. (1982). *Plant Physiol.* **70,** 1678–1682.

Fonyo, A. (1978). *J. Bioenerg. Biomembr.* **10,** 171–194.

Forde, B. G., and Leaver, C. J. (1980). *Proc. Nat. Acad. Sci. U.S.A.* **77,** 418–422.

Forde, B. G., Oliver, R. J. C., and Leaver, C. J. (1978). *Proc. Natl. Acad. Sci. U.S.A.* **75,** 3841.

Forman, H., and Boveris, A. (1982). *In* "Free Radicals in Biology" (W. A. Pryor, ed.), pp. 65–90. Academic Press, New York.

Forman, N. G., and Wilson, D. F. (1982). *J. Biol. Chem.* **257,** 12,908–12,915.

Fowler, M. W., and ap Rees, T. (1970). *Biochim. Biophys. Acta* **201,** 33–44.

Fox, T. D., and Leaver, C. J. (1981). *Cell* **26,** 315–323.

Foyer, C., Walker, D., Spencer, C., and Mann, B. (1982). *Biochem. J.* **202,** 429–434.

Frédéric, J. (1958). *Arch. Biol.* **69,** 167–349.

Frédéric, J., and Chèvremont, M. (1952). *Arch Biol.* **63,** 109–129.

Frederick, S. E., and Newcomb, E. H. (1969). *J. Cell Biol.* **43,** 343–353.

Freitag, H., Janes, M., and Neupert, W. (1982). *Eur. J. Biochem.* **126,** 197–202.

Fuchs, R., Haas, R., Wrage, K., and Heinz, E. (1981). *Hoppe-Seyler's Z. Physiol. Chem.* **362,** 1069–1078.

Furbank, R. T., Badger, M. R., and Osmond, C. B. (1982). *Plant Physiol.* **70,** 927–931.

Futaesaku, Y., Mizuhira, V., and Nakamura, H. (1972). *Int. Congr. Histo-Cytochem., [Proc.] Kyoto,* 155–156.

Futai, M., and Kanazawa, H. (1980). *Curr. Top. Bioenerg.* **10,** 181.

Galante, Y. M., and Hatefi, Y. (1978). *In* "Methods in Enzymology" (S. Fleischer and L. Packer, eds.) Vol. 53, pp. 15–21. Academic Press, New York.

Galante, Y. M., and Hatefi, Y. (1979). *Arch. Biochem. Biophys.* **192,** 559–568.

Galante, Y. M., Wong, S. Y., and Hatefi Y. (1979). *J. Biol. Chem.* **254**, 12,372–12,378.
Gardeström, P. (1981). Ph. D. Thesis. Univ. of Umeå, Umeå, Sweden.
Gardeström, P., and Edwards, G. E. (1983a). *Proc. Int. Congr. Photosynth. 6th, 1982* **1**, 112–116 (Abstr.).
Gardeström, P., and Edwards, G. E. (1983b). *Plant Physiol.* **71**, 24–29.
Gardeström, P., Ericson, I., and Larsson, C. (1978). *Plant Sci. Lett.* **13**, 231–239.
Gardeström, P., Bergman, A., and Ericson, I. (1980). *Plant Physiol.* **65**, 389–391.
Gardeström, P., Bergman, A., and Ericson, I. (1981). *Physiol. Plant.* **53**, 439–444.
Gardeström, P., Bergman, A., Sahlström, S., Edman, K-A., and Ericson, I. (1983). *Plant Sci. Lett.* **31**, 173–180.
Garland, W. J., and Dennis, D. T. (1977). *Arch. Biochem. Biophys.* **182**, 614–625.
Gasser, S. M., and Schatz, G. (1983). *J. Biol. Chem.* **258**, 3427–3430.
Gasser, S. M., Ohashi, A., Da, G., Böhni, P. C., Gibson, J., Reid, G. A., Yonetani, T., and Schatz, G. (1982). *Proc. Natl. Acad. Sci. U.S.A.* **79**, 267–271.
Gautheron, D. C., and Julliard, J. H. (1979). *In* "Methods in Enzymology" (S. Fleisher and L. Packer, eds.) Vol. 56, pp. 419–430. Academic Press, New York.
Gengenbach, B. G., Miller, R. J., Koeppe, D. E., and Arntzen, C. J. (1973). *Can. J. Bot.* **51**, 2119–2125.
Gerbaud, A., and André, M. (1980). *Plant Physiol.* **66**, 1032–1036.
Gerhardt, B. (1983). *Planta* **265**, 1–60.
Giaquinta, R. T. (1976). *Plant Physiol.* **57**, 872–875.
Giaquinta, R. T. (1979). *Plant Physiol.* **63**, 744–748.
Giersch, C. (1982). *Arch. Biochem. Biophys.* **219**, 379–387.
Gimpel, J. A., De Haan, E. J., and Tager, J. M. (1973). *Biochim. Biophys. Acta* **292**, 582–591.
Givan, C. V. (1968). *Plant Physiol.* **43**, 948–952.
Givan, C. V. (1974). *Phytochemistry* **13**, 1741–1745.
Givan, C. V. (1983). *Physiol. Plant.* **57**, 311–316.
Givan, C. V., and Hodgson, J. M. (1983). *Plant Sci. Lett.* **32**, 233–242.
Givan, C. V., and Torrey, J. G. (1968). *Plant Physiol.* **43**, 635–640.
Goddard, D. (1944). *Am. J. Bot.* **31**, 270–276.
Goldstein, A. H., Anderson, J. O., and McDaniel, R. G. (1980). *Plant Physiol.* **66**, 488–493.
Goldthwaite, J. (1974). *Plant Physiol.* **54**, 399–403.
Gonzalez-Cadavid, N. F. (1974). *Subcell. Biochem.* **3**, 275–309.
Gottlieb, L. D. (1982). *Science* **216**, 373–380.
Gounaris, K., Sen, A., Brain, A. P. R., Quinn, P. J., Williams, W. P. (1983). *Biochim. Biophys. Acta* **728**, 129–139.
Graham, D. (1980). *In* "The Biochemistry of Plants," Vol. 2: Metabolism and Respiration (D. D. Davies, ed.), pp. 525–579. Academic Press, New York.
Graham, D., and Chapman, E. A. (1979). *In* "Encyclopedia of Plant Physiology," Vol. 6: Photosynthesis II (M. Gibbs and E. Latzko, Eds.), pp. 150–162. Springer-Verlag, Berlin.
Gray, G. M. (1982). *Can. J. Biochem.* **60**, 157–171.
Gray, M. W., and Doolittle, W. F. (1982). *Microbiol. Rev.* **46**, 1–42.
Gray, M. W., and Spencer, D. F. (1983). *FEBS Lett.* **161**, 323–327.
Green, D. E. (1966). *Comp. Biochem.* **14**, 309–329.
Greenblatt, G. A., and Sarkissian, I. V. (1973). *Physiol. Plant.* **29**, 361–364.
Groen, A. K., Wanders, R. J. A., Westerhoff, H. V., Van der Meer, R., and Tager, J. M. (1982). *J. Biol. Chem.* **257**, 2754–2757.
Grover, S. D., and Laties, G. G. (1981). *Plant Physiol.* **68**, 393–400.

Grover, S. D., and Wedding, R. T. (1982). *Plant Physiol.* **70**, 1169–1172.

Gruber, P. J., Trelease, R. N., Becker, W. M., and Newcomb, E. H. (1970). *Planta* **93**, 269–288.

Grubmeyer, C., and Spencer, M. (1978). *Plant Physiol.* **61**, 567–569.

Grubmeyer, C., Duncan, I., and Spencer, M. (1977). *Can. J. Biochem.* **55**, 812–818.

Grubmeyer, C., Melanson, D., Duncan, I., and Spencer, M. (1979). *Plant Physiol.* **64**, 757–762.

Guillemaut, P., and Weil, J. H. (1975). *Biochim. Biophys. Acta* **407**, 240–248.

Guillemaut, P., Burkard, G., and Weil, J. H. (1972). *Phytochemistry* **11**, 2217–2219.

Guillemaut, P., Steinmetz, A., Burkard, G., and Weil, J. H. (1975). *Biochim. Biophys. Acta* **378**, 64–72.

Guillermond, A. (1924). *Ann. Sci. Nat. Bot. Biol. Veg.* **6**, 1–134.

Guillermond, A., Mangenot, G., and Plantefol, L. (1933). "Traité de Cytologie Végétale," Le Francois, Paris.

Guillot-Salomon, T. (1972). *C. R. Hebd. Seances Acad. Sci. Ser. D:* **274**, 869–872.

Gunning, B. E. S., and Steer, M. W. (1975). "Ultrastructure and the Biology of Plant Cells." Arnold, London.

Gutierrez, M., Gracen, V. E., and Edwards, G. E. (1974). *Planta* **119**, 279–300.

Gutman, M. (1976). *Biochemistry* **15**, 1342–1348.

Gutman, M. (1980). *Biochim. Biophys. Acta* **594**, 53–84.

Gutman, M., Kearney, E. B., and Singer, T. P. (1971a). *Biochemistry* **10**, 4763–4770.

Gutman, M., Kearney, E. B., and Singer, T. P. (1971b). *Biochemistry* **10**, 2726–2733.

Haas, R., Heinz, E., Popovici, G., and Weissenböck, G. (1979). *Z. Naturforsch. C: Biochem., Biophys., Biol., Virol.* **34**, 854–864.

Hack, E., and Leaver, C. J. (1983). *EMBO J.* **2**, 1783–1789.

Hackenbrock, C. R. (1968). *Proc. Natl. Acad. Sci. U.S.A.* **61**, 598–605.

Hackenbrock, C. R. (1975). *Arch. Biochem. Biophys.* **170**, 139–148.

Hackenbrock, C. R., and Miller, K. J. (1975). *J. Cell Biol.* **65**, 615–630.

Hackett, D. P. (1959). *Annu. Rev. Plant Physiol.* **10**, 113–116.

Hackett, D. P. (1963). *In* "Control Mechanisms in Respiration and Fermentation" (B. Wright, ed.), pp. 105–127. Ronald Press, New York.

Hackett, D. P., Rice, B., and Schmid, C. (1960a). *J. Biol. Chem.* **235**, 2140–2144.

Hackett, D. P., Hass, D. W., Griffith, S. K. and Niederpruem, D. J. (1960b). *Plant Physiol.* **35**, 8–19.

Halestrap, A. P., and Denton, R. M. (1974). *Biochem. J.* **138**, 313–316.

Hallermayer, G., and Neupert, W. (1974). *Hoppe-Seyler's Z. Physiol. Chem.* **355**, 279–288.

Hampp, R., and Ziegler, H. (1980). *Planta* **147**, 485–494.

Hampp, R., Goller, M., and Ziegler, H. (1982). *Plant Physiol.* **69**, 448–455.

Hampson, R. K., Barron, L. C., and Olson, M. S. (1983). *J. Biol. Chem.* **258**, 2993–2999.

Hannig, K., and Heidrich, H. G. (1974). *In* "Methods in Enzymology" (S. Fleisher and L. Packer, eds.), Vol. 31, pp. 746–760. Academic Press, New York.

Hansford, R. G. (1972). *FEBS Lett.* **21**, 139–141.

Hansmann, P., and Sitte, P. (1984). *Z. Naturforsch.* **39c**, 758–766.

Hanson, J. B., and Day, D. A. (1980). *In* "The Biochemistry of Plants," Vol. 1: The Plant Cell (N. E. Tolbert, ed.), pp. 315–358. Academic Press, New York.

Hanson, J. B., and Hodges, T. K. (1967). *Curr. Top. Bioenerg.* **2**, 65–98.

Hanson, J. B., Bertagnolli, B. L., and Shepherd, W. D. (1972). *Plant Physiol.* **50**, 347–354.

Harmon, H. J., Hall, J. D., and Crane, F. L. (1974). *Biochim. Biophys. Acta* **344**, 119–155.

Harrison, B. D., and Roberts, I. M. (1968). *J. Gen. Virol.* **3**, 121–131.

Harrison, B. D., and Woods, R. D. (1966). *Virology* **28**, 610–620.

Hartmann, E., Jeck, U., and Grasmück, I. (1981). *Z. Pflanzenphysiol.* **103**, 427–433.

Hartree, E. F. (1957). *Adv. Enzymol.* **18**, 1–64.

Harwood, J. L. (1980). *In* "The Biochemistry of Plants," Vol. 4: Lipids: Structure and Function (P. K. Stumpf, ed.), pp. 1–55. Academic Press, New York.

Hasson, E. P., and Laties, G. G. (1976). *Plant Physiol.* **57**, 148–152.

Hatch, M. D., and Kagawa, T. (1974). *Arch. Biochem. Biophys.* **160**, 346–349.

Hatch, M. D., and Mau, S. (1977). *Arch. Biochem. Biophys.* **179**, 361–369.

Hatch, M. D., and Osmond, C. B. (1976). *In* "Encyclopedia of Plant Physiology," Vol. 3, Transport in Plants III, Intracellular Interactions and Transport Processes (C. R. Stocking, and U. Heber, eds.), pp. 144–184, Springer-Verlag, Berlin.

Hatch, M. D., Mau, S. L., and Kagawa, T. (1974). *Arch. Biochem. Biophys.* **165**, 188–200.

Hatch, M. D., Kagawa, T., and Craig, S. (1975). *Aust. J. Plant Physiol.* **2**, 111–128.

Hatch, M. D., Tsuzuki, M., and Edwards, G. E. (1982). *Plant Physiol.* **69**, 483–491.

Hatefi, Y. (1976). *In* "Enzymes of Biological Membranes," Vol. 4: Electron Transport Systems and Receptors (A. Martonosi, ed.), pp. 3–41. Plenum, New York.

Hatefi, Y. (1978). *In* "Methods in Enzymology" (S. Fleischer and L. Packer, eds.), Vol. 53, pp. 35–40. Academic Press, New York.

Hatefi, Y., Galante, Y. M., Stiggall, D. L., and Djavadi-Ohaniance, L. (1976). *In* "The Structure Basis of Membrane Function" (Y. Hatefi and L. Djavadi-Ohaniance, eds.), pp. 169–188. Academic Press, New York.

Hatefi, Y., Galante, Y. M., Stiggall, D. L., and Ragan, C. I. (1979). *In* "Methods in Enzymology" (S. Fleischer and L. Packer, eds.) Vol. 56, pp. 577–602. Academic Press, New York.

Hattori, T., and Asahi, T. (1982). *Plant Cell Physiol.* **23**, 525–532.

Haurowitz, F., Schwerin, P., and Yenson, M. M. (1941). *J. Biol. Chem.* **140**, 353–359.

Haussman, K. (1977). *Naturwissenschaften* **64**, 95.

Hay, R., Böhni, P. C., and Gasser, S. (1984). *Biochim. Biophys. Acta* **779**, 65–87.

Heber, U., and Santarius, K. A. (1970). *Z. Naturforsch. B: Anorg. Chem., Org. Chem., Biochem., Biophys., Biol.* **25**, 718–728.

Heber, U., Egneus, H., Hanck, U. Jensen, M., and Köster, S. (1978). *Planta* **143**, 41–49.

Heber, U., Takahama, U., Neimanis, S., and Shimizu-Takahama, M. (1982). *Biochim. Biophys. Acta* **679**, 287–299.

Heinrich, R., and Rapoport, T. A. (1974). *Eur. J. Biochem.* **42**, 97–105.

Heinz, E. (1977). *In* "Lipids and Lipid Polymers in Higher Plants" M. Tevini and H. K. Lichtenthaler, eds.), pp. 102–120. Springer-Verlag, Berlin.

Heldt, H. W., Sauer, F., and Rapley, L. (1972a). *Proc. Int. Cong. Photosynth. 2nd, 1971* **2**, 1345–1355.

Heldt, H. W., Klingenberg, M., and Milovancev, M. (1972b). *Eur. J. Biochem.* **30**, 434–440.

Heldt, H. W., Chon, C. J., Maronde, D., Herold, A., and Stankovic, Z. S., Walker, D. A., Kraminer, A., Kirk, M. R., and Heber, U. (1977). *Plant Physiol.* **59**, 1146–1155.

Henry, M. F., and Nyns, E. J. (1975). *Subcell. Biochem.* **4**, 1–65.

Hensley, J. R., and Hanson, J. B. (1975). *Plant Physiol.* **56**, 13–18.

Heron, C., Ragan, C. I., and Trumpower, B. L. (1978). *Biochem. J.* **174**, 791–800.

Heslop-Harrisson, J. (1971). *In* "Pollen: Development and Physiology" (J. Heslop-Harrison, ed.), pp. 16–31. Butterworth, London.

Higgins, J. (1963). *Ann. N. Y. Acad. Sci.* **108**, 305–321.

Hiraï, M. (1978). *Phytochemistry* **17**, 1507–1510.

Hiraga, K., and Kikuchi, G. (1980a). *J. Biol. Chem.* **255**, 11,664–11,670.

Hiraga, K., and Kikuchi, G. (1980b). *J. Biol. Chem.* **255**, 11,671–11,676.

Hiraga, K., Kochi, H., Motokawa, Y., and Kikuchi, G. (1972). *J. Biochem.* **72**, 1285–1289.

Hirel, B., Perrot-Rechenmann, C., Suzuki, A., Vidal, J., and Gadal, P. (1982). *Plant Physiol.* **69**, 983–987.

Hirose, F., and Ashihara, H. (1982). *Z. Naturforsch. C: Biochem., Biophys., Biol., Virol.* **37**, 1288–1289.

Ho, L. C., and Thornley, J. H. M. (1978). *Ann. Bot.* **42**, 481–483.

Hoch, G., Von Owens, O. H., and Kok, B. (1963). *Arch. Biochem. Biophys.* **101**, 171–180.

Hoekstra, F. A., and Van Roekel, T. (1983). *Plant Physiol.* **73**, 995–1001.

Holm, R. H., and Ibers, J. A. (1977). *In* "Iron–Sulfur Protein," (W. Lovenberg, ed.), Vol. 3, pp. 205–285. Academic Press, New York.

Honda, S. I., Hongladarom, T., and Laties, G. G. (1966). *J. Exp. Bot.* **17**, 460–472.

Huang, A. H. C., and Cavalieri, A. J. (1979). *Plant Physiol.* **63**, 531–535.

Huber, S. C. (1979). *Plant Physiol.* **64**, 846–851.

Huber, S. C., and Moreland, D. E. (1979). *Plant Physiol.* **64**, 115–119.

Hulme, A. C., and Rhodes, M. J. C. (1968). *In* "Plant Cell Organelles" (J. B. Pridham, ed.), pp. 99–118. Academic Press, New York.

Hulme, A. C., Jones, J. D., and Wooltorton, L. S. C. (1964). *Phytochemistry* **3**, 173–179.

Hunt, L., and Fletcher, J. S. (1976). *Plant Physiol.* **57**, 304–307.

Huq, S., and Palmer, J. M. (1978a). *In* "Plant Mitochondria" (G. Ducet and C. Lance, eds.), pp. 225–232. Elsevier, Amsterdam.

Huq, S., and Palmer, J. M. (1978b). *FEBS Lett.* **92**, 317–320.

Huq, S., and Palmer, J. M. (1978c). *FEBS Lett.* **95**, 217–220.

Huq, S., and Palmer, J. M. (1978d). *Plant Sci. Lett.* **11**, 351–356.

Hutton, D., and Stumpf, P. K. (1969). *Plant Physiol.* **44**, 508–516.

Ikeda, T., Matsumoto, T., Kisaki, T., and Noguchi, M. (1980). *Agric. Biol. Chem.* **44**, 135–142.

Ikuma, H. (1970). *Plant Physiol.* **45**, 773–781.

Ikuma, H. (1972). *Annu. Rev. Plant Physiol.* **23**, 419–436.

Ikuma, H., and Bonner, W. D., Jr. (1967a). *Plant Physiol.* **42**, 67–75.

Ikuma, H., and Bonner, W. D., Jr. (1967b). *Plant Physiol.* **42**, 1400–1406.

Iwasaki, Y., and Asahi, T. (1983). *Arch. Biochem. Biophys.* **227**, 164–173.

Jackson, C., and Moore, A. L. (1979). *In* "Plant Organelles" (E. Reid, ed.), pp. 1–12. Ellis Horwood, Chichester, England.

Jackson, C., Dench, J. E., Hall, D. O., and Moore, A. L. (1979a). *Plant Physiol.* **64**, 150–153.

Jackson, C., Dench, J. E., Morris, P., Lui, S. C., Hall, D. O., and Moore, A. L. (1979b). *Biochem. Soc. Trans.* **7**, 1122–1124.

Jacob, F., Brenner, S., and Cuzin, F. (1963). *Cold Spring Harbor. Symp. Quant. Biol.* **28**, 329–348.

Jacobus, W. E., Moreadith, R. W., and Vandegaer, K. M. (1982). *J. Biol. Chem.* **257**, 2397–2402.

James, T. W., and Spencer, M. S. (1982). *Plant Physiol.* **69**, 1113–1115.

James, W. O. (1953). "Plant Respiration." Oxford Univ. Press (Clarendon), London and New York.

Jardetsky, O., and Jardetsdy, C. D. (1962). *In* "Methods of Biochemical Analysis" Vol. 9 (D. Glick, ed.), pp. 235–410. Wiley, New York.

Jeannin, G., Burkard, G., and Weil, J. H. (1976). *Biochim. Biophys. Acta* **442**, 24–31.

John, P., and Whatley, F. R. (1975). *Nature (London)* **254**, 495–498.

John P., and Whatley, F. R. (1977). *Adv. Bot. Res.* **4**, 51–115.

Johnson, H. M., and Wilson, R. H. (1972). *Biochim. Biophys. Acta* **267**, 398–408.

Johnston, S. P., Møller, I. M., and Palmer, J. M. (1979). *FEBS Lett.* **108**, 28–32.

Jolliot, A., Demandre, C., Kader, J. C., and Mazliak, P. (1978). *In* "Plant Mitochondria" (G. Ducet and C. Lance, eds.), pp. 445–452. Elsevier, Amsterdam.

Jordan, E. G., and Luck, B. T. (1980). *Eur. J. Cell. Biol.* **22**, 766–771.
Journet, E. P., and Douce, R. (1983). *Plant Physiol.* **72**, 802–808.
Journet, E. P., and Douce, R. (1984). *C. R. Hebd. Seances Acad. Sci. Ser.* D **13**, 365–370.
Journet, E. P., and Douce, R. (1985). *Plant Physiol.* (in press).
Journet, E. P., Neuburger, M., and Douce, R. (1980). *Plant Physiol.* **67**, 467–469.
Journet, E. P., Bonner, W. D., and Douce, R. (1982). *Arch. Biochem. Biophys.* **214**, 366–375.
Jukes, T. H. (1983). *Nature (London)* **301**, 19–20.
Jung, D. W., and Brierley, G. P. (1979). *Plant Physiol.* **64**, 948–953.
Jung, D. W., and Hanson, J. B. (1973). *Arch. Biochem. Biophys.* **158**, 139–148.
Jung, D. W., and Hanson, J. B. (1975). *Arch. Biochem. Biophys.* **168**, 358–368.
Jung, D. W., and Laties, G. G. (1976). *Plant Physiol.* **57**, 583–588.
Jung, D. W., and Laties, G. G. (1979). *Plant Physiol.* **63**, 591–597.
Junge, W., and DeVault, D. (1975). *Biochim. Biophys. Acta.* **408**, 200–214.
Kadenbach, B., Mende, P., Kolbe, H. V. J., Stipani, I., and Palmieri, F. (1982). *FEBS Lett.* **139**, 109–112.
Kader, J. C. (1977). *In* "Dynamic Aspects of Cell Surface Organization" (G. Poste and G. L. Nicholson, eds.), pp. 127–204. North-Holland Publ., Amsterdam.
Kagawa, T., and Hatch, M. D. (1975). *Arch. Biochem. Biophys.* **167**, 687–696.
Kaine, B. P., Gupta, R., and Woese, C. R. (1983). *Proc. Natl. Acad. Sci. U.S.A.* **80**, 3309–3312.
Kalina, M., and Pease, D. C. (1977). *J. Cell Biol.* **74**, 742–746.
Kaplan, R. S., and Pedersen, P. L. (1983). *Biochem. J.* **212**, 279–288.
Keilin, D., and Hartree, E. F. (1949). *Nature (London)* **164**, 254–258.
Kellems, R. E., and Butow, R. A. (1972). *J. Biol. Chem.* **247**, 8043–8050.
Kellenberger, E., Ryter, A., and Sechand, J. (1958). *J. Biophys. Biochem. Cytol.* **4**, 323–325.
Kelly, G. J. (1982). *Trends Biochem. Sci. (Pers. Ed.)* **7**, 233.
Kelly, G. J., and Gibbs, M. (1973). *Plant Physiol.* **52**, 674–676.
Kelley, G. J., and Latzko, E. (1975). *Nature (London)* **256**, 429–430.
Kelly, G. J., and Latzko, E. (1977). *Plant Physiol.* **60**, 290–294.
Kelly, G. J., and Turner, J. F. (1969). *Biochem. J.* **115**, 481–487.
Kemble, R. J., and Bedbrook, J. R. (1980). *Nature (London)* **284**, 565–566.
Kemble, R. J., Mans, R. J., Gabay-Laughnan, S., and Laughnan, J. R. (1983). *Nature (London)* **304**, 744–747.
Kennedy, E. P. (1961). *Fed. Proc., Fed. Am. Soc. Exp. Biol.* **20**, 934–940.
Kent, S. S. (1979). *Plant Physiol.* **64**, 159–161.
Keys, A. J. (1980). *In* "The Biochemistry of Plants," Vol. 5: Amino Acids and Derivatives (B. J. Miflin, ed.), pp. 359–374. Academic Press, New York.
Keys, A. J., Bird, I. F., Cornelius, M. J., Lea, P. J., Wallsgrave, R. M., and Miflin, B. J. (1978). *Nature (London)* **275**, 741–743.
Kim, B. D., Mans, R. J., Conde, M. F., Pring, D. R., and Levings, C. S., III (1982). *Plasmid* **7**, 1–8.
King, T. E. (1978). *In* "Membrane Proteins, FEBS Meeting, 1977" (P. Nicholls, J. V. Møller, P. L. Jørgensen, and A. J. Moody, eds) Vol. 45, pp. 17–31. Pergamon Press, Oxford.
Kirk, B. I., and Hanson, J. B. (1973). *Plant Physiol.* **51**, 357–362.
Kisaki, T., and Tolbert, N. E. (1969). *Plant Physiol.* **44**, 242–250.
Kisaki, T., Imai, A., and Tolbert, N. E. (1971). *Plant Cell Physiol.* **12**, 267–273.
Kislev, N., Swift, H., and Bogorad, L. (1965). *J. Cell Biol.* **25**, 327–344.
Klein, G., Satre, M., Dianoux, A. C., and Vignais, P. V. (1981). *Biochemistry* **20**, 1339–1344.
Klein, R. R., and Burke, J. J. (1984). *Plant Physiol.* **76**, 436–441.
Kleppinger-Sparace, K. F., and Moore, T. S., Jr. (1985). *Plant Physiol.* **77**, 12–15.
Klingenberg, M. (1961). *Biochem. Z.* **335**, 263–272.

Klingenberg, M. (1970). *Assay Biochem.* **6**, 119–159.

Klingenberg, M. (1980). *J. Membr. Biol.* **56**, 97–105.

Klingenberg, M. (1981). *Nature (London)* **290**, 449–453.

Klingenberg, M., and Rottenberg, H. (1977). *Eur. J. Biochem.* **73**, 125–130.

Klingenberg, M., and Shollmeyer, P. (1960). *Biochem. Z.* **333**, 335–350.

Knight, V. A., Wiggins, P. M., Harvey, J. D., and O'Brien, J. A. (1981). *Biochim. Biophys. Acta* **637**, 146–151.

Knutson, R. M. (1974). *Science* **186**, 746–747.

Kobr, M. J., and Beevers, H. (1971). *Plant Physiol.* **47**, 48–52.

Koeppe, D. E., and Miller, R. J. (1972). *Plant Physiol.* **49**, 353–357.

Kok, B. (1949). *Biochim. Biophys. Acta* **3**, 625–631.

Kolodner, R., and Tewari, K. K. (1972). *Proc. Natl. Acad. Sci. U.S.A.* **69**, 1830–1834.

Kombrink, E., Kurger, N. J., and Beevers, H. (1984). *Plant Physiol.* **74**, 395–401.

Kono, Y., Takeuchi, S., and Kawarada, A. (1980). *Tetrahedron Lett.* **21**, 1537–1540.

Korb, H., and Neupert, W. (1978). *Eur. J. Biochem.* **91**, 609–620.

Kow, Y. W., Erbes, D. L., and Gibbs, M. (1982). *Plant Physiol.* **69**, 442–447.

Kowallik, K. V., and Hermann, R. G. (1972). *J. Cell Sci.* **11**, 357–364.

Krab, K., and Wikström, M. (1978). *Biochim. Biophys. Acta* **504**, 200–214.

Krause, G. H., and Heber, U. (1976). *In* "The Intact Chloroplast" (J. Barber, ed.), pp. 171–214. Elsevier, Amsterdam.

Krebs, H. A., and Johnson, W. A. (1937). *Enzymologia* **4**, 148–155.

Kroger, A., and Klingenberg, M. (1973). *Eur. J. Biochem.* **34**, 358–368.

Kroon, A. M., and Saccone, C. (1980). "The Organisation and Expression of the Mitochondrial Genome." Elsevier, Amsterdam.

Kurger, N. J., and Beevers, H. (1984). *Plant Physiol.* **76**, 49–54.

Kruger, N. J., Kombrink, E., and Beevers, H. (1983). *FEBS Lett.* **153**, 409–412.

Ku, H. S., Pratt, H. K., Spurr, A. R., and Harris, W. M. (1968). *Plant Physiol.* **43**, 883–887.

Kurzok, H. G., and Feierabend, J. (1984). *Biochim. Biophys. Acta* **788**, 214–221.

Lagunas, R., and Gancedo, C. (1983). *Eur. J. Biochem.* **137**, 479–483.

Lambers, H. (1980). *Plant Cell Environ.* **3**, 293–302.

Lambers, H., Day, D. A., and Azcón-Bieto, J. (1983). *Physiol. Plant.* **58**, 148–154.

Lambowitz, A. M., and Bonner, W. D., Jr. (1973). *Biochem. Biophys. Res. Commun.* **52**, 703–711.

Lambowitz, A. M., and Bonner, W. D., Jr. (1974). *J. Biol. Chem.* **249**, 2428–2440.

Lambowitz, A. M., Bonner, W. D., and Wikström, M. K. F. (1974). *Proc. Natl. Acad. Sci. U.S.A.* **71**, 1183–1187.

Lambowitz, A. M., Chua, N. H., and Luck, D. J. L. (1976). *J. Mol. Biol.* **107**, 223.

Lance, C. (1972). *Ann. Sci. Nat., Bot. Biol. Veg.* **13**, 477–495.

Lance, C. (1981). *In* "Recent Advances in the Biochemistry of Fruits and Vegetables" (J. Friends and M. J. C. Rhodes, eds.), pp. 63–87. Academic Press, London.

Lance, C. and Bonner, W. D., Jr. (1968). *Plant Phsyiol.* **43**, 756–766.

Lance, C., Hobson, G. E., Young, R. E., and Biale, J. B. (1965). *Plant Physiol.* **40**, 1116–1123.

Lance, C., Hobson, G. E., Young, R. E., and Biale, J. B. (1967). *Plant Physiol.* **42**, 471–478.

Lance, C., Dizengremel, P., and Chauveau, M. (1978). *In* "Functions of Alternative Terminal Oxidases, FEBS meeting 1977" (H. Degn, D. Loyd, and G. C. Hill, eds.), Vol. 49, pp. 133–139. Pergamon Press, Oxford.

Lance-Nougarède, A. (1966). *C. R. Hebd. Seance Acad. Sci.* **263**, 246–249.

LaNoue, K. F., and Schoolwerth, A. C. (1979). *Annu. Rev. Biochem.* **48**, 871–922.

LaPolla, R. J., and Lambowitz, A. M. (1981). *J. Biol. Chem.* **256**, 7064–7067.

Larsson, C. (1979). *Physiol. Plant.* **46**, 221–226.

Larsson, C., and Anderson, B. (1979). *In* "Plant Organelles" (E. Ried, ed.), pp. 35–46. Ellis Horwood, Chichester, England.

Laties, G. G. (1953). *Physiol. Plant.* **6**, 199.

Laties, G. G. (1973). *Biochemistry* **12**, 3350–3355.

Laties, G. G. (1978). *In* "Biochemistry of Wounded Plant Tissues" (G. Kahl, ed.), pp. 421–466. de Gruyter, Berlin.

Laties, G. G. (1982). *Annu. Rev. Plant Physiol.* **33**, 519–555.

Laties, G. G. (1983). *Plant Physiol.* **72**, 953–958.

Laties, G. G., and Treffry, T. (1969). *Tissue Cell* **1**, 575–583.

Laughnan, J. R., Gabay-Laughnam, S., and Carlson, J. E. (1981). *Stadler Genet. Symp.* **13**, 93–99.

Lauquin, G. J. M., Brandolin, G., Lunardi, J., and Vignais, P. V. (1978). *Biochim. Biophys. Acta* **501**, 10–19.

Lawyer, A. L., and Zelitch, I. (1979). *Plant Physiol.* **64**, 706–711.

Lazowska, J., Jacq, C., and Slonimski, P. P. (1980). *Cell* **22**, 333–348.

Leak, L. V. (1968). *J. Ultrastruct. Res.* **24**, 102–108.

Leaver, C. J. (1979). *In* "Nucleic Acids in Plants" (T. C. Hall and J. W. Davies, eds.), Vol. 1, pp. 193–215. CRC Press, Boca Raton, Florida.

Leaver, C. J., and Forde, B. G. (1980). *In* "Genome Organization and Expression" (C. J. Leaver, ed.), pp. 407–425. Plenum, New York.

Leaver, C. J., and Gray, M. W. (1982). *Annu. Rev. Plant Physiol.* **33**, 373–402.

Leaver, C. J., and Harmey, M. A. (1973). *Biochem. Soc. Symp.* **38**, 175.

Leaver, C. J., and Harmey, M. A. (1976). *Biochem. J.* **157**, 275–277.

Leaver, C. J., Forde, B. G., Dixon, L. K., and Fox, T. D. (1982). *In* "Mitochondrial Genes" (P. Slonimsky, P. Borst, and G. Attardi, eds.), pp. 457–470. Cold Spring Harbor Lab., Cold Spring Harbor, New York.

Leaver, C. J., Hack, E., and Ford, B. G. (1983). *In* "Methods in Enzymology" Protein Synthesis by Isolated Plant Mitochondria. (S. Fleischer and B. Fleischer, eds.) Vol. 97, pp. 476–484. Academic Press, New York.

Lebacq, P., and Vedel, F. (1981). *Plant Sci. Lett.* **23**, 1–9.

Ledbetter, M. C., and Porter, K. R. (1970). "Introduction to the Fine Structure of Plant Cells." Springer-Verlag, Berlin.

Leech, R. M., and Murphy, D. J. (1976). *In* "The Intact Chloroplast" (J. Barber, ed.), pp. 365–401. Elsevier, Amsterdam.

Le Floc'h, F., and Lafleuriel, J. (1983). *Z. Pflanzenphysiol.* **113**, 61–71.

Lehninger, A. L. (1964). "The Mitochondrion." pp. 132–156. Benjamin, New York and Amsterdam.

Lemberg, R., and Barrett, J. (1973). "Cytochromes." Academic Press, London and New York.

Lendzián, K. J. (1980). *Plant Physiol.* **66**, 8–12.

Lê-quôc, K., and Lê-quôc, D. (1982). *Arch. Biochem. Biophys.* **216**, 639–651.

Levi, C., and Gibbs, M. (1976). *Plant Physiol.* **57**, 933–935.

Levings, C. S., III (1983). *Cell* **32**, 659–666.

Levings, C. S., III Shah, D. M., Hu, W. W. L., Pring, D. R., and Timothy, D. M. (1979). *ICN-UCLA Symp. Mol. Cell. Biol.* **15**, 63–73.

Lévy, M., Toury, R., and André, J. (1966). *C. R. Hebd. Seance Acad. Sci.* **262**, 1593–1599.

Lieberman, M., and Baker, J. E. (1965). *Annu. Rev. Plant Physiol.* **16**, 343–382.

Liedvogel, B., and Kleinig, H. (1980). *Planta* **150**, 170–173.

Liedvogel, B., and Stumpf, P. K. (1982). *Plant Physiol.* **69**, 897–903.

Lilley, R. McC., Stitt, M., Mader, G., and Heldt, H. W. (1982). *Plant Physiol.* **70**, 965–970.

Bibliography

Lin. W., and Hanson, J. B. (1974). *Plant Physiol.* **54,** 250–256.
Litvak, S., Keclard-Christophe, L., Echeverria, M., and Castroviejo, M, (1983). *In* "Structure and Function of Plant Genomes" (O. Cifferri and L. Dure, III, eds.), pp. 381–385. Plenum, New York.
Lloyd, D. (1974). "The Mitochondria of Microorganisms." Academic Press, London and New York.
Lonsdale, D. M. (1984). *Plant Mol. Biol.* **3,** 201–206.
Lonsdale, D. M., Hodge, T. P., Howe, C. J., and Stern, D. B. (1983a). *Cell* **34,** 1007–1014.
Lonsdale, D. M., Hodge, T. P., Fauron, C. M. R., and Flavell, R. B. (1983b). *UCLA-Symp. Mol. Cell. Biol.* **12,** 445–456.
Lonsdale, D. M., Hodge, T. P., and Fauron, C. M. R. (1984). *Nucleic Acids Res.* **12,** 9249–9261.
Loomis, W. D. (1974). *In* "Methods in Enzymology" (S. Fleischer and L. Packer, eds.), Vol. 31, pp. 528–544. Academic Press, New York.
Loomis, W. D., and Battaile, J. (1966). *Phytochemistry* **5,** 423–438.
Lord, J. M., and Beevers, H. (1972). *Plant Physiol.* **49,** 249–251.
Lord, J. M., Kagawa, T., and Beevers, H. (1972). *Proc. Natl. Acad. Sci. U.S.A.* **69,** 2429–2432.
Lord, J. M., Kagawa, T. Moore, T. S., and Beevers, H. (1973). *J. Cell Biol.* **57,** 659–674.
Lorimer, G. H., and Andrews, T. J. (1981). *In* "The Biochemistry of Plants," Vol. 8: Photosynthesis (M. D. Hatch and N. K. Boardman, eds.), pp. 329–374. Academic Press, New York.
Lovenberg, W. (1973, 1977). "Iron–Sulfur Proteins," Vols. 1–3. Academic Press, New York.
Lütke-Brinkhaus, F., Liedvogel, B., and Kleinig, H. (1984). *Eur. J. Biochem.* **141,** 537–541.
Ludwig, L. J., Charles-Edwards, D. A., and Withers, A. C. (1975). *In* "Environmental and Biological Control of Photosynthesis" (R. Marcelle, ed.), pp. 29–36. Jung, The Hague.
Lynen, F. (1963). *In* "Control Mechanisms in Respiration and Fermentation" (B. Wright, ed.), pp. 289–306. Ronald Press, New York.
Lyons, J. M. (1973). *Annu. Rev. Plant Physiol.* **24,** 445–466.
McCaig, T. N., and Hill, R. D. (1977). *Can. J. Bot.* **55,** 549–555.
McCarty, R. E., Douce, R., and Benson, A. A. (1973). *Biochim. Biophys. Acta* **316,** 266–270.
McDonald, F. D., and ap Rees, T. (1983). *Biochim. Biophys. Acta* **755,** 81–89.
Macey, M., and Stumpf, P. K. (1982/83). *Plant Sci. Lett.* **28,** 207–212.
McGivan, J. D., and Klingenberg, M. (1971). *Eur. J. Biochem.* **20,** 392–399.
Macher, B. A., and Mudd, J. B. (1974). *Plant Physiol.* **53,** 171–175.
Macino, G., and Tzagoloff, A. (1979). *J. Biol. Chem.* **254,** 4617–4623.
Mackender, R. O., and Leech, R. M. (1974). *Plant Physiol.* **53,** 496–502.
McLean, J. R., Cohn, G. L., Brandt, I. K., and Simpson, M. V. (1958). *J. Biol. Chem.* **233,** 657–663.
McMurray, W. C. (1973). *In* "Forms and Function of Phospholipids" (G. B. Ansell, J. N. Hawthorne, and R. M. C. Dawson, eds.), pp. 205–251. Elsevier, Amsterdam.
McMurray, W. C., and Jarvis, E. C. (1978). *Can. J. Biochem.* **56,** 414–419.
Macrae, A. R. (1971a). *Biochem. J.* **122,** 495–501.
Macrae, A. R. (1971b). *Phytochemistry* **10,** 1453–1458.
Macrae, A. R., and Moorhouse, R. (1970). *Eur. J. Biochem.* **16,** 96–102.
Maeshima, M., and Asahi, T. (1978). *Arch. Biochem. Biophys.* **187,** 423–430.
Maeshima, M., and Asahi, T. (1981). *J. Biochem.* **90,** 399–406.
Mahler, H. R. (1981). *Ann. N. Y. Acad. Sci.* **361,** 53–75.
Malhotra, S. K., and Eakin, R. T. (1967). *J. Cell Sci.* **2,** 205–212.

Malhotra, S. S., and Spencer, M. (1971). *J. Exp. Bot.* **22**, 70–77.

Malmström, B. G. (1979). *Biochim. Biophys. Acta* **549**, 281–303.

Malone, C., Koeppe, D. E., and Miller, R. J. (1974). *Plant Physiol.* **53**, 918–927.

Mandolino, G., De Santis, A., and Melandri, B. A. (1983). *Biochim. Biophys. Acta* **723**, 428–439.

Mangat, B. S., Levin, W. B., and Bidwell, R. G. S. (1974). *Can. J. Bot.* **52**, 673–681.

Mannella, C. A., (1974). Ph D. Thesis. Univ. of Pennsylvania, Philadelphia, U.S.A.

Mannella, C. A. (1981). *Biochim. Biophys. Acta* **645**, 33–40.

Mannella, C. A., and Bonner, W. D. Jr. (1975). *Biochim. Biophys. Acta* **413**, 226–233.

Mannella, C. A., and Parsons, D. F. (1977). *Biochim. Biophys. Acta* **460**, 375–378.

Margulis, L. (1975). *Symp. Soc. Exp. Biol.* **29**, 21–32.

Marré, E., Cornaggia, M. P., and Bianchetti, R. (1968). *Phytochemistry* **7**, 1115–1123.

Marshall, M. O., and Kates, M. (1973). *FEBS Lett.* **31**, 199–202.

Marshall, M. O., and Kates, M. (1974). *Can. J. Biochem.* **52**, 469–482.

Marsho, T. V., Behrens, P. W., and Radmer, R. J. (1979). *Plant Physiol.* **64**, 656–659.

Martin, J. B., Bligny, R., Rebeillé, F., Douce, R., Leguay, J. J., Mathieu, Y., and Guern, J. (1982). *Plant Physiol.* **70**, 1156–1161.

Martinez, G., Rochat, H., and Ducet, G. (1974). *FEBS Lett.* **47**, 212–217.

Marx, R., and Brinkmann, K. (1979). *Planta* **144**, 359–365.

Massey, V. (1960). *Biochim. Biophys. Acta* **38**, 447–460.

Matile, P. (1978). *Annu. Rev. Plant Physiol.* **29**, 193–213.

Matsuoka, M., and Asahi, T. (1982). *Agric Biol. Chem.* (in press).

Matsuoka, I., and Nakamura. T. (1979). *J. Biochem.* **86**, 675–681.

Matsuoka, M., Maeshima, M., and Asahi, T. (1981). *J. Biochem.* **90**, 649–655.

Matthews, D. E., Gregory, P., and Gracen, V. E. (1979). *Plant. Physiol.* **63**, 1149–1153.

Mayer, A. M., and Harel, E. (1979). *Phytochemistry* **18**, 193–215.

Mazliak, P., and Ben Abdelkader, A. (1971). *Phytochemistry* **10**, 2879–2890.

Medina, F. J., Risueno, M. C., and Rodriguez-Garcia, M. I. (1981). *Planta* **151**, 215–225.

Meeuse, B. J. D. (1975). *Annu. Rev. Plant Physiol.* **26**, 117–126.

Meijer, A. J., and Tager, J. M. (1969). *Biochim. Biophys. Acta* **189**, 136–139.

Meijer, A. J., Van Woerkom, G. M., and Eggelte, T. A. (1976). *Biochim. Biophys. Acta* **430**, 53–61.

Mettler, I. J., and Beevers, H. (1980). *Plant Physiol.* **66**, 555–560.

Meunier, D., and Mazliak, P. (1972). *C. R. Hebd. Seance Acad. Sci. Ser. C* **275**, 213–216.

Meunier, D., Pianeta, C., and Coulomb, P. (1971). *C. R. Hebd. Seance Acad. Sci. Ser. D* **272**, 1376–1379.

Meves, F. (1904). *Ber. Dtsch. Bot. Ges.* **22**, 284–302.

Meyer, J., and Vignais, P. M. (1973). *Biochim. Biophys. Acta* **325**, 375–384.

Miernyk, J. A., and Dennis, D. T. (1984). *Arch. Biochem. Biophys.* **233**, 643–651.

Miflin, B. J., and Beevers, H. (1974). *Plant Physiol.* **53**, 870–874.

Miflin, B. J., and Lea, P. J. (1980). *In* "The Biochemistry of Plants," Vol. 5: Amino Acids and Derivatives (B. J. Miflin, ed.), pp. 169–202. Academic Press, New York.

Mikulska, E., Odintsova, M. S., and Turischeva, M. S. (1970). *J. Ultrastr. Res.* **35**, 258–265.

Miller, M. G., and Obendorf, R. L. (1981). *Plant Physiol.* **67**, 962–964.

Miller, R. J., and Koeppe, D. E. (1971). *Science* **173**, 67–69.

Miller, R. J., Dumford, S. W., Koeppe, D. E., and Hanson, J. B. (1970). *Plant Physiol.* **45**, 649–653.

Miller, R. W., de la Roche, I. A., and Pomeroy, M. K. (1974). *Plant Physiol.* **53**, 426–433.

Millerd, A., Bonner, J., Axelrod, B., and Bandurski, P. (1951). *Proc. Natl. Acad. Sci. U.S.A.* **37**, 855–862.

Millhouse, J., Wiskich, J. T., and Beevers, H. (1983). *Aust. J. Plant Physiol.* **10**, 167–177.

Mitchell, J. A., and Moore, A. L. (1984). *Biochem. Soc. Trans.* **12**, 849–850.
Mitchell, P. (1975). *FEBS Lett.* **56**, 1.
Mitchell, P. (1976). *J. Theor. Biol.* **62**, 327–367.
Mitchell, P. (1980). *Ann. N. Y. Acad. Sci.* **341**, 564–584.
Mitchell, P., and Moyle, J. (1983). *FEBS Lett.* **151**, 167–178.
Mohamed, A. H., and Anderson, L. E. (1983). *Plant Physiol.* **71**, 248–250.
Møller, I. M., and Palmer, J. M. (1981a). *Biochem. J.* **195**, 583–588.
Møller, I. M., and Palmer, J. M. (1981b). *Physiol. Plant.* **53**, 413–420.
Møller, I. M., and Palmer, J. M. (1981c). *Biochim. Biophys. Acta* **638**, 225–233.
Møller, I. M., and Palmer, J. M. (1982). *Physiol. Plant.* **54**, 267–274.
Møller, I. M., Johnston, S. P., and Palmer, J. M. (1981a). *Biochem. J.* **194**, 487–495.
Møller, I. M., Bergman, A., Gardeström, P., Ericson, I., and Palmer, J. M. (1981b). *FEB Lett.* **126**, 13–17.
Møller, I. M., Chow, W. S., Palmer, J. M., and Barber, J. (1981c). *Biochem. J.* **193**, 37–46.
Montes, G., and Bradbeer, J. W. (1976). *Plant Sci. Lett.* **6**, 35–41.
Moore, A. L. (1978). *In* "Plant Mitochondria" (G. Ducet and C. Lance, eds.), pp. 85–92. Elsevier, Amsterdam.
Moore, A. L., and Åkerman, K. E. O. (1982). *Biochem. Biophys. Res. Commun.* **109**, 513–517.
Moore, A. L., and Bonner, W. D., Jr. (1976). *Biophys. J.* **16**, 20.
Moore, A. L., and Bonner, W. D., Jr. (1977). *Biochim. Biophys. Acta* **460**, 455–466.
Moore, A. L., and Bonner, W. D., Jr. (1981). *Biochim. Biophys. Acta* **634**, 117–128.
Moore, A. L., and Bonner, W. D., Jr. (1982). *Plant Physiol.* **70**, 1271–1276.
Moore, A. L., and Rich, P. (1980). *Trends Biochem. Sci. (Pers. Ed.)* **5**, 284–288.
Moore, A. L., and Wilson, S. B. (1977). *J. Exp. Bot.* **28**, 607–618.
Moore, A. L., and Wilson, S. B. (1978). *Planta* **141**, 297–302.
Moore, A. L., Jackson, C., Halliwell, B., Dench, J. E., and Hall, D. O. (1977). *Biochem. Biophys. Res. Commun.* **78**, 483–491.
Moore, A. L., Rich, P. R., and Bonner, W. D., Jr. (1978). *J. Exp. Bot.* **29**, 1–12.
Moore, A. L., Jackson, C., Dench, J., Morris, P., and Hall, D. O. (1979). *Plant Physiol.* **63**, S-110, Abstr. no. 613.
Moore, A. L., Dench, J. E., Jackson, C., and Hall, D. O. (1980). *FEBS Lett.* **115**, 54–58.
Moore, T. S. (1974). *Plant Physiol.* **54**, 164–168.
Moore, T. S. (1976). *Plant Physiol.* **57**, 383–386.
Moore, T. S., Jr. (1982). *Annu. Rev. Plant Physiol.* **33**, 235–359.
Moore, T. S., Jr., Lord, J. M., Kagawa, T., and Beevers, H. (1973). *Plant Physiol.* **52**, 50–53.
Moreau, F. (1976). *Plant Sci. Lett.* **6**, 215–221.
Moreau, F. (1978). Ph. D. Thesis. Univ. Pierre et Marie Curie, Paris, France.
Moreau, F., and Lance, C. (1972). *Biochimie* **54**, 1335–1348.
Moreau, F., and Romani, R. (1982). *Plant Physiol.* **70**, 1380–1384.
Moreau, F., Dupont, J., and Lance, C. (1974). *Biochim. Biophys. Acta* **345**, 295–304.
Morisset, C., Raymond, P., Mocquot, B., and Pradet, A. (1982). *Bull. Soc. Bot. Fr.* **129**, 73–92.
Morohashi, Y., and Matsushima, H. (1983). *Plant Physiol.* **73**, 82–86.
Morohashi, Y., Bewley, J. D., and Yeung, E. C. (1981). *J. Exp. Bot.* **32**, 605–613.
Morré, D. J., Merritt, W. D., and Lembi, C. A. (1971). *Protoplasma* **73**, 43–50.
Moyle, J., and Mitchell, P. (1978). *FEBS Lett.* **88**, 268–272.
Mudd, J. B. (1980). *In* "The Biochemistry of Plants," Vol. 4: Lipids: Structure and Function (P. K. Stumpf, ed.), pp. 249–282. Academic Press, New York.
Munn, E. A. (1974). "The Structure of Mitochondria." Academic Press, London and New York.

Murphy, D. J., and Leech, R. M. (1978). *FEBS Lett.* **88**, 192–196.

Muto, S., and Beevers, H. (1974). *Plant Physiol.* **54**, 23–28.

Nagel, W. O., and Sauer, L. A. (1982). *J. Biol. Chem.* **257**, 12,405–12,411.

Naik, M. S., and Singh, P. (1980). *FEBS Lett.* **111**, 277–280.

Nakayama, N., and Asahi, T. (1981). *Plant Cell Physiol.* **22**, 79–89.

Nash, D., and Wiskich, J. T. (1983). *Plant Physiol.* **71**, 627–634.

Nass, M. M. K. (1969). *Science* **165**, 25–35.

Nass, M. M. K., Nass, S., and Afzelius, B. A. (1965). *Exp. Cell. Res.* **37**, 516–539.

Nauen, W., and Hartmann, T. (1980). *Planta* **148**, 7–16.

Nawa, Y., and Asahi, T. (1973). *Plant Physiol.* **51**, 833–838.

Neal, G. H., and Beevers, H. (1959). *Biochem. J.* **74**, 409–416.

Nelson, B. D., and Gellerfors, P. (1978). *In* "Methods in Enzymology" (S. Fleischer and L. Packer, eds.) Vol. 53, pp. 80–98. Academic Press, New York.

Neuburger, M. (1980). Ph. D. Thesis, Univ. Grenoble, France.

Neuburger, M., and Douce, R. (1977). *C. R. Hebd. Seance Acad. Sci. Ser.* D **285**, 881–884.

Neuburger, M., and Douce, R. (1978). *In* "Plant Mitochondria" (G. Ducet and C. Lance, eds.), pp. 109–116. Elsevier, Amsterdam.

Neuburger, M., and Douce, R. (1980). *Biochim. Biophys. Acta* **589**, 176–189.

Neuburger, M., and Douce, R. (1983). *Biochem. J.* **216**, 443–450.

Neuburger, M., Journet, E. P., Bligny, R., Carde, J. P., and Douce, R. (1982). *Arch. Biochem. Biophys.* **217**, 312–323.

Neuburger, M., Day, D. A., and Douce, R. (1984). *Arch. Biochem. Biophys.* **229**, 253–258.

Neuburger, M., Day, D. A., and Douce, R. (1985). *Plant Physiol.* (in press).

Neupert, W., and Schatz, G. (1981). *Trends Biochem. Sci. (Pers. Ed.)* **6**, 1–5.

Nicholls, D., and Åkerman, K. (1982). *Biochim. Biophys. Acta* **683**, 57–88.

Nishimura, M., and Beevers, H. (1978). *Plant Physiol.* **62**, 40–43.

Nishimura, M., and Beevers, H. (1979). *Nature (London)* **277**, 412–413.

Nishimura, M., and Beevers, H. (1981). *Plant Physiol.* **67**, 1255–1258.

Nishimura, M., Graham, D., and Akazawa, T. (1976). *Plant Physiol.* **58**, 309–315.

Nishimura, M., Douce, R., and Akazawa, T. (1982). *Plant Physiol.* **69**, 916–920.

Noack, E. A., Crea, A. E. G., and Falsone, G. (1980). *Toxicon* **18**, 165–174.

Oestreicher, G., Hogue, P., and Singer, T. P. (1973). *Plant Physiol.* **52**, 622–626.

Ohnishi, J. I., and Kanai, R. (1983). *Plant Cell Physiol.* **24**, 1411–1420.

Ohnishi, T. (1975). *Biochim. Biophys. Acta* **387**, 475–490.

Ohnishi, T. (1976). *Eur. J. Biochem.* **64**, 91–103.

Ohnishi, T., Salerno, J. C., Winter, D. B., Lim, J., Yu, C. A. Yu, L., and King, T. E. (1976). *J. Biol. Chem.* **251**, 2094–2109.

Ohnishi, T., Blum, H., Galante, Y. M., and Hatefi, Y. (1981). *J. Biol. Chem.* **256**, 9216–9220.

Olabiyi, O., Kaiser, W. M., and Heber, U. (1983). *Z. Pflanzenphysiol.* **111**, 155–167.

Oliver, D. J. (1981). *Plant Physiol.* **68**, 1031–1034.

Oliver, J., and Walker, G. H. (1984). *Plant Physiol.* **76**, 409–413.

O'Neal, D., and Joy, K. W. (1974). *Plant Physiol.* **54**, 773–779.

Opik, H., (1968). *In* "Plant Cell Organelles" (J. B. Pridham, ed.), pp. 47–88. Academic Press, London and New York.

Öpik, H. (1973). *J. Cell Sci.* **12**, 725–739.

Öpik, H. (1974). *In* "Dynamic Aspects of Plant Ultrastructure" (A. W. Robards, ed.), pp. 52–83. McGraw-Hill, London and New York.

Oshino, N., Sugano, T., Oshino, R., and Chance, B. (1974). *In* "Dynamics of Energy Transducing Membrane" (L. Ernster, R. W. Estabrook, and E. C. Slater, eds.), pp. 201–214. Elsevier, Amsterdam.

Osmond, C. B. (1976). *In* "Encyclopedia of Plant Physiology," Vol. 2: Transport in Plants II, Part A Cells (U. Lüttge and M. G. Pitman, eds.), pp. 347–372. Springer-Verlag, Berlin.

Osmond, C. B., and ap Rees, T. (1969). *Biochim. Biophys. Acta* **184**, 35–42.

Oursel, A., Lamant, A., Salsac, L., and Mazliak, P. (1973). *Phytochemistry* **12**, 1865–1874.

Overman, A. R., Lorimer, G. H., and Miller, R. J. (1970). *Plant Physiol.* **45**, 126–132.

Packer, L., Murakami, S., and Mehard, C. W. (1970). *Annu. Rev. Plant Physiol.* **21**, 271–304.

Packer, L., Williams, M. A., and Criddle, R. S. (1973). *Biochim. Biophys. Acta* **292**, 92–104.

Paech, C., Reynolds, J. G., Singer, T. P., and Holm, R. H. (1981). *J. Biol. Chem.* **256**, 3167–3170.

Palade, G. E. (1952). *Anat. Rec.* **114**, 427–451.

Palade, G. E. (1953). *J. Histochem. Cytochem.* **1**, 188–211.

Palmer, J. D., and Shields, C. R. (1984). *Nature (London)* **307**, 437–440.

Palmer, J. D., Shields, C. R., Cohen, D. B., and Orton, T. J. (1983). *Nature (London)* **301**, 725–728.

Palmer, J. M. (1976). *Annu. Rev. Plant Physiol.* **27**, 133–157.

Palmer, J. M. (1979). *Biochem. Soc. Trans.* **7**, 246–255.

Palmer, J. M. (1980). *J. Exp. Bot.* **31**, 1497–1508.

Palmer, J. M., and Arron, G. P. (1976). *J. Exp. Bot.* **27**, 418–430.

Palmer, J. M., and Kirk, B. I. (1974). *Biochem. J.* **140**, 79–86.

Palmer, J. M., and Møller, I. M. (1982). *Trends Biochem. Sci.* **7**, 258–261.

Palmer, J. M., and Wedding, R. T. (1966). *Biochim. Biophys. Acta* **113**, 167–174.

Palmer, J. M., Schwitzguébel, J. P., and Møller, I. M. (1982). *Biochem. J.* **208**, 703–711.

Palmieri, F., and Klingenberg, M. (1979). *In* "Methods in Enzymology" (S. Fleischer and L. Packer, eds.) Vol. 56, pp. 279–301. Academic Press, New York.

Pande, S. V., and Parvin, R. (1978). *J. Biol. Chem.* **253**, 1565–1573.

Papa, S., Francavilla, A., Paradies, G., and Meduri, B. (1971). *FEBS Lett.* **12**, 285–288.

Papa, S., Lorusso, M., Izzo, G., and Capuano, F. (1981). *Biochem. J.* **194**, 395–406.

Papa, S., Guerrieri, F., and Izzo, G. (1983). *Biochem. J.* **216**, 259–272.

Parrish, D. J., and Leopold, A. C. (1978). *Plant Physiol.* **62**, 470–472.

Parsons, D. F. (1967). *In* "Methods in Enzymology" (R. W. Estabrook and M. E. Pullman, eds.) Vol. 10, pp. 655–667. Academic Press, New York.

Parsons, D. F., Bonner, W. D., Jr., and Verboon, J. G. (1965). *Can. J. Bot.* **43**, 647–655.

Parsons, D. F., Williams, G. R., and Chance, B. (1966). *Ann. N. Y. Acad. Sci.* **137**, 643–666.

Passam, H. C., and Coleman, J. O. D. (1975). *J. Exp. Bot.* **26**, 536–543.

Passam, H. C., and Palmer, J. M. (1971). *J. Exp. Bot.* **22**, 304–313.

Passam, H. C., and Palmer, J. M. (1972). *J. Exp. Bot.* **23**, 366–374.

Passam, H. C., Souverijn, J. H. M., and Kemp, A., Jr. (1973). *Biochim. Biophys. Acta* **305**, 88–94.

Passarella, S., Palmieri, F., and Quagliariello, E. (1977). *Arch. Biochem. Biophys.* **180**, 160–168.

Pauly, G., Douce, R., and Carde, J. P. (1981). *Z. Pflanzenphysiol.* **104**, 199–206.

Pertoft, H., and Laurent, T. C. (1977). *In* "Methods of Cell Separation" (N. Catsimpoolas, ed.), pp. 25–65. Plenum, New York.

Peterman, T. K., and Siedow, J. N. (1983). *Plant Physiol.* **71**, 55–58.

Pfaff, E., Klingenberg, M., Ritt, E., and Vogell, W. (1968). *Eur. J. Biochem.* **5**, 222–232.

Pharmacia Fine Chemicals (1980). "Percoll Methodology and Applications," Uppsala, Sweden.

Phillips, M. L., and Williams, G. R. (1973). *Plant Physiol.* **51**, 667–670.

Pichersky, E., and Gottlieb, L. D. (1984). *Plant Physiol.* **74**, 340–347.

Plantefol, L. (1922). *C. R. Hebd. Seances Acad. Sci.* **174,** 123–127.
Plantefol, L. (1932). *Ann. Physiol. Physicochim. Biol.* **8,** 124–153.
Platt, S. G., Plaut, Z., and Bassham, J. A. (1977). *Plant Physiol.* **60,** 230–234.
Plaut, G. W. E. (1970). *Curr. Top. Cell. Regul.* **2,** 1–32.
Pollak, J. K., and Sutton, R. (1980). *Biochem. J.* **192,** 75–83.
Pollard, C. J., Stemler, A., and Blaydes, D. F. (1966). *Plant Physiol.* **41,** 1323–1329.
Pomeroy, M. K. (1977). *Plant Physiol.* **59,** 250–251.
Pomeroy, M. K., and Raison, J. K. (1981). *Plant Physiol.* **68,** 382–385.
Popp, M., Osmond, C. B., and Summons, R. E. (1982). *Plant Physiol.* **69,** 1289–1292.
Poulsen, L. L., and Wedding, R. T. (1970). *J. Biol. Chem.* **245,** 5709–5717.
Powling, A. (1981). *Mol. Gen. Genet.* **183,** 82–84.
Pradet, A. (1982). *In* "The Physiology and Biochemistry of Seed Development, Dormancy and Germination" (A. A. Khan, ed.), pp. 347–369. Elsevier, Amsterdam.
Pradet, A., and Raymond, P. (1985). *In* "Physiology and Biochemistry of Plant Respiration" (J. M. Palmer, ed.), pp. 184–188. Cambridge Univ. Press, London and New York.
Preiss, J. (1982). *Annu. Rev. Plant Physiol.* **33,** 431–454.
Preiss, J. (1984). *Trends Biochem. Sci. (Pers. Ed.)* **8,** 24.
Pressman, B. C., and Lardy, H. A. (1955). *Biochim. Biophys. Acta* **18,** 482–487.
Prince, R. C., Bonner, W. D., Jr., and Bershak, P. A. (1981). *Fed. Proc., Fed. Am. Soc. Exp. Biol.* **40,** 1667 Abstr.
Pring, D. R. (1974). *Plant Physiol.* **53,** 677–683.
Pring, D. R., and Levings, C. S., III (1978). *Genetics* **89,** 121–136.
Pring, D. R., and Thornbury, D. W., (1975). *Biochim. Biophys. Acta* **383,** 140–146.
Pring, D. R., Levings, C. S., III, Hu, W. W. L., and Timothy, D. H. (1977). *Proc. Natl. Acad. Sci. U.S.A.* **74,** 2904–2908.
Pring, D. R., Conde, M. F., Schertz, K. F., and Levings, C. S., III (1982). *Mol. Gen. Genet.* **186,** 180–184.
Proudlove, M. O., and Moore, A. L. (1982). *FEBS Lett.* **147,** 26–30.
Proudlove, M. O., and Thurman, D. A. (1980). *New Phytol.* **88,** 255–264.
Pumphrey, A. M., and Redfearn, E. R. (1960). *Biochem. J.* **76,** 61–64.
Quetier, F., and Vedel, F. (1974). *FEBS Lett.* **42,** 305–308.
Quetier, F., and Vedel, F. (1977). *Nature (London)* **268,** 365–368.
Quinn, P. J., and Williams, W. P. (1983). *Biochim. Biophys. Acta* **737,** 223–266.
Racker, E. (1974). *Mol. Cell. Biochem.* **5,** 17–23.
Radzali, B. M., and Givan, C. V. (1981). *Biochem. Physiol. Pflanz.* **176,** 490–494.
Raff, R. A., and Mahler, H. R. (1972). *Science* **177,** 575–582.
Ragan, C. I. (1978). *Biochem. J.* **172,** 539–547.
Raison, J. K. (1973). *Symp. Soc. Exp. Biol.* **27,** 485–512.
Raison, J. K. (1980). *In* "The Biochemistry of Plants," Vol. 2: Metabolism and Respiration (D. D. Davies, ed.), pp. 613–626. Academic Press, New York.
Raison, J. K., and Chapman, E. A. (1976). *Aust. J. Plant Physiol.* **3,** 291–299.
Raison, J. K., Lyons, J. M., and Campbell, L. C. (1973). *J. Bioenerg.* **4,** 397–408.
Ralph, R. K., and Wojcik, S. J. (1978). *Plant Sci. Lett.* **12,** 227–232.
Ramirez-Mitchell, R., Johnson, H. M., and Wilson, R. H. (1973). *Exp. Cell Res.* **76,** 449–460.
Randall, D. D., Williams, M., and Rapp, D. J. (1981). *Arch. Biochem. Biophys.* **207,** 437–444.
Ranson, S. L. (1965). *In* "Plant Biochemistry" (J. Bonner and J. E. Varner, eds.), pp. 493–525. Academic Press, New York.
Rathnam, C. K. M. (1978). *FEBS Lett.* **96,** 367–372.
Ravanel, P., Tissut, M., and Douce, R. (1981). *Phytochemistry* **20,** 2101–2103.

Ravanel, P., Tissut, M., and Douce, R. (1982). *Plant Physiol.* **69**, 375–378.
Ravanel, P., Tissut, M., and Douce, R. (1984). *Plant Physiol.* **75**, 414–420.
Raven, J. A. (1972a). *New Phytol.* **71**, 227–247.
Raven, J. A. (1972b). *New Phytol.* **71**, 995–1014.
Raymond, P., and Pradet, A. (1980). *Biochem. J.* **190**, 39–44.
Rayner, J. R., and Wiskich, J. T. (1983). *Aust. J. Plant Physiol.* **10**, 55–63.
Rebeillé, F., Bligny, R., and Douce, R. (1980). *Biochim. Biophys. Acta* **620**, 1–9.
Rebeillé, F., Bligny, R., and Douce, R. (1983). *Arch. Biochem. Biophys.* **225**, 143–148.
Rebeillé, F., Bligny, R., and Douce, R. (1984a). *Plant Physiol.* **74**, 355–359.
Rebeillé, F., Bligny, R., and Douce, R. (1984b). *Plant Physiol.* (in press).
Rebeillé, F., Bligny, R., Martin, J. B., and Douce, R. (1985). *Biochem. J.* (in press).
Reed, L. J. (1981). *In* "Current Topics in Cellular Regulation," Vol. 18, Biological Cycles, (R. W. Estabrook and P. Srere, eds.), pp. 95–106. Academic Press, New York.
Reed, L. J., and Cox, T. J. (1966). *Annu. Rev. Biochem.* **35**, 57.
Reed, L. J., Pettit, F., and Yeaman, S. (1978). *In* "Microenvironments and Metabolic Compartmentation" (P. A. Srere and R. W. Estabrook, eds.), pp. 305–321. Academic Press, New York.
Reich, J. G., and Sel'kov, E. E. (1981). "Energy Metabolism of the Cell–A Theoretical Treatise." Academic Press, London and New York.
Reid, E. E., Lyttle, C. E., Canvin, D. T., and Dennis, D. T. (1975). *Biochem. Biophys. Res. Commun.* **62**, 42–47.
Ricard, B., Echeverria, M., Christophe, L., and Litvak, S. (1983). *Plant Mol. Biol.* **2**, 167–175.
Rich, P. R. (1978). *FEBS Lett.* **96**, 252–256.
Rich, P. R. (1981). *FEBS Lett.* **130**, 173–178.
Rich, P. R. (1984). *Biochim. Biophys. Acta* **768**, 53–79.
Rich, P. R., and Bonner, W. D., Jr. (1978a). *Arch. Biochem. Biophys.* **188**, 206–213.
Rich, P. R., and Bonner, W. D., Jr. (1978b). *Biochim. Biophys. Acta* **501**, 381–395.
Rich, P. R., and Moore, A. L. (1976). *FEBS Lett.* **65**, 339–341.
Rich, P. R., Moore, A. L., and Bonner, W. D., Jr. (1977a). *Biochem. J.* **162**, 205–208.
Rich, P. R., Moore, A. L., Ingledew, W. J., and Bonner, W. D., Jr. (1977b). *Biochim. Biophys. Acta* **462**, 501–514.
Rich, P. R., Wiegand, N. K., Blum, H., Moore, A. L., and Bonner, W. D., Jr. (1978). *Biochim. Biophys. Acta* **525**, 325–337.
Richaud, P., and Denis, M. (1984). *Arch. Biochem. Biophys.* (in press).
Rieske, J. S. (1976). *Biochim. Biophys. Acta* **456**, 195–247.
Rieske, J. S., Hansen, R. E., and Zaugg, W. S. (1964). *J. Biol. Chem.* **239**, 3017–3022.
Robertson, J. D. (1960). *Prog. Biophys. Biophys. Chem.* **10**, 343–356.
Robinson, B. H., Williams, G. R., Halperin, M. L., and Leznoff, C. C. (1972). *J. Membr. Biol.* **7**, 391–401.
Robinson, S. P., and Walker, D. A. (1979). *Arch. Biochem. Biophys.* **196**, 319–323.
Robinson, S. P., and Wiskich, J. T. (1977). *Plant Physiol.* **59**, 422–427.
Romani, R. J., Tuskes, S. E., and Özelkök, S. (1974). *Arch. Biochem. Biophys.* **164**, 743–751.
Rosamond, J. (1982). *Biochem. J.* **202**, 1–8.
Ross, E., and Schatz, G. (1976). *J. Biol. Chem.* **251**, 1991–1996.
Rothman, J. E., and Kennedy, E. P. (1977). *J. Mol. Biol.* **110**, 603–618.
Rothman, J. E., and Lenard, J. (1977). *Science* **195**, 743–753.
Roughan, P. G., Holland, R., and Slack, C. R. (1979). *Biochem. J.* **184**, 193–202.
Rubin, P. M., and Randall, D. D. (1977). *Plant Physiol.* **60**, 34–39.
Rubin, P. M., Zahler, W. L., and Randall, D. D. (1978). *Arch. Biochem. Biophys.* **188**, 70–77.

Ruigrok, TH. J. C., Van Zaane, D., Wirtz, K. W. A., and Scherphof, G. L. (1972). *Cytobiologie* **5**, 412.

Rupp, H., and Moore, A. L. (1979). *Biochim. Biophys. Acta* **548**, 16–25.

Russo, M., and Martelli, G. P. (1969). *Phytopathol. Mediterr.* **8**, 65–73.

Rustin, P., and Alin, M. F. (1983). *Physiol. Veg.* **22**, 93–101.

Rustin, P., Julienne, M., and Kader, J. C. (1980a). *C. R. Hebd. Seance Acad. Sci. Ser. D* **291**, 105–108.

Rustin, P., Moreau, F., and Lance, C. (1980b). *Plant Physiol.* **66**, 457–462.

Rustin, P., Dupont, J., and Lance, C. (1983a). *Trends Biochem. Sci. (Pers. Ed.)* **8**, 155–157.

Rustin, P., Dupont, J., and Lance, C. (1983b). *Arch. Biochem. Biophys.* **225**, 630–639.

Ruzicka, F. J., Beinert, H., Schepler, K. L., Dunham, W. R., and Sands, R. H. (1975). *Proc. Natl. Acad. Sci. U.S.A.* **72**, 2886–2890.

Rychter, A., Janes, H. W., and Frenkel, C. (1979). *Plant Physiol.* **63**, 149–151.

Sabularse, D. C., and Anderson, R. L. (1981). *Biochem. Biophys. Res. Commun.* **103**, 848–855.

Sager, R. (1972). "Cytoplasmic Genes and Organelles." Academic Press, New York.

Saglio, P. H., and Pradet, A. (1980). *Plant Physiol.* **66**, 516–519.

Saglio, P. H., Raymond, P., and Pradet, A. (1980). *Plant Physiol.* **66**, 1053–1057.

Saglio, P. H., Raymond, P., and Pradet, A. (1983). *Plant Physiol.* **72**, 1035–1039.

Salemme, F. R. (1977). *Annu. Rev. Biochem.* **46**, 299–329.

Salerno, J. C., Harmon, H. J., Blum, H., Leigh, J. S., and Ohnishi, T. (1977). *FEBS Lett.* **82**, 179–181.

Santarius, K. A., and Heber, U. (1965). *Biochim. Biophys. Acta* **102**, 39–54.

Santora, G. T., Gee, R., and Tolbert, N. E. (1979). *Arch. Biochem. Biophys.* **196**, 403–411.

Sarojini, G., and Oliver, D. J. (1983). *Plant Physiol.* **72**, 194–199.

Sato, S., and Asahi, T. (1975). *Plant Physiol.* **56**, 816–820.

Satre, M., Klein, G., and Vignais, P. V. (1979). *In* "Methods in Enzymology" (S. Fleischer and L. Packer, eds.) Vol. 55, pp. 421–426. Academic Press, New York.

Sawhney, S. K., Naik, M. S., and Nicholas, D. J. D. (1978). *Nature (London)* **272**, 647–648.

Sawhney, S. K., Nicholas, D. J. D., and Naik, M. S. (1979). *Indian J. Biochem. Biophys.* **16**, 37–38.

Schaefer, J., Kier, L. D., and Stejskal, E. O. (1980). *Plant Physiol.* **65**, 254–259.

Schardl, C. L., Lonsdale, D. M., Pring, D. R., and Rose, K. (1984). *Nature* **310**, 292–296.

Schatz, G., and Mason, T. L. (1974). *Annu. Rev. Biochem.* **43**, 51–87.

Schell, J., Van Montagu, M., de Beuckeleer, M., de Block, M., Depicker, A., de Wilde, M., Engler, G. Genetello, C., Hernalsteens, J. P., Holsters, M., Seurinck, J., Silva, B., Van Vliet, F., and Villarroel, R. (1979). *Proc. R. Soc. London Ser. B* **204**, 251–266.

Schindler, S., Lichtenthaler, H. K., Dizengremel, P., Rustin, P., and Lance, C. (1984). *In* "Structure, Function and Metabolism of Plant Lipids" (P. A. Siegenthaler and W. Eichenberger, eds.) pp. 267–272. Elsevier, Amsterdam.

Schirch, L., and Peterson, D. (1980). *J. Biol. Chem.* **255**, 7801–7806.

Schleyer, M., Schmidt, B., and Neupert, W. (1982). *Eur. J. Biochem.* **125**, 109–116.

Schmitt, M. R., and Edwards, G. E. (1983). *Plant Physiol.* **72**, 728–734.

Schnaitman, C. A. (1971). *J. Bacteriol.* **108**, 553–563.

Schnaitman, C. A., and Greenawalt, J. W. (1968). *J. Cell Biol.* **38**, 158–165.

Schnare, M. N., and Gray, M. W. (1982). *Nucleic Acids Res.* **10**, 2085–2092.

Schnarrenberger, C., Oeser, A., and Tolbert, N. E. (1973). *Arch. Biochem. Biophys.* **154**, 438–448.

Schneider, H., Lemasters, J. J., Höchlim, M., and Hackenbrook, C. R. (1980). *Proc. Natl. Acad. Sci. U.S.A.* **77**, 442–446.

Schonbaum, G. R., Bonner, W. D., Jr., Storey, B. T., and Bahr, J. T. (1971). *Plant Physiol.* **47**, 124–128.

Schwertner, H. A., and Biale, J. B. (1973). *J. Lipid Res.* **14**, 235–242.

Schwitzguebel, J. P., and Siegenthaler, P. A. (1984). *Plant Physiol.* **75**, 670–674.

Sells, G. D., and Koeppe, D. E. (1981). *Plant Physiol.* **68**, 1058–1068.

Sha'afi, R. I. (1981). *In* "New Comprehensive Biochemistry," Vol. 2: Membrane Transport (S. L. Bonting and J. J. H. H. M. de Pont, eds.), pp. 29–60. Elsevier, Amsterdam.

Shimomura, Y., and Ozawa, T. (1981). Biochem. Int. **2**, 313–318.

Siedow, J. N. (1982). *Recent Adv. Phytochem.* **16**, 47–83.

Siedow, J. N., and Bickett, D. M. (1981). *Arch. Biochem. Biophys.* **207**, 32–39.

Siedow, J. N., and Girvin, M. E. (1980). *Plant Physiol.* **65**, 669–674.

Siess, E. A., Brocks, D. G., and Wieland, O. H. (1982). *In* "Metabolic Compartmentation" (H. Sies, ed.), pp. 235–257. Academic Press, New York.

Silva-Lima, M., and Denslow, N. D. (1979). *Arch. Biochem. Biophys.* **193**, 368–372.

Silva-Lima, M., and Pinheiro, P. A. (1975). *Biochimie* **57**, 1401–1403.

Simcox, P. D., Reid, E. E., Canvin, D. T., and Dennis, D. T. (1977). *Plant Physiol.* **59**, 1128–1132.

Sinclair, D., and Pillay, D. T. N. (1981). *Z. Pflanzenphysiol.* **104**, 299–310.

Singer, S. J. (1974). *Annu. Rev. Biochem.* **43**, 805.

Singer, S. J. and Nicholson, G. L. (1972). *Science* **175**, 720–721.

Singer, T. P., and Gutman, M. (1974). *Horiz. Biochem. Biophys.* **1**, 261–302.

Singer, T. P., Kearney, E. B., and Kenney, W. C. (1972). *Adv. Enzymol. Relat. Subj. Biochem.* **37**, 189–272.

Singh, P., and Naik, M. S. (1984). *FEBS Lett.* **165**, 145–150.

Sitaramam, V., and Janardana Sarma, M. K. (1981). *Proc. Natl. Acad. Sci. U.S.A.* **78**, 3441–3445.

Sjöstrand, F. S. (1953). *Nature (London)* **171**, 30–31.

Sjöstrand, F. S., and Barajas, L. (1968). *J. Ultrastruct. Res.* **25**, 121–155.

Slater, E. C., Rosing, J., and Mol, A. (1973). *Biochim. Biophys. Acta* **292**, 534–553.

Sluse, F. E., Duyckaerts, C., Sluse-Goffart, C. M., Fux, J. P., and Liébecq, C. (1980). *FEBS Lett.* **120**, 94–98.

Smith, B. N., and Meeuse, B. J. D. (1966). *Plant Physiol.* **41**, 343–347.

Smith, S., and Ragan, C. I. (1980). *Biochem. J.* **185**, 315–326.

Smyth, D. A., and Black, C. C. (1984). *Plant Physiol.* **75**, 862–864.

Smyth, D. A., Wu, M. X., and Black, C. C. (1984a). *Plant Physiol.* **76**, 316–320.

Smyth, D. A., Wu, M. X., and Black, C. C. (1984b). *Plant Sci. Lett.* **33**, 61.

Solomos, T. (1977). *Annu. Rev. Plant Physiol.* **28**, 279–297.

Solomos, T., and Laties, G. G. (1975). *Plant Physiol.* **55**, 73–78.

Soloway, S., and Rosen, P. (1955). *Science* **121**, 99–103.

Sottocasa, G. L., Kuylenstierna, B., Ernster, L., and Bergstrand, A. (1967). *In* "Methods in Enzymology" (R. W. Estabrook and M. E. Pullman, eds.) Vol. 10, pp. 448–463. Academic Press, New York.

Spahr, P. F., and Edsall, J. T. (1964). *J. Biol. Chem.* **239**, 850–854.

Spalding, M. H., Schmitt, M. R., Ku, S. B., and Edwards, G. E. (1979). *Plant Physiol.* **63**, 738–743.

Sparace, S., and Moore, T. S., Jr. (1979). *Plant Physiol.* **63**, 963–972.

Sparace, S., and Moore, T. S., Jr. (1981). *Plant Physiol.* **67**, 261–265.

Sparks, R. B., Jr., and Dale, R. M. K. (1980). *Mol. Gen. Genet.* **180**, 351–355.

Spencer, D. F., Bonen, L., and Gray, M. W. (1981). *Biochemistry* **20**, 4022–4029.

Sperk, G., and Tuppy, H. (1977). *Plant Physiol.* **59**, 155–157.

Srere, P. A. (1980). *Trends Biochem. Sci. (Pers. Ed.)* **5**, 120–121.

Staehelin, L. A., and Probine, M. C. (1970). *Adv. Bot. Res.* **3**, 1–52.

Stenlid, G. (1970). *Phytochemistry* **9**, 2251–2256.

Stern, D. B., and Lonsdale, D. M. (1982). *Nature (London)* **299**, 698–702.

Stern, D. B., Dyer, T. A., and Lonsdale, D. M. (1982). *Nucleic Acids Res.* **10**, 3333–3340.

Stewart, C. R., and Lai, E. Y. (1974). *Plant Sci. Lett.* **3**, 173–181.

Stewart, C. R., Boggess, S. F., Aspinall, D., and Paleg, L. G. (1977). *Plant Physiol.* **59**, 930–932.

Stewart, G. R., Mann, A. F., and Fentem, P. A. (1980). *In* "The Biochemistry of Plants," Vol. 5: Amino Acids and Derivatives (B. J. Miflin, ed.), pp. 271–327. Academic Press, New York.

Stitt, M., and ap Rees, T. (1979). *Phytochemistry* **18**, 1905–1911.

Stitt, M., and Heldt, H. W. (1981a). *Biochim. Biophys. Acta* **638**, 1–11.

Stitt, M., and Heldt, H. W. (1981b). *Plant Physiol.* **68**, 755–761.

Stitt, M., Wirtz, W., and Heldt, H. W. (1980). *Biochim. Biophys. Acta* **593**, 85–102.

Stitt, M., Lilley, McC. R., and Heldt, H. W. (1982a). *Plant Physiol.* **70**, 971–977.

Stitt, M., Mieskes, G., Söling, H. D., and Heldt, H. W. (1982b). *FEBS Lett.* **145**, 217–221.

Stitt, M., Gerhardt, R., Kürzel, B., and Heldt, H. W. (1983). *Plant Physiol.* **72**, 1139–1141.

Stitt, M., Herzog, B., and Heldt, H. W. (1984a). *Plant Physiol.* **75**, 548–553.

Stitt, M., Kürzel, B., and Heldt, H. W. (1984b). *Plant Physiol.* **75**, 554–560.

Stoner, C. D., and Hanson, J. B. (1966). *Plant Physiol.* **41**, 255–266.

Storey, B. T. (1969). *Plant Physiol.* **44**, 413–421.

Storey, B. T. (1970a). *Plant Physiol.* **45**, 455–460.

Storey, B. T. (1970b). *Plant Physiol.* **45**, 447–454.

Storey, B. T. (1970c). *Plant Physiol.* **46**, 13–20.

Storey, B. T. (1971a). *Plant Physiol.* **48**, 493–497.

Storey, B. T. (1972). *Biochim. Biophys. Acta* **267**, 48–64.

Storey, B. T. (1973). *Biochim. Biophys. Acta* **292**, 592–602.

Storey, B. T. (1976). *Plant Physiol.* **58**, 521–525.

Storey, B. T. (1980). *In* "The Biochemistry of Plants," Vol. 2: Metabolism and Respiration (D. D. Davies, ed.), pp. 125–195. Academic Press, New York.

Storey, B. T., and Bahr, J. T. (1969). *Plant Physiol.* **44**, 115–125.

Storey, B. T., and Bahr, J. T. (1972). *Plant Physiol.* **50**, 95–102.

Stubbs, M., Veech, R. L., and Krebs, H. A. (1972). *Biochem. J.* **126**, 59–65.

Stubbs, M., Vignais, P. V., and Krebs, H. A. (1978). *Biochem. J.* **172**, 333–342.

Stumpf, P. K. (1980). *In* "The Biochemistry of Plants," Vol. 4: Lipids: Structure and Function (P. K. Stumpf ed.), pp. 177–204. Academic Press, New York.

Sumida, S., and Mudd, J. B. (1968). *Plant Physiol.* **43**, 1162–1164.

Sumida, S., and Mudd, J. B. (1970). *Plant Physiol.* **45**, 719–722.

Suyama, Y., and Bonner, W. D., Jr. (1966). *Plant Physiol.* **41**, 383–388.

Swamy, G. S., and Pillay, D. T. N. (1982). *Plant Sci. Lett.* **25**, 73–84.

Swift, H., and Wolstenholme, D. R. (1969). *In* "Handbook of Molecular Cyctology" (A. Lima-de-Faria, ed.), pp. 972–1046. North-Holland Publ. Amsterdam.

Swift, H., Sinclair, J. H., Stevens, B. J., Rabinowitz, M., and Gross, N. (1968). *In* "Biochemical Aspects of the Biogenesis of Mitochondria" (E. C. Slater, J. M. Tager, S. Papa, and E. Quagliariello, eds.), pp. 71–77. Adriatica Editrice, Bari.

Szaniawski, R. K. (1981). *Z. Pflanzenphysiol.* **101**, 391–398.

Szarkowska, L., and Klingenberg, M. (1963). *Biochem. Z.* **338**, 674.

Tager, J. M., Wanders, R. J. A., Groen, A. K., Kunz, W. Bohnensack, R., Küster, U., Letko, G., Böhme, G., Duszynski, J., and Wojtczak, L. (1983). *FEBS Lett.* **151**, 1–9.

Takeda, Y., Hizukuri, S., and Nikumi, Z. (1967). *Biochim. Biophys. Acta* **146**, 568–575.

Tanner, G. J., Copeland, L., and Turner, J. F. (1983). *Plant Physiol.* **72**, 659–663.

Tassi, F., Restivo, F. M., Ferrari, C., and Puglisi, P. P. (1983). *Plant Sci. Lett.* **29**, 215–225.

Terada, H. (1981). *Biochim. Biophys. Acta* **639**, 225–242.

Tetley, R. M., and Thimann, K. V. (1974). *Plant Physiol.* **54,** 294–303.
Tezuka, T., and Laties, G. G. (1983). *Plant Physiol.* **72,** 959–963.
Thebud, R., and Santarius, K. A. (1981). *Planta* **152,** 242–247.
Theologis, A., and Laties, G. G. (1978). *Plant Physiol.* **62,** 232–237.
Thimann, K. V., Yocum, C. S., and Hackett, D. P. (1954). *Arch. Biochem. Biophys.* **53,** 239–257.
Thomas, D. R., and McNeil, P. H. (1976). *Planta* **132,** 61–63.
Thomas, S. M., and ap Rees, T. (1972). *Phytochemistry* **11,** 2177–2185.
Thompson, E. W., Notton, B. A., Richardson, M., and Boulter, D. (1971). *Biochem. J.* **124,** 787–791.
Ting, I. P., and Dugger, W. M., Jr. (1966). *Plant Physiol.* **41,** 500–505.
Ting, I. P., and Dugger, W. M., Jr. (1967). *Plant Physiol.* **42,** 712–718.
Ting, I. P., and Gibbs, M. (1982). "Crassulacean Acid Metabolism." Amer. Soc. Plant Physiologists, Rockville, Maryland.
Tischler, M. E., Pachence, J., Williamson, J. R., and La Noue, K. F. (1976). *Arch. Biochem. Biophys.* **173,** 448–462.
Tobin, A., Djerdjour, B., Journet, E., Neuburger, M., and Douce, R. (1980). *Plant Physiol.* **66,** 225–229.
Tolbert, N. E. (1971). *In* "Photosynthesis and Photorespiration" (M. D. Hatch, C. B. Osmond, and R. C. Slatyer, eds.), pp. 458–471. Wiley (Interscience), New York.
Tolbert, N. E. (1980). *In* "The Biochemistry of Plants," Vol. 2: Metabolism and Respiration (D. D. Davies, ed.), pp. 487–523. Academic Press, New York.
Tolbert, N. E., and Cohan, M. S. (1953). *J. Biol. Chem.* **204,** 649–654.
Tomomatsu, A., and Asahi, T. (1980). *Plant Cell Physiol.* **21,** 689–698.
Tourte, Y. (1975). *J. Microsc. Biol. Cell.* **23,** 301–315.
Trumpower, B. L. (1981a). *J. Bioenerg. Biomembr.* **13,** 1–24.
Trumpower, B. L. (1981b). *Biochim. Biophys. Acta* **639,** 129–155.
Tuppy, H., and Sperk, G. (1976). *Eur. J. Biochem.* **68,** 13–19.
Turner, J. F., and Turner, D. H. (1980). *In* "The Biochemistry of Plants," Vol. 2: Metabolism and Respiration (D. D. Davies, ed.), pp. 279–316. Academic Press, New York.
Tzagoloff, A., and Meagher, P. (1972). *J. Biol. Chem.* **247,** 594–603.
Tzagoloff, A., Macino, G., and Sebald, W. (1979). *Annu. Rev. Biochem.* **48,** 419–441.
Usuda, H., Arron, G., and Edwards, G. E. (1980). *J. Exp. Bot.* **31,** 1477–1483.
Uzzell, T., and Spolsky, C. (1974). *Am. Sci.* **62,** 334–343.
Valenti, V., and Pupillo, P. (1981). *Plant Physiol.* **68,** 1191–1196.
Vanderleyden, J., Vanden Eynde, E., and Verachtert, H. (1980). *Biochem. J.* **186,** 309–316.
Van Schaftingen, E., Lederer, B., Bartrons, R., and Hers, H. G. (1982). *Eur. J. Biochem.* **129,** 191–195.
Van Sumere, C. F., Cottenic, J., De Greef, J., and Kint, J. (1972). *Recent Adv. Phytochem.* **4,** 165–186.
Vartapetian, B. B., Andreeva, I. N., and Kursanov, A. L. (1974). *Nature (London)* **248,** 258–259.
Vartapetian, B. B., Andreeva, I. N., Kozlova, G. I., and Agapova, L. P. (1977). *Protoplasma* **91,** 243–256.
Vartapetian, B. B., Andreeva, I. N., and Nuritdinov, N. (1978). *In* "Plant Life in Anaerobic Environments" (D. D. Hook and R. M. M. Crawford, eds.), pp. 13–88. Ann Arbor Science, Ann Arbor, Michigan.
Vasconcelos, A. C. L., and Bogorad, L. (1971). *Biochim. Biophys. Acta* **228,** 492–502.
Vedel, F., and Quetier, F. (1974). *Biochim. Biophys. Acta* **340,** 374–387.
Verhaeren, E. H. C. (1980). *Phytochemistry* **19,** 501–503.

Vignais, P. V. (1976). *Biochim. Biophys. Acta* **456,** 1–38.

Vignais, P. V., and Lauquin, G. J. M. (1979). *Trends Biochem. Sci. (Pers. Ed.)* **4,** 90–92.

Vignais, P. V., Douce, R., Lauquin, G. J. M., and Vignais, P. M. (1976). *Biochim. Biophys. Acta* **440,** 688–696.

Vignais, P. V., and Satre, M. (1984). *Mol. Cell. Biochem.* **60,** 33–70.

Von Jagow, G., and Engel, W. D. (1980). *Angew. Chem. Int. Ed. Engl.* **19,** 659–748.

Von Jagow, G., and Klingenberg, M. (1970). *Eur. J. Biochem.* **12,** 583–592.

Wagner, G. J., Mulready, P., and Cutt, J. (1981). *Plant Physiol.* **68,** 1081–1089.

Wainio, W. W. (1970). "The Mammalian Mitochondrial Respiratory Chain." Academic Press, London and New York.

Walker, D. A., and Beevers, H. (1956). *Biochem. J.* **62,** 120–127.

Walker, G. H., Sarojini, G., and Oliver, D. J. (1982). *Biochem. Biophys. Res. Commun.* **107,** 856–861.

Walker, W. H., Singer, T. P., Ghisla, S., Hemmerich, P., Hartmann, U., and Zeszotek, E. (1972). *Eur. J. Biochem.* **26,** 279–289.

Wallsgrove, R. M., Lea, P. J., and Miflin, B. J. (1979). *Plant Physiol.* **63,** 232–236.

Wallsgrove, R. M., Keys, A. J., Bird, I. F., Cornelius, M. J., Lea, P. J., and Miflin, B. J. (1980). *J. Exp. Bot.* **31,** 1005–1017.

Walsh, C. (1978). *Annu. Rev. Biochem.* **47,** 881–931.

Ward, B. L., Anderson, R. S., and Bendich, A. J. (1981). *Cell* **25,** 793–803.

Waterton, J. C., Bridges, I. G., and Irving, M. P. (1983). *Biochim. Biophys. Acta* **763,** 315–320.

Wattiaux-De Coninck, S., Dubois, F., and Wattiaux, P. (1977). *Biochim. Biophys. Acta* **471,** 421–435.

Waidyanatha, V. P., Keys, A. J., and Whittingham, C. P. (1975). *J. Exp. Bot.* **26,** 15–26.

Weil, J. H. (1979). *In* "Nucleic Acids in Plants," (T. C. Hall and J. W. Davies), Vol. 1, pp. 143–192. CRC Press, Boca Raton, Florida.

Weinbach, E. C., and Garbus, J. (1966). *J. Biol. Chem.* **241,** 169–175.

Wells, R., and Ingle, J. (1970). *Plant Physiol.* **46,** 178–179.

Whatley, F. R. (1981). *Ann. N. Y. Acad. Sci.* **361,** 330.

Whatley, J. M., John, P., and Whatley, F. R. (1979). *Proc. R. Soc. London, Ser. B* **204,** 165–187.

Wickner, W. (1979). *Annu. Rev. Biochem.* **48,** 23–45.

Widger, W. R., Cramer, W. A., Herrmann, R. G., and Trebst, A. (1984). *Proc. Natl. Acad. Sci. U.S.A.* **81,** 674–678.

Wikström, M. K. F., and Berden, J. A. (1972). *Biochim. Biophys. Acta* **283,** 403–420.

Wikström, M., and Lambowitz, A. M. (1974). *FEBS Lett.* **40,** 149–153.

Wikström, M., Krab, K., and Saraste, M. (1981). *Ann. Rev. Biochem.* **50,** 623–655.

Wildman, S. G., Hongladarum, T. and Honda, S. I. (1966). "Organelles in Living Plant Cells," 16 mm sound film. Educational Film Sales and Rentals, Univ. Ext. Univ. Ca., Berkeley.

Wildman, S. G., Jope, C., and Atchison, B. A. (1974). *Plant Physiol.* **54,** 231–237.

Williams, J. F. (1980). *Trends Biochem. Sci. (Pers. Ed.)* **5,** 315–319.

Williams, R. J. P. (1978). *Biochim. Biophys. Acta* **505,** 1–44.

Williamson, J. R. (1966). *J. Biol. Chem.* **241,** 5026–5036.

Williamson, J. R., and La Noue, K. F. (1975). *PAABS Symp. Ser.* **4,** 153.

Wilson, D. F., Erecinska, M., and Schramm, V. L. (1983). *J. Biol. Chem.* **258,** 10,464–10,473.

Wilson, R. H., and Graesser, R. J. (1976). *In* "Encyclopedia of Plant Physiology," Vol. 3: Transport in Plants III. Intracellular Interactions and Transport Processes (C. R. Stocking and U. Heber, eds.), pp. 377–397. Springer-Verlag, Berlin.

Wilson, R. H., and Hanson, J. B. (1969). *Plant Physiol.* **44**, 1335–1341.
Wilson, S. B. (1978). *Biochem. J.* **176**, 129–136.
Wilson, S. B. (1980). *Biochem. J.* **190**, 349–360.
Wilson, S. B., and Bonner, W. D., Jr. (1970a). *Plant Physiol.* **46**, 25–30.
Wilson, S. B., and Bonner, W. D., Jr. (1970b). *Plant Physiol.* **46**, 31–35.
Wingfield, P., Arad, T., Leonard, K., and Weiss, H. (1979). *Nature (London)* **280**, 696–697.
Wirtz, K. W. A., Devaux, P. F., and Bienvenue, A. (1980). *Biochemistry* **19**, 3395–3399.
Wirtz, W., Stitt, M., and Heldt, H. W. (1980). *Plant Physiol.* **66**, 187–193.
Wiskich, J. T. (1974). *Aust. J. Plant. Physiol.* **1**, 177–181.
Wiskich, J. T. (1977). *Annu. Rev. Plant Physiol.* **28**, 45–69.
Wiskich, J. T. (1980). *In* "The Biochemistry of Plants," Vol. 2: Metabolism and Respiration (D. D. Davies, ed.), pp. 243–278. Academic Press, New York.
Wiskich, J. T., and Bonner, W. D., Jr. (1963). *Plant Physiol.* **38**, 594–603.
Wiskich, J. T., and Day, D. A. (1979). *J. Exp. Bot.* **30**, 99–107.
Wiskich, J. T., and Day, D. A. (1982). *Plant Physiol.* **70**, 959–964.
Wittenbach, V. A., Lin, W., and Hebert, R. R. (1982). *Plant Physiol.* **69**, 98–102.
Woese, C. R., and Fox, G. E. (1977). *Proc. Natl. Acad. Sci. U.S.A.* **74**, 5088–5091.
Wohlrab, H. (1980). *J. Biol. Chem.* **255**, 8170–8173.
Wojtczak, L., and Nalecz, M. J. (1979). *Eur. J. Biochem.* **94**, 99–107.
Wojtczak, L., and Zaluska, H. (1969). *Biochim. Biophys. Acta* **193**, 64–72.
Wojtczak, L., Wojtczak, A. B., and Ernster, L. (1969). *Biochim. Biophys. Acta* **191**, 10–21.
Woo, K. C. (1979). *Plant Physiol.* **63**, 783–787.
Woo, K. C., and Osmond, C. B. (1976). *Aust. J. Plant Physiol.* **3**, 771–785.
Woo, K. C., and Osmond, C. B. (1977). *In* "Special Issue of Plant and Cell Physiol." (S. Miyachi, S. Katoh, Y. Fujita, and K. Shibata, eds.), no. 3, pp. 315–323. Center Academic Publ., Tokyo.
Woo, K. C., Jokinen, M., and Canvin, D. T. (1980). *Plant Physiol.* **65**, 433–436.
Wrigglesworth, J. M., Packer, L., and Branton, D. (1970). *Biochim. Biophys. Acta* **205**, 125–135.
Wu, M. X., Smyth, D. A., and Black, C. C., Jr. (1983). *Plant Physiol.* **73**, 188–191.
Yonetani, T. (1960). *J. Biol. Chem.* **235**, 845–852.
Yoshida, K., and Takeuchi, Y. (1970). *Plant Cell Physiol.* **11**, 403–409.
Yoshida, S., and Tagawa, F. (1979). *Plant Cell Physiol.* **20**, 1243–1250.
Yu, C. A., and Yu, L. (1981). *Biochim. Biophys. Acta* **639**, 99–128.
Yu, C. A., Yu, L., and King, T. E. (1972). *J. Biol. Chem.* **247**, 1012–1019.
Yu, C. A., Yu, L., and King, T. E. (1979). *Arch. Biochem. Biophys.* **198**, 314–322.
Yu, C. A., Claybrook, D. L., and Huang, A. H. (1983). *Arch. Biochem. Biophys* **227**, 180–187.
Zalman, L. S., Nikaido, H., and Kagawa, Y. (1980). *J. Biol. Chem.* **255**, 1771–1774.
Zimmermann, R., and Neupert, W. (1980). *Eur. J. Biochem.* **109**, 217–229.
Zimmermann, R., Hennig, B., and Neupert, W. (1981). *Eur. J. Biochem.* **116**, 455–460.
Zoratti, M., Pietrobon, D., and Azzone, G. F. (1983). *Biochim. Biophys. Acta* **723**, 59–70.
Zwizinski, C., and Neupert, W. (1983). *J. Biol. Chem.* **258**, 13,340–13,346.
Zwizinski, C., Schleyer, M., and Neupert, W. (1983). *J. Biol. Chem.* **258**, 4071–4074.

Index